Finite Element Analysis

Procedures in Engineering

H V Lakshminarayana

AF520227

Universities Press

Universities Press (India) Private Limited

Registered Office
3-6-747/1/A and 3-6-754/A Himayatnagar,Hyderabad 500 029 (A.P.), India
e-mail: info@universitiespress.com

Distributed by
Orient Blackswan Private Limited

Registered Office
3-6-752 Himayatnagar, Hyderabad 500 029 (A.P.), INDIA

Other Offices
Bangalore, Bhopal, Bhubaneshwar, Chennai,Ernakulam,Guwahati,Hyderabad, Jaipur, Kolkata, Lucknow, Mumbai, New Delhi, Noida, Patna

© Universities Press (India) Private Limited 2004

First published 2004
Reprinted 2008, 2012, 2013

ISBN: 978 81 7371 476 4

Typeset by
AICRA Info Mates Pvt. Ltd.
Hyderabad

Printed in India at
Glorious Printers, New Delhi

Published by
Universities Press (India) Private Limited
3-6-747/1/A and 3-6-754/A Himayatnagar,
Hyderabad 500 029 (A.P.), India

To our grand daughter

Greeshma

Contents

Preface

Every engineering curriculum should include a course on finite element method (FEM). This textbook has emerged from over a decade of learning how to teach a first level course on FEM to undergraduate and postgraduate students of engineering. The basic concepts, mathematical models and computational algorithms supporting FEM are clearly and concisely developed herein. Our intent is that this book be complete and self-contained, so that a novice can both understand FEM as a numerical method for engineering analysis, and learn to use commercial finite element analysis (FEA) software to solve practical problems. By the time a student completes the study of the book and runs the test problems, adequate knowledge of FEM and skills to use a commercial FEA program should be at hand. The application problems demonstrate the capability of FEM in general and commercial FEA codes in particular, to handle the complexities of practical problems. The book will also be useful for teaching short-term courses to professional engineers who use a general purpose FEM system as a computer aided engineering (CAE) tool.

The goal of analysis in engineering is to verify a design by predicting its behaviour prior to building a prototype. While there are several methods of engineering analysis, the most comprehensive is the FEA. The strength of a commercial FEA program is that it is not written to solve just a specific problem for the novice. It is a powerful tool, which once understood, can be applied to almost any type of problem. Commercial FEA programs provide a number of material models, a library of finite elements, a set of analysis procedures and adequate data processing capabilities for problems in solid mechanics, heat transfer, fracture mechanics, electromagnetic fields, etc. In this book, solid and structural mechanics applications are introduced first, to provide a firm basis for the finite element procedure.

The mathematical theory underlying FEM employs a variational or a weighted residual formulation, valid for the exact solution as well as any approximation. It then introduces a discretization of the solution domain into finite elements (of different types, shapes, and orders). Approximation of field variable(s) over each element domain using interpolation functions produces a set of element equations which are collected together using a procedure called assembly to form the global matrix statement. The method then uses the various linear algebra techniques to numerically solve the governing matrix equations, after enforcing appropriate boundary and constraint conditions.

This text does not present a survey of finite elements. Instead, we concentrate only on certain finite elements and analysis procedures which are very useful in engineering.

Several commercial FEA programs are available for use on personal computers. A standard course on FEM should therefore include computational work in which the test problems and computational exercises presented in the book are solved by means of a commercial FEA code, and the computed results critically examined. The instructor may sit with the students at the computer centre to offer advice, monitor their skills in FE modelling, and assess the computed results. This book is not tailored to any particular FEA program and therefore does not discuss the formalisms of input data preparation.

Chapter 1 provides a brief history of FEM; the mathematical preliminaries and numerical methods for FEA are also introduced. Chapter 2 is devoted to one-dimensional elements. The stress is on applications to linear static analysis of bars, plane- and space-trusses. Two-dimensional continuum elements, both triangular and quadrilateral in shape, are presented in Chapter 3. Creation of C^0 continuous interpolation functions and use of Gauss quadrature formulae to compute element matrices and vectors are emphasized. Application to plane stress and plane strain problems featuring complex geometries subjected to body forces, surface tractions and initial strains is demonstrated. Higher order isoparametric elements are introduced and their design identified as a topic for home assignment. Finite elements for modelling and analysis of axisymmetric solid bodies subjected to thermal and mechanical loads are described in Chapter 4. Three-dimensional continuum elements are presented in Chapter 5. A family of 3-D solid elements of different shapes (tetrahedra, hexahedra, and pentahedra) and increasing orders (linear, quadratic, and cubic) are described. Applications to stress analysis of complex solid bodies are demonstrated.

Finite elements for modelling and analysis of structures, namely beams, plates, shells and their combinations are presented in Chapters 6, 7, and 8 respectively. Analysis of beams, plane frames, and space frames subjected to static and dynamic loads is addressed in Chapter 6. Plate elements implemented in commercial FEA programs and their application to bending, buckling, vibration and transient response analysis of plates are documented in Chapter 7. Finite element modelling of shell structures for reliable prediction of their behaviour subject to static and dynamic loads is outlined in Chapter 8.

A brief description of general-purpose FEA programs with focus on their organization and capabilities is given in Chapter 9. Advanced applications of the FEM using commercial FEA codes are identified and presented in Chapter 10. Today, it is impossible for one person to write knowledgeably about all aspects of the topics listed here. A specialist is to be invited to do so for each topic!

During a first level course on the FEM, undergraduate students who are exposed to finite element modelling for stress analysis of solids. Therefore the focus of Chapters 1 through 5 is on linear elastic static analysis. Previous exposure to the continuum mechanics of solids is desirable.

A second level course on the FEM is taken by postgraduate students who are exposed to structural analysis of beams, plates and shells. We therefore discuss these in Chapters 6 through 9. Previous exposure to the theories of beams, plates and shells is not necessary since the essential physical behaviour of such structures is easily grasped. The focus is on bending, buckling, vibration and transient response analysis, both linear and nonlinear.

In addition to serving as the primary text for the courses listed above, the book should be useful as an adjunct text in a course for practicing engineers where the role of FEA in CAE is stressed. This is done in Chapters 1 and 9; Chapter 10 has an in-depth discussion on advanced applications of FEM and finite element solutions to practical problems in engineering.

Many researchers, authors and colleagues have contributed to the technical contents of this book. This is gratefully acknowledged. I am thankful to Dr. S.R. Shankapal, Director, M.S.Ramaiah School of Advanced Studies for his support. Special thanks are due to Ms. Roopa and Mr. K. Brahmananda Reddy, the word processors, who cheerfully endured the numerous revisions and editing sessions required to produce the manuscript. Finally, truly unbounded thanks to Rathna, my beloved wife, who with her love and patience lend me support while writing this book.

Bangalore, India

Hulikal V. Lakshminarayana

1 Introduction to Finite Element Method

The digital computer has exerted a profound impact on engineering education, research, and practice. A versatile numerical method in the hands of engineers is the Finite Element Method (FEM). A general-purpose program based on FEM implemented on a computer provides a universal tool for engineering analysis, design optimization, and simulation. This chapter sets the stage for the study of finite elements and solution procedures that are described in detail in the subsequent chapters. Applications to solid mechanics and structural mechanics problems are stressed in these chapters. More advanced applications of FEM are identified in the last chapter as topics for further study.

1.1 Engineering Analysis

1.1.1 Objectives of engineering analysis

During the design and development of a product (as an assemblage of parts), the analyst is quite often required to: (i) calculate the displacements at certain points; (ii) calculate the entire distribution or displacement field; (iii) determine the stress distribution and hence predict strength; (iv) determine the natural frequencies and associated modes of vibration; (v) determine the critical buckling loads and the associated mode shapes; (vi) predict and plot forced vibration response; (vii) predict and plot transient response; (viii) predict temperature distribution and hence thermal stress distribution; (ix) predict crack growth, residual strength and fatigue life; (x) predict velocity, pressure and temperature distribution in fluids; (xi) study fluid–structure interactions (hydro-elasticity, aero-elasticity, etc.); (xii) study nonlinear effects (geometric and material nonlinearities); (xiii) determine electric and magnetic fields, and many more!

1.1.2 Methods of engineering analysis

To achieve the above objectives, the analyst has at his disposal three distinct approaches: (i) analytical methods; (ii) experimental techniques and (iii) numerical methods.

Analytical methods [1.1] provide quick closed form solutions. But, they treat only simple geometries and idealized support and loading conditions. Using experimental techniques, scaled models or prototypes can be tested. This approach is costly both in terms of the model, instrumentation, test facilities and the actual test itself. Numerical methods require very few restrictive assumptions; it can treat complex geometries and realistic support and loading conditions. They are far more cost effective than experimental techniques. The current interest in the engineering community on the development and application of computational tools based on numerical methods is thereby justified. This in fact was the motivation to develop the most versatile numerical method, namely, the finite element method (FEM).

The goal of analysis is to verify a design prior to manufacture. While there are several methods of engineering analysis, the most comprehensive is the finite element analysis (FEA).

The finite element method is essentially dependent for its success on the skillful use of digital computers. The method is put in the hands of professional engineers in the form of general-purpose programs.

It is in general possible to use the FEM to provide accurate numerical solutions to almost any mathematical problem or mathematically modelled physical problem in diverse fields like solid mechanics, mechanics of composites, fluid dynamics, heat transfer, etc. Many continuum mechanics problems arise in engineering and these are usually posed by appropriate differential equations and boundary conditions to be imposed on unknown functions. All such problems can also be dealt with by FEM.

1.1.3 History of finite element method

Finite element technology has emerged as a new discipline combining continuum mechanics with approximation theory, numerical analysis, and computer science. It draws from recent advances in each of these disciplines and is nourished by them, but it also stands as a viable branch of engineering in its own right.

In the pre-computer era, the finite element concept was used in the analysis of naturally discrete systems such as trusses, frames, electrical circuits, etc. The modern version of FEM was first used in engineering practice by discretizing a continuum using simple sub-domains with multiple connecting points (nodes) by Turner et al., [1.2], Argyris [1.3], and Clough [1.4]. The name finite element was first coined by Clough [1.4] in the year 1960. Parallel developments were also reported in applied mathematics literature.

The development of modern finite element technology therefore spans a period of forty years and may be divided into four stages, with each stage spanning eight to ten years.

The first stage is characterized by development of simple two- and three-dimensional continuum elements, mostly for solid mechanics applications. The variational approach for deriving element equations was introduced and all the elements developed belonged to the stiffness (displacement) model.

In the second stage, the variational approach was expanded to include multi-field and generalized variational principles with relaxed inter-element displacement continuity requirements leading to the mixed models and hybrid models. Weighted-residual methods were introduced to extend the finite element method for the analysis of field problems. The mathematical foundations of FEM received a great deal of attention during this stage. Efficient numerical methods for the solution of algebraic equations and for the extraction of eigen values were developed. Sub-structuring (super elements) and modal synthesis techniques for solving very large problems were introduced. A number of commercial FEM systems were developed and released for public use. Also, FEM penetrated other areas of engineering analysis such as heat transfer, fluid dynamics, biomechanics, geomechanics, aeromechanics, fracture mechanics, mechanics of laminated composite materials and structures, electromagnetism, etc. Further applications to nonlinear and time-dependent (transient) response and coupled-field problems (soil–structure, fluid–structure, etc.) were also made.

The third stage involved the development and applications of special elements. Singular elements for computational fracture mechanics, boundary-layer elements for viscous fluid flow analysis, infinite elements for modelling unbounded domains, rigid links, and gap elements for contact problems are some of the examples. Other activities included the development of the boundary element method, coupling of FEM with continuum mechanics methods such as Rayleigh–Ritz and Bubnov–Galerkin methods, and establishing equivalence and similarities between various finite element methods.

The fourth stage is characterized by new application fields, development of efficient algorithms and computational strategies for new computing systems (e.g., vector multi-processor and massively-parallel processors), widespread availability of commercial FEA software on personal computers, workstations and supercomputers. Also increasing attention was focused on quality assessment and control of finite element solutions. Strategies were proposed for adaptive refinement of finite element approximation in order to achieve optimal solutions.

The success of FEM is mainly attributed to its generality, versatility, ability to model and analyze complex geometries and robustness. To date there are approximately five-hundred user-friendly, widely distributed, well documented general purpose FEA programs and over two hundred pre and post processor packages [1.5].

The literature on finite element technology is nearly over-whelming. The first textbook on the FEM was published in 1967. Since then over 387 textbooks and monographs and over 338 conference proceedings have been published on the subject [1.6].

The Finite Element Method by O.C. Zienkiewicz and R.L. Taylor, now in its fifth edition, (in volumes 1–3) is the pre-eminent reference work [1.7].

1.1.4 Advantages of finite element method

In initiating the prediction of displacements, stresses, vibration frequencies, buckling loads, etc., for a given product or its parts, the analyst must first derive the governing equations. A basic difficulty in this approach, quite apart from the solvability of the derived equations is the ability of these equations to represent the design conditions. Complexities in geometry, applied loads, support conditions and material properties enter into this condition. A basic promise of the finite element method is that a system of matrix equations governing the behaviour can be formed automatically and solved efficiently, irrespective of the complexities of practical design conditions.

The underlying mathematics of FEM is simple to understand and the procedure easy to use. However, successful application of FEM in practice depends on the availability of a general-purpose finite element analysis software implemented on a digital computer. Commercial FEM systems abound. A handbook on the topic [1.5] lists and describes over fifty programs by name along with their availability. For example, these programs can solve limitless variety of problems in solid mechanics and structural mechanics, whether linear or nonlinear, static or dynamic, elastic or plastic.

A more subtle attribute of FEM is its ability to deal with complex material models. For example, the heterogeneous, anisotropic, nonlinear, inelastic models of laminated composite materials and sandwich construction are handled without any significant expansion of the cost or complexity of the numerical simulation process. FEM brings a number of special advantages to coupled thermal–structural analysis. A consistent methodology of finite element heat transfer analysis is available [1.8] for the computation of temperature distribution in solids and structures. It is possible to use the same general-purpose FEA program to predict both temperature distribution due to thermal input and thermal stresses arising from these temperatures. Also, in cases where the material properties are a function of temperature, it is possible to assign material properties to each finite element consistent with the temperature level of that element. It is a revolution you must be familiar with – the marriage of personal computers and finite element analysis programs. Over thirty-five PC-based FEM systems are currently available in the market [1.9]. Of these, some stand out as market leaders based on their performance, popularity and advertising. What makes FEA on a PC so successful is its affordability. This will change the face of engineering design in general and structural and mechanical design in particular. Technically, the size of a finite element model that a PC can handle is limited only by the capacity of the hard disk. Best of all, a PC-based FEM system is an excellent training tool for teaching FEM.

1.1.5 Variational principles and finite element methods

The finite element method applied to the numerical solution of solid mechanics problems can be regarded as applications of the known variational principles. There are several advantages: (i) the method is thereby put on a sound theoretical foundation; (ii) the requirements for compatibility are clarified as these are quite explicit in the statement of the energy principles; and (iii) greater flexibility in the design of finite elements comes about because of the variety of alternatives at hand.

In addition to the minimum potential energy and minimum complementary energy principles, more general mixed variational principles due to Hellinger–Reissner and Hu–Washizu are available. Also, modified forms of these with relaxed inter-element displacement or stress continuity are also available. Table 1.1 is particularly useful in classifying the many different avenues for developing finite element methods. Among the finite element methods identified in Table 1.1, the displacement method is certainly the best understood and most widely used. The subsequent discussion is therefore confined to the displacement method. Finite element analysis of solids and structures based on the principle of minimum potential energy employs a piece-wise Rayleigh–Ritz procedure. Note that the assumed displacement functions must satisfy sufficient continuity conditions within the domain under consideration and the kinematic boundary conditions. There is, however, no requirement that the force boundary conditions be a priori satisfied.

Table 1.1 *Variational principles and finite element methods*

Variational principle	Finite element method
Principle of minimum potential energy/Principle of virtual work (PMPE)	Stiffness method/Displacement method
Principle of minimum complementary energy (PMCE)	Equilibrium method/Force method
Hellinger–Reissner mixed variational principle	Mixed method I
Modified PMPE	Hybrid displacement method
Modified PMCE	Hybrid stress method
Modified mixed variational principle	Mixed method II
Hu–Washizu principle	Displacement method

1.1.6 Basic steps in finite element analysis

The first step is the *discretization* of a given domain using finite elements. The domain can be a solid, a liquid, a gas, or their combinations. A library of finite elements of different types, shapes and orders is available for this purpose. Each element has a finite number of nodes and each node a finite number of degrees of freedom, which are the fundamental unknowns. The elements are inter-connected at their nodes only and the finite element mesh (see Fig. 1.1) is generated using a pre-processor. Commercial pre-processors have the capability of automated mesh generation and adaptive mesh refinement. The second step is to *approximate the field variable(s)* over each element domain in terms of their nodal values using *interpolation functions* (also called *shape functions*). *Derivation of element equations*, which are necessary and sufficient to determine the vector of nodal degrees of freedom for each element, is the objective of the third step. Variational and weighted residual approaches are widely used in this step. *Computation of element matrices and vectors* involves numerical integration over each element domain. The fourth step involves *assembly of element equations*. In the fifth step, the governing matrix equations are appropriately modified to enforce *boundary, support, symmetry* and *constraint* conditions. Special elements such as springs, rigid links, and gap elements are made use of for this purpose. *Solution of the governing matrix equations* is accomplished in the sixth step. This will provide the vector of nodal degrees of freedom for the assemblage as well as for the individual elements. The final step is called *post-processing* where the numerical results are printed, plotted, displayed, and animated graphically. Interactive computer graphics are used for this purpose.

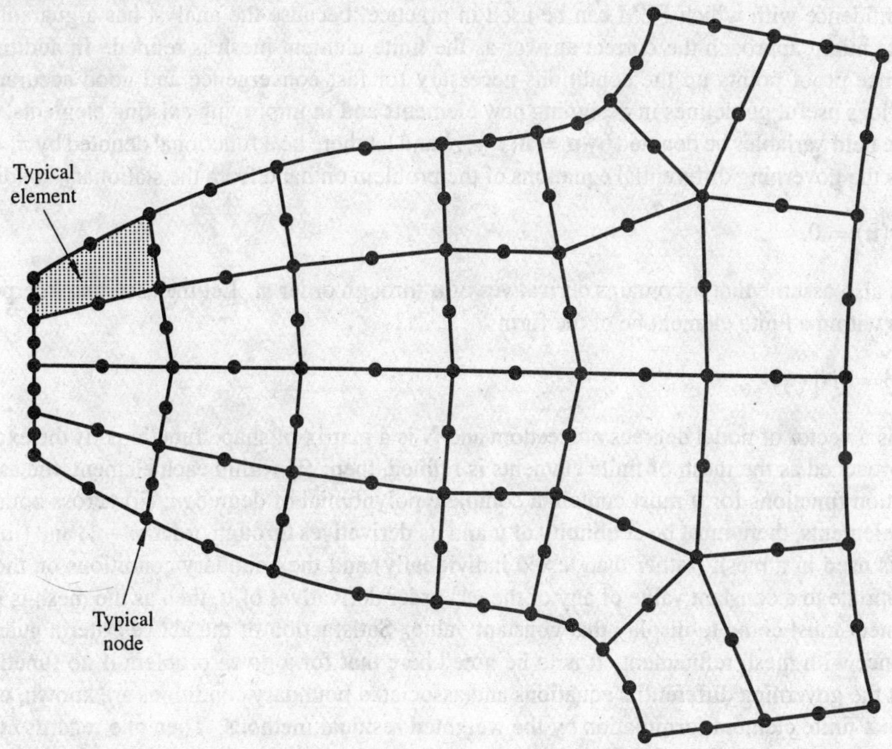

Fig. 1.1 *A coarse-mesh, two-dimensional finite element model of a spur gear teeth*

The above finite element procedure of artificially subdividing a given domain into convenient subdomains and assuming separate interpolation functions for each subdomain can be termed piece-wise Rayleigh–Ritz method.

Two approaches are available for refining a finite element model as the one shown in Fig. 1.1. The first approach provides more number of elements of the same type. This leads to the so-called h-convergence. In the second approach, the number of elements remains fixed, but the order of the interpolation functions used within each element is increased successively. This leads to p-convergence. In practice, a combination of h- and p-convergence is also used. Does an approximate numerical solution, obtained by the finite element method converge to the exact solution as the finite element mesh is uniformly refined? To ensure convergence, some criteria have to be satisfied. These are discussed in the next section.

1.1.7 Convergence criteria

If a particular problem in solid or structural mechanics is repeatedly analyzed, each time using a finer mesh of finite elements, we generate a sequence of approximate numerical solutions. How can we be assured that the sequence converges to the theoretically correct results? For conforming displacement model, a relatively simple proof of convergence can be given, based on the minimum property of the potential energy. A convergence proof is of more than academic interest. For one thing, it contributes to the confidence with which FEM can be used in practice, because the analyst has a guarantee that the results might approach the correct answer as the finite element mesh is refined. In addition, the convergence proof points up the conditions necessary for fast convergence and good accuracy, and thus provides useful guidelines in designing new elements and in improving existing elements.

Let the field variables be denoted by $\mathbf{u} = \mathbf{u}(x, y, z)$ and let there be a functional denoted by $\pi = \pi(\mathbf{u})$ that gives the governing differential equations of the problem on hand from the stationary condition

$$\partial\pi(\mathbf{u}) = 0.$$

Let us also assume that π contains derivatives of $\mathbf{u}$ through order m. Let the assumed interpolation functions within a finite element be of the form

$$\{\mathbf{u}\} = [\mathrm{N}]\,\{\mathbf{q}\}$$

where $\mathbf{q}$ is a vector of nodal degrees of freedom and $\mathbf{N}$ is a matrix of shape functions. If the exact $\mathbf{u}$ is to be approached as the mesh of finite elements is refined, then: (i) within each element, the assumed interpolation functions for $\mathbf{u}$ must contain a complete polynomial of degree m; (ii) across boundaries between elements, there must be continuity of $\mathbf{u}$ and its derivatives through order $m - 1$; and (iii) if the element is used in a mesh (rather than tested individually) and the boundary conditions on the mesh are appropriate to a constant value of any of the mth order derivatives of $\mathbf{u}$, then as the mesh is refined each element must come to display that constant value. Satisfaction of the above criteria guarantees convergence with mesh refinement. It is to be noted here that for a given problem if no functional π exists but the governing differential equations and associated boundary conditions are known, one can yet obtain a finite element formulation by the weighted residual methods. Then one regards m as the highest order derivative of the field variables $\mathbf{u}$ to be found in an integral expression used to generate the finite element equations.

The patch test [1.10] was originally created by Irons as a simple test that can be performed on a computer, so as to check the validity of a finite element formulation and its programmed implementation. If a particular element passes the patch test, we have the assurance that all convergence criteria noted above are met. Therefore, when this element is used to model any other structure or machine component, successive mesh refinement will produce a sequence of approximate solutions that converge to the exact solution. In other words, the patch test serves as a necessary and sufficient condition for convergence. An element that fails the patch test cannot be trusted.

1.1.8 Role of finite element analysis in computer-aided design

Figure 1.2 identifies the modules and their functions in a typical computer-aided design procedure.

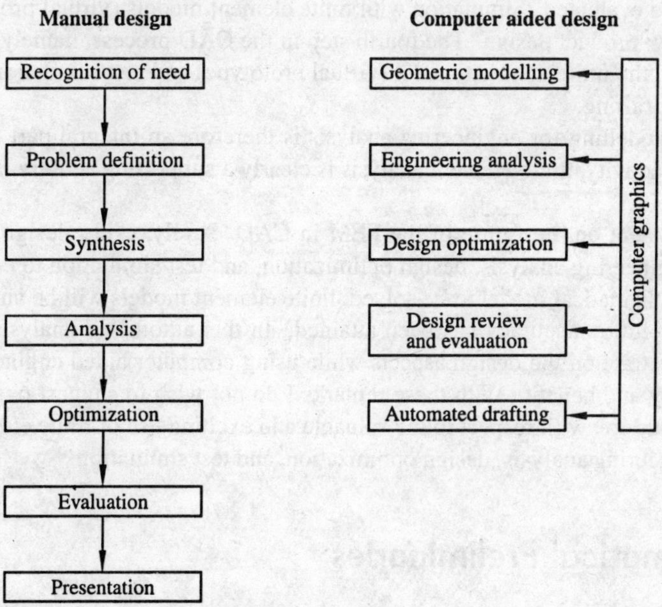

Fig. 1.2 *Computer-aided design: Modules*

We note that the first step is the geometric modelling of the product (an assemblage of parts). A large number of commercial CAD systems can be employed to perform this step. They are: Catia, UniGraphics, Pro/Engineer, I-Deas, Mechanical Desk Top, etc. In this step, for each part a mathematical model (the collection of all equations and data required to define the geometry) is generated and stored in the database. Given this information, the second step, the engineering analysis, may proceed.

In the second step, use of a general-purpose finite element analysis program (pre-processor, solver, post-processor) is now an accepted practice. A large number of commercial FEM systems are available for this purpose. They are NASTRAN, ANSYS, MARC, NISA, ALGOR, etc. This step generally is and should be performed by design engineers and not necessarily by engineering analysis specialists. Therefore, we recognize that the finite element method used must be robust, reliable and efficient. A basic pre-requisite to perform engineering analysis by FEM is a tractable mathematical model (the collection of all equations and data that can be used to predict the behaviour) of the product. FEM provides numerical solutions to the chosen mathematical model, which may be changed depending on the objectives of the analysis.

It is no longer sufficient to design a workable product which performs the desired functions. It has become essential to optimize the product design in order to maximize or minimize chosen variable(s) called objective function(s). This in fact is the third step in the CAD process. Keeping this in mind, commercial CAD and FEM systems have implemented FEM-based optimization methods as a module.

Simulation is the process of subjecting the part/product to various inputs, such as loads and environments, to determine how it behaves and thus predict the characteristics of the physical system. Though simulation may be carried out with scale models and prototypes, the cost and effort involved make it impossible to use these for product design since many feasible designs and operating conditions need to be considered and evaluated. Simulation with finite element models (virtual prototypes) is particularly valuable in new product design. The fourth step in the CAD process, namely design review and evaluation, relies on the finite element model (virtual prototype) and test simulation. The final step in CAD is automated drafting.

Finite element modelling for engineering analysis is therefore an integral part of CAD. Although an exciting field of activity, finite element analysis is clearly a supporting activity in the larger field of CAD.

Finally, we comment on the future of the FEM in CAD. Surely, every design engineer wants to use FEM-based engineering analysis, design optimization, and test simulation to enhance the product design. Given a mathematical model to be solved, finite element models will be automatically refined until the required solution accuracy has been attained. In this automated analysis environment, the engineer can concentrate on the design aspects while using computer aided engineering (CAE) tools with great efficiency and benefit. With these remarks I do not wish to suggest overconfidence but to express a realistic outlook with respect to the valuable and exciting use of finite elements and solution procedures in engineering analysis, design optimization, and test simulation.

1.2 Mathematical Preliminaries

1.2.1 Physical problems, mathematical models and finite element solutions

The finite element method is widely used to perform engineering analysis of physical systems. The derivation of an appropriate mathematical model of the physical problem is an important pre-requisite for engineering analysis to proceed. The finite element method provides numerical solutions to this mathematical model and therefore it is necessary to assess the accuracy of the numerical solution. If the accuracy criteria are not met, the finite element model has to be refined until sufficient accuracy is reached. Hence the development of an appropriate finite element model is crucial.

Once a mathematical model has been solved accurately by finite element analysis and the results interpreted with respect to the physical system, we may well decide to consider next a refined mathematical model in order to improve our understanding of the response of the physical system. Furthermore, a change in the physical problem statement itself may be necessary, and this in turn will also lead to additional mathematical models and their finite element solutions. This iterative solution process is shown in Fig. 1.3.

In summary, we should keep in mind that the crucial step in finite element analysis of physical systems is always choosing an appropriate mathematical model. Furthermore, the chosen mathematical model must be reliable and effective. In the process of analysis, the analyst has to judge not only the accuracy of the finite element solution but also its validity. Choosing the mathematical model,

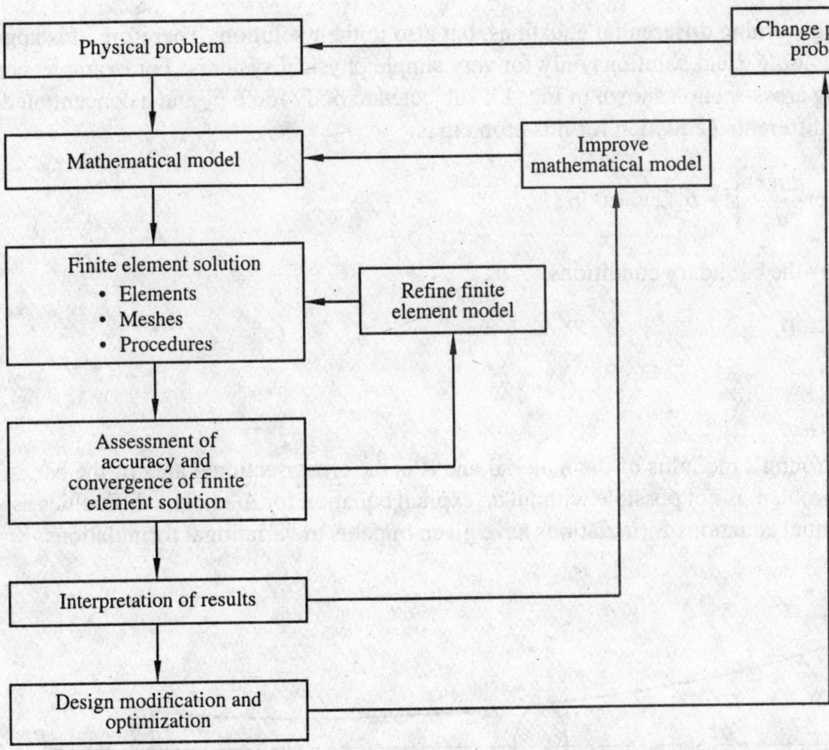

Fig. 1.3 *Physical problems, mathematical models, finite element solutions*

solving this model by appropriate finite element procedures, and judging the results are the fundamental ingredients of an engineering analysis.

Some classical techniques used for the formulation and solution of mathematical models of engineering systems is well documented (see Bathe [1.11]). Two categories of mathematical models are considered: lumped-parameter models and continuum-mechanics-based models. We also refer to these as discrete-systems and continuous-systems mathematical models. Discrete-system mathematical models lead to steady state problems, propagation problems and eigen value problems. Continuous-system mathematical models lead to differential equation formulations, variational formulations and weighted residual methods.

1.2.2 Differential equations formulations

Continuum-mechanics-based mathematical modelling of systems lead us to differential equations. The governing differential equations must be satisfied throughout the domain of the physical system, and before their solution can be attempted, they must be supplemented by boundary conditions and also by initial conditions. In initiating engineering analysis of a part/product, the analyst must first derive the governing differential equations. A basic difficulty in the approach, quite apart from the solvability of the derived equations, is the ability of these equations to represent the complexities in geometry, applied loads, support conditions and material properties. In summary, we face difficulties not only in deriving the governing differential equations, but also in their solution. Therefore, this approach is suitable for providing exact solutions only for very simple physical systems. For example, consider a bar with varying cross-section shown in Fig. 1.4 subjected to body force b_x and a concentrated load R. The governing differential equation for this problem is

$$\frac{d}{dx}\left[EA(x)\frac{du(x)}{dx}\right] + b_x(x) = 0 \text{ in } \Omega, \tag{1.1}$$

supplemented by the boundary conditions,

$$u(x=0) = 0, \tag{1.2}$$

$$EA(x)\left.\frac{du}{dx}\right|_{x=L} = R, \tag{1.3}$$

where E is the Young's modulus of the material and A is the cross-sectional area of the bar. An exact solution to the problem is not possible without an explicit equation for $A(x)$. The difficulties associated with the differential equations formulations have given impetus to variational formulations.

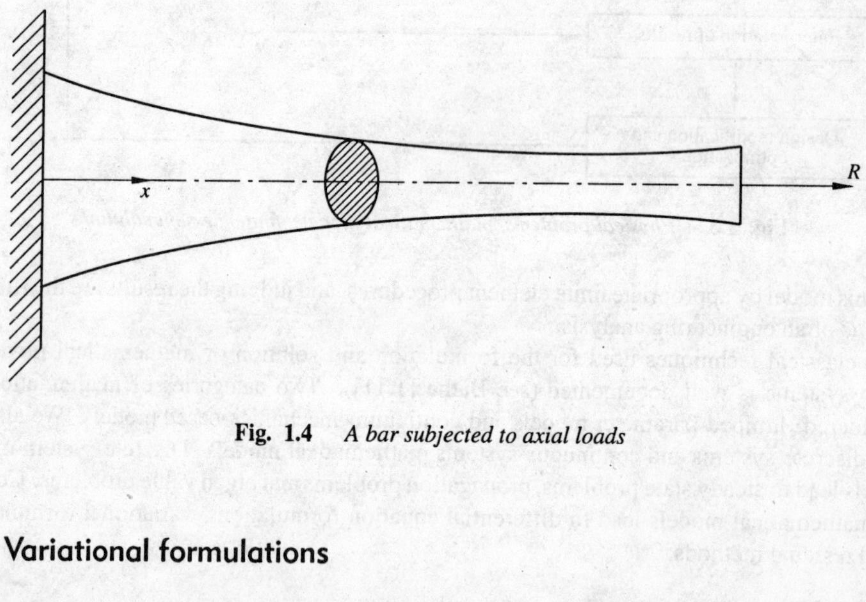

Fig. 1.4 *A bar subjected to axial loads*

1.2.3 Variational formulations

Continuum-mechanics-based mathematical modelling of solids and structures lead us to variational formulations. The essence of the approach is to calculate the total potential π of the system and to invoke the stationarity of π, i.e., $\partial\pi = 0$, with respect to the state variables. The variational formulations are effective for the solution of solid mechanics and structural mechanics problems by the Rayleigh–Ritz method. Indeed, the FEM for solid/structural mechanics problems can be regarded as a piecewise Rayleigh–Ritz method!

The total potential π is also called the functional of the problem. An important question then arises: How can we establish an appropriate functional corresponding to a physical problem?

For solid mechanics and structural mechanic problems, a number of functionals are applicable. For instance, we can employ the potential energy functional π_p, the complimentary energy functional π_c, the Hellinger–Reissner functional π_{HR}, the Hu–Washizu functional π_{HW}, and a number of modified forms of these with relaxed requirements on inter-element displacement or stress continuity which are also available. Table 1.1 is particularly useful in identifying the many different possibilities for developing finite element methods based on these functionals.

For example, the potential energy functional governing static buckling analysis of the problem shown in Fig. 1.5 is

$$\pi_p(w) = \frac{1}{2}\int_0^L EI\left(\frac{d^2w}{dx^2}\right)dx - \frac{P}{2}\int_0^L\left(\frac{dw}{dx}\right)^2 dx + \frac{1}{2}kw_L^2, \tag{1.4}$$

and the essential boundary conditions are

$$w(x=0) = 0, \tag{1.5}$$

$$\left.\frac{dw}{dx}\right|_{x=0} = 0. \tag{1.6}$$

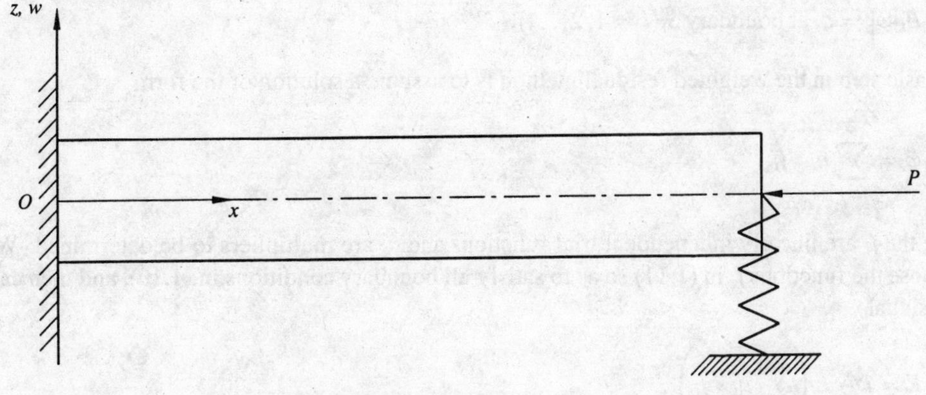

Fig. 1.5 *A column subjected to compressive load*

The basic step in the Rayleigh–Ritz method is to assume a solution of the form

$$w(x) = \sum_{i=1}^{n} a_i * f_i(x), \tag{1.7}$$

where $f_i(x)$ are linearly independent trial functions and a_i are multipliers to be determined. We have to assume the trial functions $f_i(x)$ such that the essential boundary conditions are a priori satisfied. The parameter a_i is determined from the equations

$$\frac{\partial \pi_p}{\partial a_i} = 0, i = 1, 2, 3, \ldots, n. \tag{1.8}$$

There are some classes of problems for which variational formulations are not available. This has given impetus to the development of weighted residual methods.

1.2.4 Weighted residual methods

In a previous section we have discussed the differential equations formulation of physical problems. For thermal analysis, fluid flow analysis, electromagnetic fields, etc., the governing differential equations, appropriate boundary and initial conditions are well known. However, closed form solutions are possible for simple systems only. For more complex systems, weighted residual methods must be employed. Indeed, the finite element method for field problems can be regarded as an extension of these.

Consider the analysis of a steady state field problem using its differential equations formulation,

$$L[\varphi] = r \text{ in domain } \Omega, \tag{1.9}$$

in which L is a linear differential operator, φ is the state variable to be calculated, and r is the forcing function. The solution to the problem must also satisfy the boundary conditions

$$B_i[\varphi] = q_i \text{ at boundary } S_i (i = 1, 2, \ldots). \tag{1.10}$$

The basic step in the weighted residual method is to assume a solution of the form

$$\bar{\varphi} = \sum_{i=1}^{n} a_i * f_i, \tag{1.11}$$

where the f_i are linearly independent trial functions and a_i are multipliers to be determined. We have to choose the functions f_i in (1.11) so as to satisfy all boundary conditions in (1.10), and then calculate the residual

$$R = r - L\left(\sum_{i=1}^{n} a_i * f_i\right). \tag{1.12}$$

For the exact solution the residual is of course zero. A good approximation to the exact solution would imply that R is small at all points of the solution domain Ω. The various weighted residual methods differ in the criteria that they employ to calculate a_i such that R is small. However, in all the methods we determine the a_i so as to make a weighted average of R vanish. In the Galerkin method, the parameters a_i are determined from the equation

$$\int_{\Omega} f_i * R d\Omega = 0 (i = 1, 2, 3, \ldots, n), \tag{1.13}$$

where Ω is the solution domain.

An important step in using a weighted residual method is the solution of the simultaneous equations for the parameters a_i. We note that since L is a linear operator, a linear set of equations in the parameters a_i is generated. In addition, the coefficient matrix is symmetric (and also positive definite) if L is a symmetric and also positive definite operator. The fundamental difficulty in using the weighted residual method in practice is because the trial functions must be $2m$ times differentiable and must satisfy all essential and natural boundary conditions, where $2m$ is the order of the highest derivative present in the problem governing the differential equations. Therefore, the weighted residual method is used in the context of the FEM in a different form, namely, in a form that allows the use of trial functions which are to be m times differentiable only and do not need to satisfy a priori the natural boundary conditions. The procedure would be to weigh the governing differential equations(s) in the domain with suitable weight function(s); integrate the resulting equation(s) with a transformation using integration by parts (or more generally using the divergence theorem); and substituting the natural boundary conditions. This leads us to a weak statement.

1.2.5 Matrix algebra

The use of vectors and matrices is of fundamental importance in engineering analysis by finite element method. The objective of this section is to present the fundamentals, with emphasis on those aspects that are important in finite element analysis.

Matrices are an ordered array of numbers that are subjected to specific rules of addition, multiplication, and inversion. Whenever the elements of a matrix obey a certain law, we can consider the matrix to be of special form symmetric, diagonal, banded, etc.

Two matrices $\mathbf{A}$ and $\mathbf{B}$ can be multiplied to obtain $\mathbf{C} = \mathbf{AB}$ if and only if the number of columns in $\mathbf{A}$ is equal to the number of rows in $\mathbf{B}$. If $\mathbf{A}$ is of order $p \times m$ and $\mathbf{B}$ is of order $m \times q$, then for each element of matrix $\mathbf{C}$, we have,

$$\mathbf{C}_{ij} = \sum_{k=1}^{m} a_{ik} b_{kj}, \tag{1.14}$$

and $\mathbf{C}$ is of the order $p \times q$.

With regard to matrix division, it strictly does not exist. Instead, an inverse matrix is defined. The inverse of a matrix $\mathbf{A}$ is denoted by $\mathbf{A}^{-1}$. Assuming that the inverse exists, the elements of $\mathbf{A}^{-1}$ are such that $\mathbf{A}^{-1}\mathbf{A} = \mathbf{I}$ and $\mathbf{A}\mathbf{A}^{-1} = \mathbf{I}$, where $\mathbf{I}$ is the identity matrix.

A matrix that possesses an inverse is said to be nonsingular. A matrix without an inverse is a singular matrix. To obtain the inverse of a general matrix, we need to have a general algorithm.

A practical way of calculating the inverse of a matrix $\mathbf{A}$ of order $n \times n$ is to solve the n systems of equation

$$\mathbf{AX} = \mathbf{I}, \tag{1.15}$$

where $\mathbf{I}$ is the identity matrix of order n and we have $\mathbf{X} = \mathbf{A}^{-1}$. For the solution of equations in (1.15), one can use the well known Gaussian elimination algorithm.

The trace and determinant of a matrix are defined only if the matrix is square. The trace of the matrix $\mathbf{A}$ is denoted as $\mathrm{tr}(A)$ and equal to

$$\mathrm{tr} = \sum_{i=1}^{n} a_{ii},$$

where n is the order of $\mathbf{A}$.

The determinant of an $n \times n$ matrix **A** is denoted as $\det A$ and is given by the recurrence relation

$$\det A = \sum_{j=1}^{n} (-1)^{1+j} a_{1j} \det A_{1j},$$

where A_{ij} is the $(n-1)(n-1)$ matrix obtained by eliminating the 1st row and jth column from the matrix **A**.

A set of simultaneous linear algebraic equations may be symbolized as

$$\mathbf{AX} = \mathbf{b}. \tag{1.16}$$

Solution for **X** may be symbolized as

$$\mathbf{X} = \mathbf{A}^{-1}\mathbf{b}.$$

A square matrix is called singular if its determinant is zero. If **A** in (1.16) is singular, there is no unique solution vector **X**, and standard equation solvers will fail. Let **A** be an $n \times n$ matrix and **X** an $n \times l$ column vector. Also let $\mathbf{X} \neq 0$, which means that at least one coefficient $\mathbf{X}_i$ is nonzero. Then for all **X**,

if $\mathbf{X}^T\mathbf{AX} > 0$, **A** is called positive definite.
if $\mathbf{X}^T\mathbf{AX} \geq 0$, **A** is called positive semi-definite.

1.3 Numerical Methods in Finite Element Analysis

1.3.1 Interpolation functions

In the finite element method, there is a need to approximate the field variables over each element domain in terms of their nodal values using interpolation functions (also called shape functions). However, to ensure monotonic convergence of finite element solutions, the interpolation functions should a priori satisfy the so-called convergence criteria. The two requirements for monotonic convergence are that the elements (or the mesh) must be compatible and complete. For one-dimensional elements, C^0 continuous Lagrange and C^1 continuous Hermite polynomial functions, well known in the mathematics literature, are used as interpolation functions. For two- and three-dimensional continuum elements and for structural elements (beam, plate, shell), procedures to derive appropriate interpolation functions are outlined in the subsequent chapters.

1.3.2 Numerical integration techniques

An important aspect of finite element analysis is the use of numerical integration techniques to compute element matrices and vectors. The required integrals in the one-, two- and three-dimensional cases respectively, can be written as

$$I = \int F(\xi) d\xi;$$

$$I = \int F(\xi,\eta)d\xi d\eta; \text{ and}$$

$$I = \int F(\xi,\eta,\zeta)d\xi d\eta d\zeta.$$

These integrals are evaluated numerically using

$$I = \int F(\xi)\, d\xi = \sum_{i=1}^{n} W_i F(\xi_i);$$

$$I = \int F(\xi,\eta)\, d\xi d\eta = \sum_{i=1}^{n}\sum_{j=1}^{n} W_i W_j F(\xi_i,\eta_j);$$

$$I = \int F(\xi,\eta,\zeta) d\xi d\eta d\zeta = \sum_{i=1}^{n}\sum_{j=1}^{n}\sum_{k=1}^{n} W_i W_j W_k F(\xi_i,\eta_j,\zeta_k),$$

where the W_i, W_j, and W_k are the weighting factors, and (ξ_i, η_j, ζ_k) denote sampling point locations and n denotes number of sampling points.

In finite element analysis we integrate matrices, which means that each element of the matrix considered is integrated individually. Hence for the derivation of numerical integration formulas, we need to consider a typical element of a matrix, which we denote as F. A very important numerical integration technique in which both the positions of the sampling points and the weights have been optimized is the Gauss quadrature formula. For the one-dimensional case, we have

$$I = \int F(\xi)d\xi = \sum_{i=1}^{n} W_i F(\xi_i).$$

We require n sampling points (also called Gauss points) to integrate exactly a polynomial of order at most $(2n - 1)$. Polynomials of orders less than $(2n - 1)$ are also integrated exactly.

A great deal of research has been done on the development of suitable numerical integration formulas for quadrilateral and triangular domains in two-dimensions, hexahedral and tetrahedral domains in three-dimensions.

1.3.3 Static analysis—solution of equilibrium equations

The computational efficiency of finite element analysis depends on the numerical methods used for the solution of the governing matrix equations. In this section, we are concerned with the solution of the simultaneous equations that arise in static analysis of solids and structures. Specifically, we discuss the solution of the equations that arise in linear static analysis

$$[\mathbf{K}]\{\mathbf{Q}\} = \{\mathbf{F}\}, \tag{1.17}$$

where **K** is the stiffness matrix, **Q** the vector of nodal degrees of freedom, and **F** the vector of nodal forces.

There are two different approaches for the solution of the equations in (1.17) – direct solution techniques and iterative solution techniques.

The most effective direct solution techniques currently used are basically applications of Gauss elimination algorithm. The basic algorithm can be applied to any set of simultaneous linear equations. Its effectiveness in FEA depends on the specific properties of the assembled matrix: symmetry, bandedness, sparsity and positive definiteness.

The mathematical operations of Gauss elimination reduce the matrix **K** to upper triangular form, i.e., a form in which all elements below the leading diagonal are zero. Starting with the last equation, it is then possible to calculate by back substution all unknowns in the order $Q_n, Q_{n-1}, \ldots, Q_1$. A very important aspect of the computer implementation of the Gauss elimination procedure is referred to as the active column solution or the skyline reduction method. The use of bandwidth minimization procedures can be very important in practice because the mean half-bandwidth of a stiffness matrix may initially be rather large as a result of the element and nodal point generation schemes used. Furthermore, the active columns of the matrix **K** are stored in a one-dimensional array. Cholesky factorization, static condensation, sub structuring, and frontal solver are some other schemes that are in principle, applications of the basic Gauss elimination procedure.

It is informative to note that during the initial developments of the finite element method, iterative solution algorithms have been employed. A basic disadvantage of an iterative solution is that the time of solution can be estimated only very approximately because the number of iterations required for convergence depends on the condition number of the stiffness matrix **K**. It is primarily for this reason that the use of iterative solvers in finite element analysis was largely abandoned, while the direct solvers have been refined and rendered extremely effective. The Gauss-Seidel iterative procedure continues to find use. However, the conjugate gradient method is particularly attractive.

The finite element equations to be solved in nonlinear analysis of solids and structures are, at time $t + \Delta t$,

$$ {}^{t+\Delta t}\mathbf{R} - {}^{t+\Delta t}\mathbf{F} = 0, \tag{1.18} $$

where the vector ${}^{t+\Delta t}\mathbf{F}$ stores the externally applied nodal loads and ${}^{t+\Delta t}\mathbf{R}$ is the vector of nodal forces that are equivalent to the element stresses. Both vectors in (1.18) are evaluated using the principle of virtual displacements.

Since the nodal point forces ${}^{t+\Delta t}\mathbf{R}$ depend nonlinearly on the nodal point displacements, it is necessary to iterate in the solution of (1.18). The most frequently used iteration scheme for the solution of nonlinear finite element equations are the Newton-Raphson and other closely related techniques. An important requirement of nonlinear finite element analysis is frequently the calculation of the collapse load of a structure. A load-displacement constraint method proposed by E.Riks can be used for this purpose.

1.3.4 Vibration analysis—solution of eigen problems

The finite element equations to be solved in vibration analysis of solids and structures are

$$ \mathbf{K}\phi = \lambda \mathbf{M}\phi, \tag{1.19} $$

where **K** and **M** are, respectively, the stiffness matrix and mass matrix of the finite element model. The eigen values λ_i and the eigen vectors ϕ_r are the natural frequencies (radians/s) squared (ω_i^2) and the corresponding mode shape vectors, respectively. We concentrate in particular on the calculation of the smallest eigen values $\lambda_1, \lambda_2, \ldots, \lambda_p$ and corresponding eigen vectors $\phi_1, \phi_2, \ldots, \phi_p$.

The solution methods considered here can be subdivided into four groups: the vector iteration methods; the transformation methods; polynomial iteration techniques; and sturm sequence methods. A number of solution algorithms have been developed within each of these four groups. In addition, the Lanc'zos method and the subspace iteration methods are also available.

1.3.5 Dynamic response analysis—solution of equations of motion

The finite element equations for dynamic response analysis are

$$\mathbf{M}\ddot{\mathbf{Q}}(\mathbf{t}) + \mathbf{C}\dot{\mathbf{Q}}(\mathbf{t}) + \mathbf{K}\mathbf{Q}(\mathbf{t}) = \mathbf{F}(\mathbf{t}), \tag{1.20}$$

where **M**, **C**, and **K** are the mass, damping and stiffness matrices; **F** is the vector of externally applied, time-dependent, nodal point loads, **Q**, $\dot{\mathbf{Q}}$, and $\ddot{\mathbf{Q}}$ are the displacement, nodal point velocity and nodal point acceleration vectors of the finite element model, all of them being time-dependent.

Mathematically, (1.20) represents a system of linear differential equations of second order with constant coefficients. However, the procedures proposed for the solution of general systems of differential equations can become very expensive if the order of the matrices is large, unless specific advantage is taken of the special characteristics of the coefficient matrices **K**, **C** and **M**. In practical finite element analysis, the procedures considered are divided into two groups: direct integration and mode superposition. The central different method, the Houbolt method, the Wilson θ-method, the new-mark method, belong to the direct integration methods group.

The dynamic response analysis by mode superposition requires, first the solution of the eigen values and eigen vectors of the problem, then the solution of the decoupled equilibrium equations, and finally the superposition of the response in each eigen vector. In practice, the eigen vectors are the free vibration mode shapes of the finite element model.

The choice between mode superposition analysis and direct integration methods is merely one of numerical effectiveness.

Nonlinear dynamic response analysis of a finite element model is in essence, performed using the incremental formulations, the iterative solution methods, and the time integration algorithms (explicit integration and implicit integration). The application of mode superposition in nonlinear dynamic response analysis can be effective if only a relatively few modes shapes need to be considered.

1.3.6 Linear buckling analysis—solution methods

The matrix equation for linear buckling analysis of structures by the finite element method is

$$[\mathbf{K} + \lambda\mathbf{K}_\sigma]\,\{\mathbf{dQ}\} = \{\mathbf{0}\}, \tag{1.21}$$

where the initial stress stiffness matrix $\mathbf{K}_\sigma$ is calculated from an arbitrarily chosen level of membrane stress state, and λ is the factor by which this level must be increased or decreased in order to produce buckling. At the critical (buckling) load, there is a bifurcation in a load versus displacement plot. Two infinitesimally close equilibrium states are possible — the unbuckled state and the buckled state, without any change in the applied loads F. Nodal displacement increments $\{\mathrm{d}\mathbf{Q}\}$ are departures from the configuration **Q** that exist just before buckling. The right hand side of (1.21) is the corresponding change in applied nodal point loads and is therefore a null vector. Equation (1.21) is an eigen problem.

The computed value of λ may be positive or negative, depending on the state of membrane stress used to construct $\mathbf{K}_\sigma$. Membrane stresses may be known at the outset, or they may have to be computed. Linear buckling analysis uses $\mathbf{K}$ and $\mathbf{K}_\sigma$ based on the original, undeformed geometry of the structure and often overestimates the actual buckling load. Most practical buckling problems are nonlinear, and buckling analysis should be based on the tangent stiffness that prevails at the instant of buckling. These considerations are automatically incorporated in the nonlinear finite element analysis procedures.

The reader is advised to refer Bathe [1.11] for algorithmic details. However, computer implementation of these numerical methods and its integration into a commercial FEM system that can be executed on any computer, personal to supercomputer, is a challenge and involves many human years' effort.

1.4 References

[1.1] Wang, C.T. *Applied Elasticity*. McGraw Hill, New York, 1953.

[1.2] Turner, M.J., Clough, R.W., Martin, H.C., and Topp, L.J. Stiffness and defection analysis of complex structures. *Journal of Aeronautical Sciences*, 23:805–823, 1956.

[1.3] Argyris, J.J. *Energy Theorems and Structural Analysis*. Butterworth, London, 1960.

[1.4] Clough, R.W. The finite element method in plane stress analysis. In *Proceedings of the Second ASC Conference on Electronic Computation*, Pittsburgh, PA, Sept. 1960.

[1.5] Brebbia, C.A., ed., *Finite Element Systems*. Springer Verlag, Berlin, 1985.

[1.6] Noor, A.K. Bibliography of books and monographs on finite element technology. *Applied Mechanics Reviews*, 4(6), June 1991.

[1.7] Zienkiewicz, O.C. and Taylor, R.L. *The Finite Element Method*, volume 1–3. Arnold Publishers, 5th ed., 2000.

[1.8] Baker, A.J. and Pepper, O.W. *Finite Elements 1–2–3*. McGraw Hill, New York, 1991.

[1.9] McLean, D.M. An overview of PC-based finite element programs in the USA. *Finite Element News*, June 1988.

[1.10] Taylor, R.L., Simo, J.C., Zienkiewicz, O.C., and Chan, C.E. The patch tests—condition for assessing FEM convergence. *International Journal For Numerical Methods in Engineering*, 22:39–62, 1986.

[1.11] Bathe, K.J. *Finite Element Procedures*. Prentice Hall of India, New Delhi, 1996.

1.5 Computational Problems

No specific computational problems are suggested in this chapter. However, students may wish to get acquainted with the FEA software chosen for the course by using its pre-processor to create simple geometries and discretize them using finite elements.

2 One-Dimensional Elements

Analysis of Bars and Trusses

2.1 Overview

A bar is a member subjected to axial loads only. A truss is an assemblage of bars with pin joints. Solid mechanics [2.1] provides the mathematical model for finite element analysis of bars and trusses. In this chapter, one-dimensional elements and step-by-step procedure for the analysis of bars and trusses is presented. The assumed displacement function, the minimum potential energy functional approach, is used to derive element equations. Computational aspects of element matrices and vectors are included. Procedures for the assembly of element matrices, imposing boundary and constraint conditions, and solution of a system of linear algebraic equations are presented. Test problems to assess element accuracy are described. Applications to linear static analysis of practical plane and space trusses are included. The chapter concludes with identification of topics for advanced study.

2.2 Basic Equations and Potential Energy Functional

Let us consider a bar (Fig. 2.1) whose cross-section is either constant or gradually varying in the axial direction, subjected to axial loads symmetrically positioned with respect to the geometric centroid of the cross-section. The deformation of the bar is characterized by the axial displacement $u(x)$. The axial strain ε_{xx}, assuming small deformation is

$$\varepsilon_{xx} = \frac{du}{dx}. \tag{2.1}$$

The nonzero stress component is the axial stress σ_{xx}. Assuming the material of the bar to be homogeneous, isotropic, linear, elastic, the stress σ_{xx}, is related to the strain ε_{xx} by Hook's Law,

$$\sigma_{xx} = E\left(\varepsilon_{xx} - \varepsilon_{xx}^{\text{init}}\right), \tag{2.2}$$

where E is the Young's modulus of elasticity and $\varepsilon_{xx}^{\text{init}}$, denotes initial strain if any.

The potential energy functional π_p is

$$\pi_p = \frac{1}{2}\int_0^L \varepsilon_{xx}\sigma_{xx}A(x)dx - \int_0^L u(x)b_x(x)A(x)dx - \sum_{i=1}^{n} u_i P_i, \tag{2.3}$$

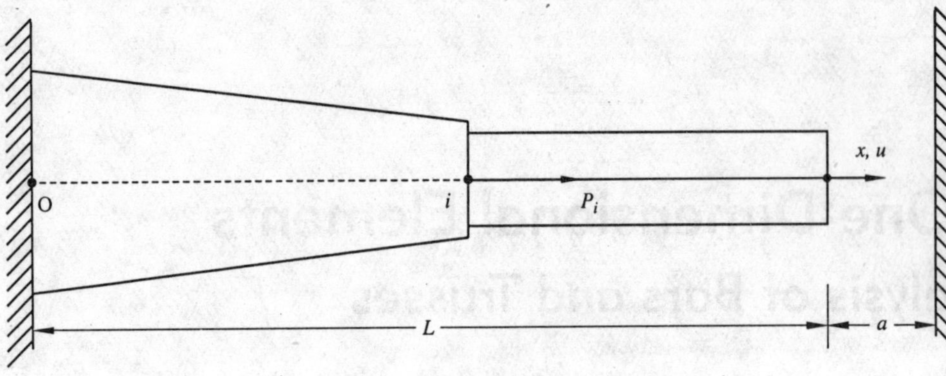

Fig. 2.1 *A bar subjected to axial load*

where $A(x)$ is the cross-sectional area, P_i is the axial load at any point i; u_i is the axial displacement at the point i; $b_x(x)$ is the body force per unit volume in the axial direction. We invoke the variational principle of minimum potential energy to satisfy the equilibrium conditions for the bar.

In order to evaluate the potential energy functional π_p, the displacement $u(x)$ should be kinematically admissible. That is, the displacement function $u(x)$ should be finite, single-valued, and continuous, satisfying a priori the displacement boundary conditions. Creating such mathematical functions for complex problems is in fact impossible. Admitting the availability of such functions, evaluation of the integrals in (2.3) is equally challenging. Both the hurdles are crossed by the introduction of the finite element discretization. The body in Fig. 2.1 is now considered as an assemblage of finite elements (see Fig. 2.2.). A typical element isolated from the mesh is shown in Fig. 2.3 and is called 1-D bar element.

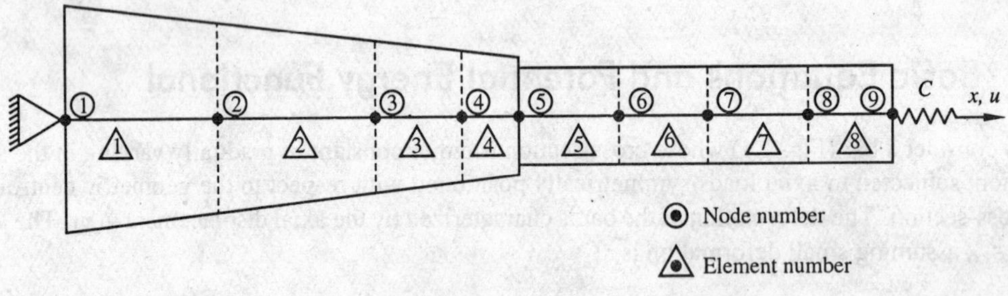

Fig. 2.2 *Finite element discretization of the bar shown in Fig. 2.1*

2.3 1-D Bar Element

The element shown in Fig. 2.3 is mathematically a line between nodes (1) and (2) and carries one degree of freedom at each node. The nodal displacement vector is

$$\{\mathbf{q}\}^T = (u_1, u_2)$$

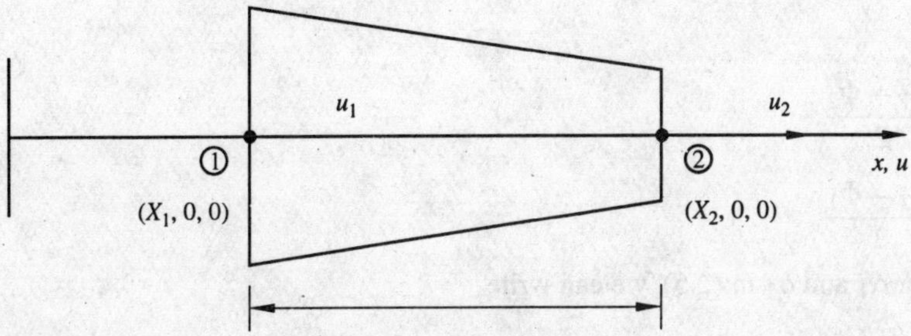

Fig. 2.3 *1-D bar element*

and the corresponding nodal force vector is

$$\{\mathbf{f}\}^T = (F_{x1}, F_{x2})\,.$$

The derivation of element equations, i.e., algebraic equations that relate the nodal forces to the nodal displacements by the assumed displacement function minimum potential energy functional approach is described below.

2.3.1 Admissible displacement function

The axial displacement **u** at any point within the element has to be uniquely defined by the nodal displacements. The simplest representation is given by

$$\mathbf{u} = [\mathbf{N}]\,\{\mathbf{q}\}\,, \tag{2.4}$$

where the shape function matrix is

$$[\mathbf{N}] = [\mathbf{N}_1 \mathbf{N}_2]\,.$$

In order to derive explicit expressions for the shape functions $\mathbf{N}_1$ and $\mathbf{N}_2$, it is convenient to work in the covariant coordinate space ξ, and to use the concept of a parent element (a bi-unit line) as shown in Fig. 2.4.

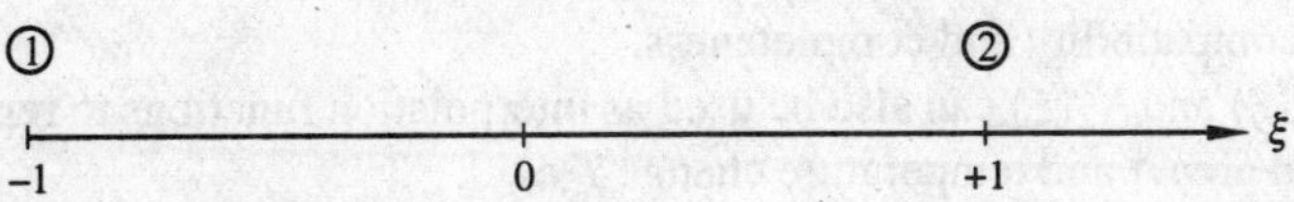

Fig. 2.4 *Parent element and natural coordinate for the 1-D bar element*

Any variable Φ is approximated over the parent element domain using a linear polynomial with respect to ξ. That is,

$$\Phi = \alpha_1 + \alpha_2 \xi. \tag{2.5}$$

Let $\Phi = \Phi_1$ at node (1), i.e., at $\xi = -1$, and

$\Phi = \Phi_2$ at node (2), i.e., at $\xi = +1$.

Then,

$$\alpha_1 = \frac{\Phi_1 + \Phi_2}{2}.$$

$$\alpha_2 = \frac{\Phi_2 - \Phi_1}{2}.$$

Substituting for α_1 and α_2 in (2.5), we can write

$$\Phi = N_1(\xi)\Phi_1 + N_2(\xi)\Phi_2, \tag{2.6}$$

with

$$N_1(\xi) = \frac{1}{2}(1 - \xi),$$

$$N_2(\xi) = \frac{1}{2}(1 + \xi).$$

A plot of these functions for $-1 \leq \xi \leq 1$ shows that $N_i = 1$ at node (i); $N_i = 0$ at the other node; and $N_1(\xi) + N_2(\xi) = 1.0$.

Let us now map, uniquely, this parent element on to the 1-D bar element. The required relationship between the Cartesian and natural coordinates to accomplish this is

$$X = \sum_{i=1}^{2} N_i(\xi) X_i, \tag{2.7}$$

where X_i denotes the x-coordinate of node i of the 1-D bar element.

With the geometry of the element now defined by the parametric equation (2.7), it is also possible to write the displacement function as

$$u = \sum_{i=1}^{2} N_i(\xi) u_i. \tag{2.8}$$

It is informative to note that (2.8)is admissible in the sense that it satisfies the convergence criteria, namely, continuity, compatibility and completeness.

The functions $N_1(\xi)$ and $N_2(\xi)$ can also be used as interpolation functions to represent the variation of the cross-sectional area A and temperature change T as

$$A = \sum_{i=1}^{2} N_i(\xi) A_i, \tag{2.9}$$

$$T = \sum_{i=1}^{2} N_i(\xi) T_i, \tag{2.10}$$

where A_i and T_i denote the value of A and T at node $i (i = 1, 2)$.

2.3.2 Strain matrix

The axial strain at any point within the 1-D bar element is evaluated using (2.1) as

$$\varepsilon_{xx} = \frac{du}{dx} = \frac{du}{d\xi}\frac{d\xi}{dx},$$

where

$$\frac{d\xi}{dx} = \frac{2}{X_2 - X_1} = \frac{2}{\ell} \tag{2.11}$$

and $\ell = (X_2 - X_1)$ is the length of the element.
Therefore,

$$\varepsilon_{xx} = [\mathbf{B}]\{\mathbf{q}\}, \tag{2.12}$$

$$\text{where} \quad [\mathbf{B}] = \left[\frac{-1}{\ell} \quad \frac{1}{\ell}\right]. \tag{2.13}$$

Equation (2.13) defines the element strain matrix **B**. The initial strain due to temperature change is evaluated using

$$\varepsilon_{xx}^{\text{init}} = \alpha T = \alpha \sum_{i=1}^{2} N_i(\xi)\, T_i, \tag{2.14}$$

where α is the thermal expansion coefficient of the material.

2.3.3 Stress recovery

The axial stress at any point within the 1-D bar element is evaluated using (2.2) as

$$\sigma_{xx} = E[\mathbf{B}]\{\mathbf{q}\} - E\alpha \sum_{i=1}^{2} N_i(\xi) T_i. \tag{2.15}$$

It is a standard practice to recover the stress either at the center of the element ($\xi = 0.0$) or at the Gauss point locations ($\xi = \pm 1/\sqrt{3}$), which are optimal locations for accurate stress recovery. However, if nodal point stresses are required, extrapolation of the Gauss point stresses to the nodes and averaging of contributions from all elements connected to a node is recommended.

2.3.4 Element equations

Combining (2.4), (2.12) and (2.15), the potential energy functional for the 1-D bar element can be expressed as

$$\pi_p(\mathbf{q}) = \frac{1}{2}\{\mathbf{q}\}^T[\mathbf{K}]\{\mathbf{q}\} - \{\mathbf{q}\}^T\{\mathbf{f}^b\} - \{\mathbf{q}\}^T\{\mathbf{f}^{\text{init}}\} - \{\mathbf{q}\}^T\{\mathbf{f}^c\}, \tag{2.16}$$

where,

$$[\mathbf{K}] = \int_{X_1}^{X_2} \mathbf{B}^T E \mathbf{B} A(x)\, dx, \tag{2.17}$$

$$\left\{\mathbf{f}^b\right\} = \int_{X_1}^{X_2} \mathbf{N}^T b_x(x) A(x)\, dx, \tag{2.18}$$

$$\left\{\mathbf{f}^{\text{init}}\right\} = \int_{X_1}^{X_2} \mathbf{B}^T E \varepsilon_{xx}^{\text{init}} A(x)\, dx, \tag{2.19}$$

$$\{\mathbf{f}^c\}^T = \left(F_{x1}^c, F_{x2}^c\right). \tag{2.20}$$

In (2.16), $\mathbf{K}$ is the element stiffness matrix, $\mathbf{f}^b$ is the consistent (work equivalent) nodal force vector due to body force; $\mathbf{f}^{\text{init}}$ is the consistent nodal force vector due to initial strain and $\mathbf{f}^c$ denotes the concentrated forces (if any) applied at the nodes.

We now invoke the principle of minimum potential energy to satisfy the equilibrium condition. The necessary condition is

$$\frac{\partial \pi_p}{\partial \{\mathbf{q}\}} = 0$$

leading to the element equations

$$[\mathbf{K}]\{\mathbf{q}\} = \left\{\mathbf{f}^b\right\} + \left\{\mathbf{f}^{\text{init}}\right\} + \{\mathbf{f}^c\}. \tag{2.21}$$

2.3.5 Stiffness matrix

Using (2.17), the formula to compute the stiffness matrix is obtained as

$$[\mathbf{K}] = \int_{-1}^{+1} \mathbf{B}^T E \mathbf{B} \left(\sum_{i=1}^{2} N_i(\xi) A_i\right) \frac{\ell}{2} d\xi. \tag{2.22}$$

After evaluating the integral we can write

$$[\mathbf{K}] \approx \frac{E}{\ell} \frac{(A_1 + A_2)}{2} \begin{bmatrix} 1 & -1 \\ -1 & 1 \end{bmatrix}. \tag{2.23}$$

In particular, if $A_1 = A_2 = A$,

$$[\mathbf{K}] = \frac{EA}{\ell} \begin{bmatrix} 1 & -1 \\ -1 & 1 \end{bmatrix}. \tag{2.24}$$

It is important to note that the stiffness matrix is symmetric. In practice, it is necessary to interpret "*EA*" as the axial stiffness of the bar so that measured value of the axial stiffness can be used in the computation of the element stiffness matrix.

2.3.6 Consistent nodal force vector: Body force

Using Eqn. (2.18), the formula to compute the nodal force vector due to body force is

$$\left\{\mathbf{f}^b\right\} = \int_{-1}^{+1} \mathbf{N}^T b_x(x) \left(\sum_{i=1}^{2} N_i(\xi) A_i\right) \frac{\ell}{2} d\xi. \tag{2.25}$$

For the particular case of self-weight,

$$b_x(x) = \rho g, \tag{2.26}$$

where ρ is the mass density of the bar material and g is the acceleration due to gravity. After evaluating the integral in (2.25) we get

$$\left\{\mathbf{f}^b\right\} \approx \rho g \ell \frac{(A_1 + A_2)}{2} \begin{Bmatrix} 1/2 \\ 1/2 \end{Bmatrix}. \tag{2.27}$$

This in fact is equivalent to lumping of the weight of the element equally at the nodes.

For the important case of centrifugal loading caused by rotation of the element about an axis perpendicular to the x-axis with an angular velocity ω (rad/s), the body force is

$$b_x(x) = \rho \omega^2 x. \tag{2.28}$$

Then,

$$\left\{\mathbf{f}^b\right\} = \int_{-1}^{+1} \mathbf{N}^T \left(\rho\omega^2 \sum_{i=1}^{2} N_i(\xi) x_i\right) \left(\sum_{i=1}^{2} N_i(\xi) A_i\right) \frac{\ell}{2} d\xi. \tag{2.29}$$

After evaluating the integral in (2.29), we get

$$\left\{\mathbf{f}^b\right\} \approx \rho\omega^2 \ell \frac{(x_1 + x_2)}{2} \frac{(A_1 + A_2)}{2} \begin{Bmatrix} 1/2 \\ 1/2 \end{Bmatrix}. \tag{2.30}$$

which is equivalent to lumping, equally at the nodes the average centrifugal force acting on the element.

2.3.7 Consistent nodal force vector: Initial strain

Using (2.19), the formula to compute the nodal force vector due to thermal expansion is

$$\left\{\mathbf{f}^{\text{init}}\right\} = \int_{-1}^{+1} \mathbf{B}^T E \left(\sum_{i=1}^{2} \alpha N_i(\xi) T_i\right) \left(\sum_{i=1}^{2} N_i(\xi) A_i\right) \frac{\ell}{2} d\xi. \tag{2.31}$$

After evaluating the integral in (2.31) we get

$$\left\{\mathbf{f}^{\mathrm{init}}\right\} \approx E\alpha \frac{(T_1 + T_2)}{2} \frac{(A_1 + A_2)}{2} \begin{Bmatrix} -1 \\ +1 \end{Bmatrix}. \tag{2.32}$$

It is now a standard practice to evaluate the integrals on the right hand sides of (2.22), (2.25), (2.29) and (2.31) accurately using numerical integration techniques. The Gaussian quadrature formula is ideally suited and hence exclusively used for this purpose. For the 1-D bar element a two-point formula with Gauss points located at $(\pm 1/\sqrt{3})$ is recommended.

2.4 Assembly Procedure

The summation of the element potential energy functionals

$$\pi_p = \sum_i \pi_p^{(i)} \tag{2.33}$$

where the summation goes over all elements in the assemblage, expresses the direct addition of the element stiffness matrices $\mathbf{K}^{(i)}$ to obtain the stiffness matrix of the element assemblage $\mathbf{K}$. In the same way, the nodal force vector of the element assemblage $\mathbf{F}$ is obtained by directly adding the element nodal force vectors $\mathbf{f}^{(i)}$. The process of assembling the element matrices by direct addition is called the direct stiffness method. Taking the connectivity of the elements into account, (2.33) can be written as

$$\pi_p(\mathbf{Q}) = \frac{1}{2}[\mathbf{Q}]^T[\mathbf{K}]\{\mathbf{Q}\} - \{\mathbf{Q}\}^T\{\mathbf{F}\}, \tag{2.34}$$

where $\mathbf{Q}$ is the vector of nodal degrees of freedom for the finite element assemblage.

In practice, the element matrices may first be calculated corresponding to element local degrees of freedom not aligned with the global degrees of freedom, in which case a transformation is necessary prior to assembly.

In addition, concentrated stiffnesses, masses, and loads corresponding to specific nodal degrees of freedom can also be added.

2.5 Boundary and Constraint Conditions

The equilibrium conditions are now satisfied by invoking the principle of minimum potential energy. Consequently, we minimize the potential energy functional π_p with respect to $\mathbf{Q}$ leading to the equilibrium equations

$$[\mathbf{K}]\{\mathbf{Q}\} = \{\mathbf{F}\}. \tag{2.35}$$

We now have to solve (2.35) subject to some constraints. Such constraints are usually of the type

$$Q_r = a \tag{2.36}$$

and

$$\beta_1 Q_r + \beta_2 Q_s = \beta_0. \tag{2.37}$$

That is, the nodal degree of freedom Q_r is specified to be equal to a in (2.36). Q_r is to be considered as an identified degree of freedom at a particular node, and hence (2.36) is called a single point constant (SPC). Equation (2.37) specifies a multi-point constraint (MPC), where β_1, β_2 and β_0 are known constants. The MPC is encountered in simulating inclined supports, rigid connections between nodes, shrink fit assemblies, etc. Two approaches are available to enforce these constraints—the elimination approach and the penalty approach [2.2]. The penalty approach is easy to implement in a finite element analysis program and retains its simplicity to enforce both SPC and MPC.

2.5.1 Single point constraint

A discrete spring with a large stiffness **C** is introduced into the finite element model. The magnitude of **C** is fixed subsequently. One end of the spring is attached to the node, whose degree of freedom Q_r is to be constrained and the other end is displaced by an amount a. The net extension of the spring is equal to $Q_r - a$. The strain energy stored in the spring is equal to

$$U_s = \frac{1}{2}\mathbf{C}(Q_r - a)^2.$$

This strain energy adds to the potential energy of the system. As a result, the modified potential energy functional is

$$\pi_{\text{mp}} = \frac{1}{2}\{\mathbf{Q}\}^T[\mathbf{K}]\{\mathbf{Q}\} + \frac{1}{2}\mathbf{C}(Q_r - a)^2 - \{\mathbf{Q}\}^T\{\mathbf{F}\}. \tag{2.38}$$

The minimization of π_{mp} is achieved by setting

$$\frac{\partial \pi_{\text{mp}}}{\partial\{\mathbf{Q}\}} = 0,$$

The resulting equilibrium equation is written as

$$\begin{bmatrix} K_{11} & K_{12} & \cdots & K_{1r} & \cdots & K_{1n} \\ K_{21} & K_{22} & \cdots & K_{2r} & \cdots & K_{2n} \\ \cdots & \cdots & \cdots & \cdots & \cdots & \cdots \\ \cdots & \cdots & \cdots & \cdots & \cdots & \cdots \\ K_{r1} & K_{r2} & \cdots & K_{rr}+C & \cdots & K_{r9} \\ \cdots & \cdots & \cdots & \cdots & \cdots & \cdots \\ \cdots & \cdots & \cdots & \cdots & \cdots & \cdots \\ K_{n1} & K_{n2} & \cdots & K_{nr} & \cdots & K_{nn} \end{bmatrix} \begin{bmatrix} Q_1 \\ Q_2 \\ - \\ - \\ Q_r \\ - \\ - \\ Q_n \end{bmatrix} = \begin{bmatrix} F_1 \\ F_2 \\ - \\ - \\ F_r + Ca \\ - \\ - \\ F_n \end{bmatrix}. \tag{2.39}$$

Therefore, the appropriate modifications to enforce the constraint $Q_r = a$ are that the spring stiffness **C** gets added to the diagonal element $\mathbf{K}_{rr}$ and **C** a gets added to F_r.

The reaction force at the node is equal to the force exerted by the spring on the structure and is equal to

$$R_r = -\mathbf{C}(Q_r - a). \tag{2.40}$$

It is to be noted that the penalty approach is an approximate method. The accuracy of the solution, the reaction force in particular, depends on the choice of **C**.

Let us consider the rth equation in (2.39)

$$\mathbf{K}_{r1}Q_1 + \mathbf{K}_{r2}Q_2 + s + (\mathbf{K}_{rr} + \mathbf{C})Q_r + s + \mathbf{K}_{rn}Q_n = \mathbf{F}_r + \mathbf{C}a.$$

Dividing by **C**, we get

$$\frac{\mathbf{K}_{r1}}{\mathbf{C}}Q_1 + \frac{\mathbf{K}_{r2}}{\mathbf{C}}Q_2 + \ldots + \left(\frac{\mathbf{K}_{rr}}{\mathbf{C}} + 1\right)Q_r + \ldots + \frac{\mathbf{K}_{rn}}{\mathbf{C}}Q_n = \frac{\mathbf{F}_r}{\mathbf{C}} + a.$$

From this equation we see that if **C** is chosen large enough, then $\mathbf{Q}_r \approx a$.

A widely used scheme for choosing the magnitude of **C** is

$$\mathbf{C} = \max|\mathbf{K}_{ij}|10^4 \qquad (1 \le i \le n, 1 \le j \le n).$$

The use of 10^4 is found to be satisfactory on most computers.

2.5.2 Multi-point constraint

The penalty approach will now be shown to apply to enforce the MPC equation (2.37).

Let us consider the modified potential energy functional

$$\pi_{\text{mp}} = \frac{1}{2}\{\mathbf{Q}\}^T[\mathbf{K}]\{\mathbf{Q}\} + \frac{1}{2}\mathbf{C}(\beta_1\mathbf{Q}_r + \beta_2\mathbf{Q}_s - \beta_0)^2 - \{\mathbf{Q}\}^T\{\mathbf{F}\}. \tag{2.41}$$

If **C** is large, π_{mp} takes on a minimum value only when $(\beta_1 Q_r + \beta_2 Qs - \beta_0)$ is very small, that is, when $\beta_1 Q_r + \beta_2 Q_s \approx \beta_0$, as desired. Setting $\dfrac{\partial \pi_{\text{mp}}}{\partial\{\mathbf{Q}\}} = 0$ yields the appropriate modifications to the stiffness matrix and nodal force vector.

These modifications are

$$\begin{bmatrix} \mathbf{K}_{rr} & \mathbf{K}_{rs} \\ \mathbf{K}_{sr} & \mathbf{K}_{ss} \end{bmatrix} \Rightarrow \begin{bmatrix} (\mathbf{K}_{rr} + \mathbf{C}\beta_1^2) & (\mathbf{K}_{rs} + \mathbf{C}\beta_1\beta_2) \\ (\mathbf{K}_{sr} + \mathbf{C}\beta_1\beta_2) & (\mathbf{K}_{ss} + \mathbf{C}\beta_2^2) \end{bmatrix} \tag{2.42}$$

and

$$\begin{Bmatrix} \mathbf{F}_r \\ \mathbf{F}_s \end{Bmatrix} \Rightarrow \begin{Bmatrix} \mathbf{F}_r + \mathbf{C}\beta_0\beta_1 \\ \mathbf{F}_s + \mathbf{C}\beta_0\beta_2 \end{Bmatrix}. \tag{2.43}$$

The reaction forces corresponding to the nodal degrees of freedom Q_r and Q_s are

$$R_r = -C\beta_1\left(\beta_1 Q_r + \beta_2 Q_s - \beta_0\right). \tag{2.44}$$

$$R_s = -C\beta_2 (\beta_1 Q_r + \beta_2 Q_s - \beta_0)\,. \tag{2.45}$$

2.6 Solution of Linear Simultaneous Equations

Consider a system of linear simultaneous equations written in matrix form as

$$[\mathbf{A}]\{\mathbf{X}\} = \{\mathbf{b}\}. \tag{2.46}$$

If the determinant of the matrix **A** is not equal to zero, we can pre-multiply both sides of (2.46) by $[\mathbf{A}]^{-1}$ to get the unique solution for $\{\mathbf{X}\}$ as

$$\{\mathbf{X}\} = [\mathbf{A}]^{-1}\{\mathbf{b}\}. \tag{2.47}$$

However, (2.47) is never used in finite element analysis programs since the arithmetic operation count for forming the inverse of **A** is prohibitive. Gaussian elimination or its variants are the most widely used algorithms to solve (2.46). These are known as direct methods. Incidentally, Gauss–Seidel and successive over relaxation are the most popular iterative methods for solving (2.46).

No other algorithm for solution of (2.46) involves fewer arithmetic operations count than Gaussian elimination, which is a two-step process. The first involves factorization and the second is back substitution.

2.6.1 Gaussian elimination algorithm

The system of equations as given by (2.46) can be restated as

$$\begin{bmatrix} a_{11} & a_{12} & a_{13} & \cdots & a_{1j} & \cdots & a_{1n} \\ a_{21} & a_{22} & a_{23} & \cdots & a_{2j} & \cdots & a_{2n} \\ a_{31} & a_{32} & a_{33} & \cdots & a_{3j} & \cdots & a_{3n} \\ \cdots & \cdots & \cdots & \cdots & \cdots & \cdots & \cdots \\ \cdots & \cdots & \cdots & \cdots & \cdots & \cdots & \cdots \\ a_{i1} & a_{i2} & a_{i3} & \cdots & a_{ij} & \cdots & a_{in} \\ \cdots & \cdots & \cdots & \cdots & \cdots & \cdots & \cdots \\ \cdots & \cdots & \cdots & \cdots & \cdots & \cdots & \cdots \\ a_{n1} & a_{n2} & a_{n3} & \cdots & a_{nj} & \cdots & a_{nn} \end{bmatrix} \begin{bmatrix} X_1 \\ X_2 \\ X_3 \\ - \\ - \\ X_i \\ - \\ - \\ X_n \end{bmatrix} = \begin{bmatrix} b_1 \\ b_2 \\ b_3 \\ - \\ - \\ b_i \\ - \\ - \\ b_n \end{bmatrix}. \tag{2.48}$$

Gaussian elimination is a systematic approach to successively eliminate variables $x_1, x_2, x_3, \ldots, x_i, \ldots, x_{n-1}$, until only variable X_n is left. This results in an upper triangular matrix with reduced coefficients and reduced right hand side. This process is called forward elimination. It is then possible to evaluate $X_n, X_{n-1}, \ldots, X_3, X_2, X_1$ successively by the process of back-substitution.

Let us start with step 1, in which the first equation is used to eliminate X_1 from the remaining equations.

The reduction process is

$$a_{ij}^{(1)} = a_{ij} - \frac{a_{i1}}{a_{11}} a_{1j} \qquad (i,j = 2,3,\ldots,n) \tag{2.49}$$

and

$$b_i^{(1)} = b_i - \frac{a_{i1}}{a_{11}} b_1 \qquad (i,j = 2,3,\ldots,n). \tag{2.50}$$

At the end of step 1, the result is

$$\begin{bmatrix} a_{11} & a_{12} & a_{13} & \ldots & a_{1j} & \ldots & a_{1n} \\ 0 & a_{22}^{(1)} & a_{23}^{(1)} & \ldots & a_{2j}^{(1)} & \ldots & a_{2n}^{(1)} \\ 0 & a_{32}^{(1)} & a_{33}^{(1)} & \ldots & a_{3j}^{(1)} & \ldots & a_{3n}^{(1)} \\ \ldots & \ldots & \ldots & \ldots & \ldots & \ldots & \ldots \\ \ldots & \ldots & \ldots & \ldots & \ldots & \ldots & \ldots \\ 0 & a_{i2}^{(1)} & a_{i3}^{(1)} & \ldots & a_{ij}^{(1)} & \ldots & a_{in}^{(1)} \\ \ldots & \ldots & \ldots & \ldots & \ldots & \ldots & \ldots \\ \ldots & \ldots & \ldots & \ldots & \ldots & \ldots & \ldots \\ 0 & a_{n2}^{(1)} & a_{n3}^{(1)} & \ldots & a_{nj}^{(1)} & \ldots & a_{nn}^{(1)} \end{bmatrix} \begin{bmatrix} X_1 \\ X_2 \\ X_3 \\ - \\ - \\ X_i \\ - \\ - \\ X_n \end{bmatrix} = \begin{bmatrix} b_1 \\ b_2^{(1)} \\ b_3^{(1)} \\ - \\ - \\ b_i^{(1)} \\ - \\ - \\ b_n^{(1)} \end{bmatrix}, \tag{2.51}$$

where the step number appears as a superscript set in parenthesis.

Let us now consider step k, for which the algorithm is

$$a_{ij}^{(k)} = a_{ij}^{(k-1)} - \frac{a_{ik}^{(k-1)}}{a_{kk}^{(k-1)}} a_{kj}^{(k-1)} \qquad (i,j = k+1, k+2, \ldots, n), \tag{2.52}$$

$$b_i^{(k)} = b_i^{(k-1)} - \frac{a_{ik}^{(k-1)}}{a_{kk}^{(k-1)}} b_k \qquad (i,j = k+1, k+2, \ldots, n). \tag{2.53}$$

After $(n-1)$ steps the result is

$$\begin{bmatrix} a_{11} & a_{12} & a_{13} & \ldots & a_{1j} & \ldots & a_{1n} \\ 0 & a_{22}^{(1)} & a_{23}^{(1)} & \ldots & a_{2j}^{(1)} & \ldots & a_{2n}^{(1)} \\ 0 & 0 & a_{33}^{(2)} & \ldots & a_{3j}^{(2)} & \ldots & a_{3n}^{(2)} \\ \ldots & \ldots & \ldots & \ldots & \ldots & \ldots & \ldots \\ \ldots & \ldots & \ldots & \ldots & \ldots & \ldots & \ldots \\ \ldots & \ldots & \ldots & \ldots & \ldots & \ldots & \ldots \\ \ldots & \ldots & \ldots & \ldots & \ldots & \ldots & \ldots \\ \ldots & \ldots & \ldots & \ldots & \ldots & \ldots & \ldots \\ 0 & 0 & 0 & \ldots & 0 & \ldots & a_{nn}^{(n-1)} \end{bmatrix} \begin{bmatrix} X_1 \\ X_2 \\ X_3 \\ - \\ - \\ X_i \\ - \\ - \\ X_n \end{bmatrix} = \begin{bmatrix} b_1 \\ b_2^{(1)} \\ b_3^{(2)} \\ - \\ - \\ b_i^{(1)} \\ - \\ - \\ b_n^{(n-1)} \end{bmatrix}. \tag{2.54}$$

The back substitution process determines $X_n, X_{n-1}, X_{n-2}, \ldots, X_1$ using (2.54).

It is to be noted that this algorithm can be applied to a solution of equilibrium equations, (2.35). However, one has to significantly specialize the above algorithm to exploit the special properties of **K**, namely, symmetry, bandedness, and sparsity. Also, multiple right hand vectors corresponding to multiple load cases need to be handled.

In such cases, it is useful to use Cholesky factorization where the coefficient matrix gets factored into a product of a lower and upper triangular matrix. The solution of global matrix equation, (2.35) for complex practical problems does present a challenge in that the matrix **K** can become so large that its direct solution becomes a challenging problem, especially in a PC environment. This in fact, is the motivation to develop and use the frontal technique and iterative methods. Practical applications of algorithms based on Gauss elimination in static condensation and multi-level sub-structuring (super elements) are noteworthy. See Bathe [2.3] for details.

2.7 2-D Bar Element

Finite element modelling and analysis of pin-jointed plane trusses is addressed in this section. A typical plane truss is shown in Fig. 2.5. The joints transmit forces only and the members are either in tension or compression. In such a case, it is required that all loads be applied only at the joints. Finite element modelling of trusses identifies joints as nodes and members as elements as illustrated in Fig. 2.5.

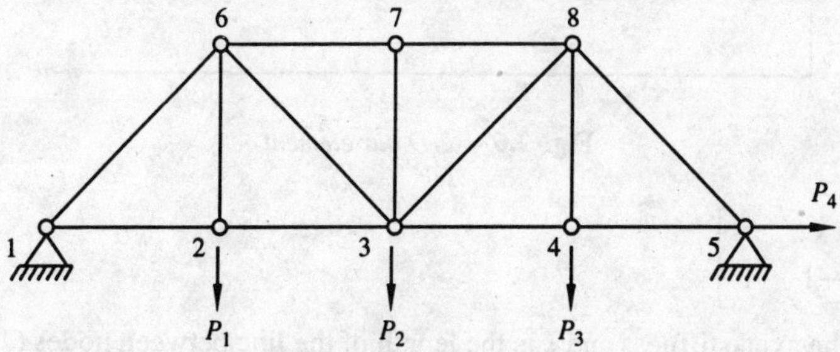

Fig. 2.5 *A plane truss*

The 2-D bar element (Fig. 2.6) is mathematically a line between two nodes whose Cartesian coordinates are (X_1, Y_1, 0) and (X_2, Y_2, 0). Each node i now carries two degrees of freedom denoted by u_i and v_i. The nodal degrees of freedom vector is

$$\{\mathbf{q}\}^T = (u_1, v_1, u_2, v_2). \tag{2.55}$$

The corresponding vector of nodal forces is

$$\{\mathbf{f}\}^T = (F_{x1}, F_{y1}, F_{x2}, F_{y2}). \tag{2.56}$$

The element equations are

$$[\mathbf{K}]\{\mathbf{q}\} = \{\mathbf{f}\} = \{\mathbf{f}^b\} + \{\mathbf{f}^{\text{init}}\} + \{\mathbf{f}^c\} \tag{2.57}$$

where **K** is the element stiffness matrix.

In order to compute the element matrices, it is convenient to define the local coordinate system $(\bar{x}, \bar{u})$ as shown in Fig. 2.6. In this new coordinate system, the 2-D bar element is nothing but the 1-D bar whose nodal degrees of freedom vector and stiffness matrix are:

$$\{\bar{\mathbf{q}}\}^T = (\bar{u}_1, \bar{u}_2). \tag{2.58}$$

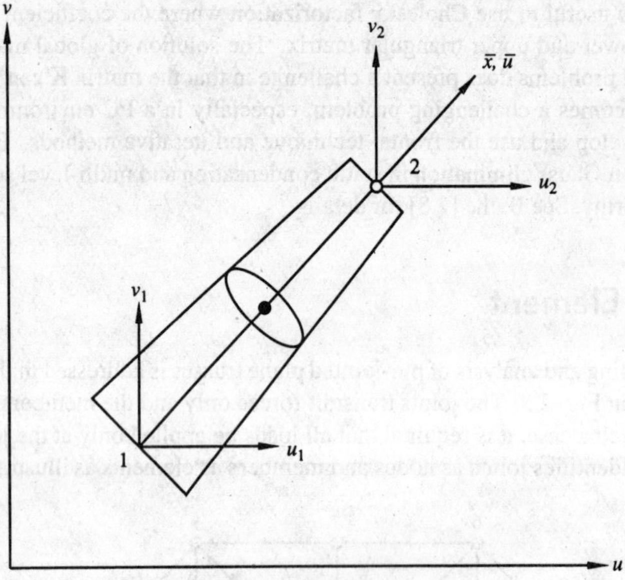

Fig. 2.6 *2-D bar element*

$$[\bar{\mathbf{K}}] = \frac{EA}{\ell}\begin{bmatrix} 1 & -1 \\ -1 & 1 \end{bmatrix}. \tag{2.59}$$

where EA denotes the axial stiffness and ℓ is the length of the line between nodes (1) and (2).

The strain energy (SE) stored in the element is evaluated as

$$SE = \frac{1}{2}\{\bar{\mathbf{q}}\}^T[\bar{\mathbf{K}}]\{\bar{\mathbf{q}}\}. \tag{2.60}$$

The relation between $\{\bar{\mathbf{q}}\}$ and $\{\mathbf{q}\}$ is given by the local–global transformation

$$\{\bar{\mathbf{q}}\} = [\mathbf{L}]\{\mathbf{q}\}, \tag{2.61}$$

where the transformation matrix $\mathbf{L}$ is given by

$$[\mathbf{L}] = \begin{bmatrix} \cos(\bar{x}, x) & \cos(\bar{x}, y) & 0 & 0 \\ 0 & 0 & \cos(\bar{x}, x) & \cos(\bar{x}, y) \end{bmatrix}. \tag{2.62}$$

In (2.62), $\cos(\bar{x}, x)$ and $\cos(\bar{x}, y)$ are the direction cosines of the $\bar{x}$-axis

$$\cos(\bar{x}, x) = \frac{(X_2 - X_1)}{\ell}, \tag{2.63}$$

$$\cos(\bar{x}, y) = \frac{(Y_2 - Y_1)}{\ell}, \tag{2.64}$$

$$\text{and} \quad \ell = \sqrt{(X_2 - X_1)^2 + (Y_2 - Y_1)^2}. \tag{2.65}$$

Substituting (2.61) into (2.60), we get

$$SE = \frac{1}{2}\{\mathbf{q}\}^T[\mathbf{K}]\{\mathbf{q}\}, \tag{2.66}$$

where

$$[\mathbf{K}] = [\mathbf{L}]^T[\bar{\mathbf{K}}][\mathbf{L}]. \tag{2.67}$$

Equation (2.67) gives the stiffness matrix for the 2-D bar element.

Similarly the nodal force vector in the local coordinate system $\{\bar{\mathbf{f}}\}^T = (F_{\bar{x}1}, F_{\bar{x}2})$ is transformed to the global coordinates using

$$\{\mathbf{f}\} = [\mathbf{L}]^T\{\bar{\mathbf{f}}\}. \tag{2.68}$$

For example, the consistent nodal force vector due to thermal strain in the local coordinate system $(\bar{\mathbf{x}}, \bar{\mathbf{u}})$ is

$$\{\bar{\mathbf{f}}^{\text{init}}\} = E\alpha\frac{(T_1+T_2)}{2}\frac{(A_1+A_2)}{2}\begin{Bmatrix}-1\\+1\end{Bmatrix}, \tag{2.69}$$

where E is the Young's modulus of elasticity, α is the coefficient of thermal expansion, A_1 and A_2 are the cross-sectional areas at nodes 1 and 2, and T_1 and T_2 denote temperature changes at nodes 1 and 2. Combining (2.69) and (2.68) and performing the indicated matrix multiplication, we get the consistent nodal force vector due to thermal strain $\mathbf{f}^{\text{init}}$.

$$\{\mathbf{f}^{\text{init}}\} = E\alpha\frac{(T_1+T_2)}{2}\frac{(A_1+A_2)}{2}\begin{Bmatrix}-\cos(\bar{x},x)\\-\cos(\bar{x},y)\\+\cos(\bar{x},x)\\+\cos(\bar{x},y)\end{Bmatrix}. \tag{2.70}$$

The element matrices can be assembled in the usual manner since the nodal displacements are referred to a common global Cartesian coordinate system.

It is important to note that element strains and stresses have to be recovered in the local system only. For example, the axial stress $\sigma_{\bar{x}\bar{x}}$ is therefore evaluated using

$$\sigma_{\bar{x}\bar{x}} = E[\mathbf{B}][\mathbf{L}]\{\mathbf{q}\}, \tag{2.71}$$

$$\text{where} \quad [\mathbf{B}] = \begin{bmatrix}\frac{-1}{\ell} & \frac{+1}{\ell}\end{bmatrix}. \tag{2.72}$$

Note that a positive stress implies that the member is in tension and a negative value implies compression.

2.8 3-D Bar Element

Finite element modelling and analysis of space trusses subjected to joint forces, body forces, and thermal loads is addressed in this section. A typical space truss is shown in Fig. 2.7. Finite element modelling of such a truss identifies each joint as a node and each member as an element.

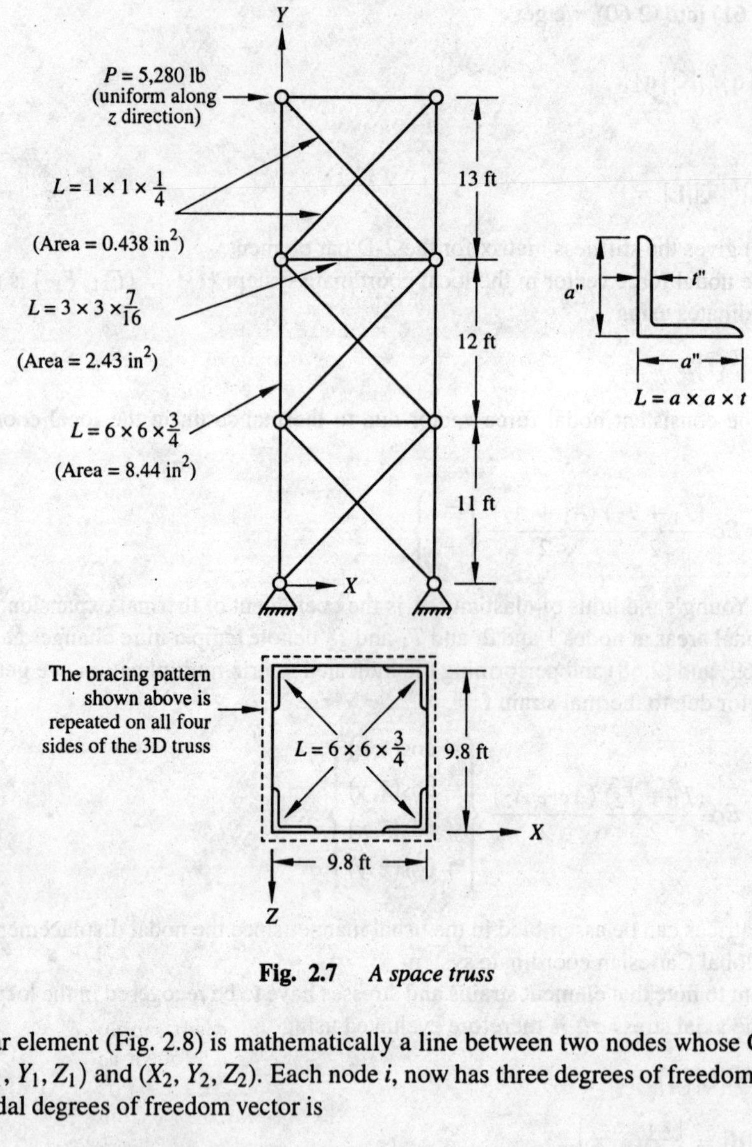

Fig. 2.7 *A space truss*

The 3-D bar element (Fig. 2.8) is mathematically a line between two nodes whose Cartesian coordinates are (X_1, Y_1, Z_1) and (X_2, Y_2, Z_2). Each node i, now has three degrees of freedom denoted by u_i, v_i, w_i. The nodal degrees of freedom vector is

$$\{\mathbf{q}\}^T = (u_1, v_1, w_1, u_2, v_2, w_2). \tag{2.73}$$

The corresponding vector of nodal forces is

$$\{\mathbf{f}\}^T = (F_{x1}, F_{y1}, F_{z1}, F_{x2}, F_{y2}, F_{z2}). \tag{2.74}$$

The element equations are

$$[\mathbf{K}]\{\mathbf{q}\} = \{\mathbf{f}\} = \{\mathbf{f}^c\} + \{\mathbf{f}^b\} + \{\mathbf{f}^{\text{init}}\}. \tag{2.75}$$

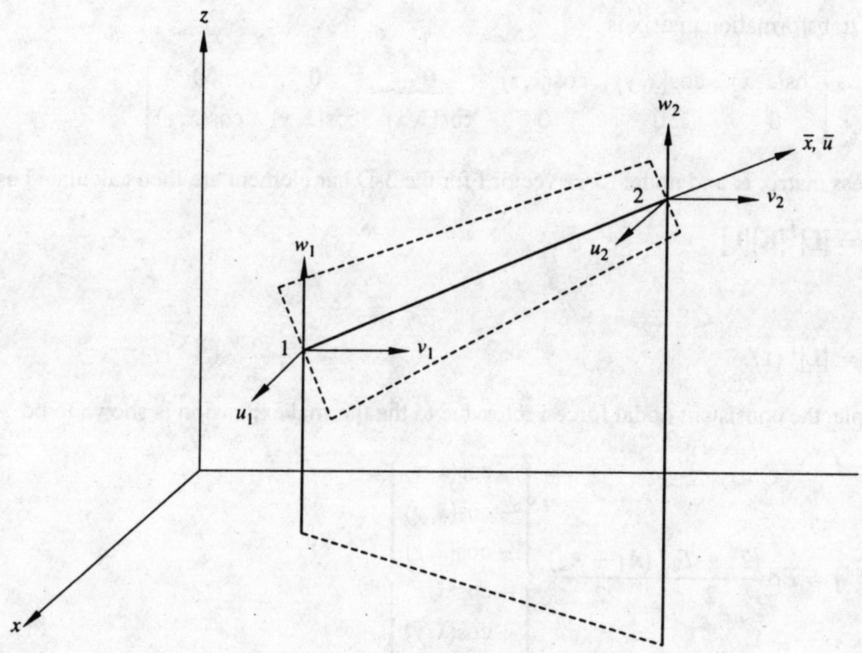

Fig. 2.8 *3-D bar element*

In order to compute the 3-D bar element matrices, it is useful to define a local coordinates system $(\bar{x}, \bar{u})$ as shown in Fig. 2.8. The direction cosines of the $\bar{x}$-axis are

$$\cos(\bar{x}, x) = \frac{(X_2 - X_1)}{\ell}, \tag{2.76}$$

$$\cos(\bar{x}, y) = \frac{(Y_2 - Y_1)}{\ell}, \tag{2.77}$$

$$\cos(\bar{x}, z) = \frac{(Z_2 - Z_1)}{\ell}, \tag{2.78}$$

$$\text{where} \quad \ell = \sqrt{(X_2 - X_1)^2 + (Y_2 - Y_1)^2 + (Z_2 - Z_1)^2}. \tag{2.79}$$

In the local coordinate system $(\bar{x}, \bar{u})$, the 3-D bar element is nothing but the 1-D bar element whose nodal degrees of freedom vector $\{\bar{\mathbf{q}}\}$, nodal force vector $\{\bar{\mathbf{f}}\}$, and stiffness matrix $[\bar{\mathbf{K}}]$ are

$$\{\bar{\mathbf{q}}\}^T = (\bar{u}_1, \bar{u}_2), \tag{2.80}$$

$$\{\bar{\mathbf{f}}\}^T = (F_{\bar{x}1}, F_{\bar{x}2}), \tag{2.81}$$

$$[\bar{\mathbf{K}}] = \frac{EA}{\ell}, \begin{bmatrix} 1 & -1 \\ -1 & 1 \end{bmatrix}, \tag{2.82}$$

where EA denotes the axial stiffness of the bar and ℓ is the length of the line between nodes (1) and (2). Transformation to the global Cartesian coordinate system (x, y, z) gives

$$\{\bar{\mathbf{q}}\}^T = [\mathbf{L}]\{\mathbf{q}\}, \tag{2.83}$$

where the transformation matrix is

$$[\mathbf{L}] = \begin{bmatrix} \cos(\bar{x},x) & \cos(\bar{x},y) & \cos(\bar{x},z) & 0 & 0 & 0 \\ 0 & 0 & 0 & \cos(\bar{x},x) & \cos(\bar{x},y) & \cos(\bar{x},z) \end{bmatrix}. \tag{2.84}$$

The stiffness matrix **K** and nodal force vector **f** for the 3-D bar element are then calculated using

$$[\mathbf{K}] = [\mathbf{L}]^T[\bar{\mathbf{K}}][\mathbf{L}] \tag{2.85}$$

and

$$\{\mathbf{f}\} = [\mathbf{L}]^T\{\bar{\mathbf{f}}\}. \tag{2.86}$$

For example, the consistent nodal force vector due to the thermal expansion is shown to be

$$\{\mathbf{f}^{\text{init}}\} = E\alpha\frac{(T_1+T_2)}{2}\frac{(A_1+A_2)}{2}\begin{Bmatrix} -\cos(\bar{x},x) \\ -\cos(\bar{x},y) \\ -\cos(\bar{x},z) \\ +\cos(\bar{x},x) \\ +\cos(\bar{x},y) \\ +\cos(\bar{x},z) \end{Bmatrix}, \tag{2.87}$$

where E is the Young's modulus, α is the coefficient of thermal expansion, A_1 and A_2 are the cross-sectional areas at nodes 1 and 2, and T_1 and T_2 denote temperature changes at nodes 1 and 2. The element matrices and vectors can be assembled in the usual manner. The assembled matrix equation can then be solved after enforcing support conditions (to suppress all possible rigid body displacements), symmetry and constraint conditions (if any) to determine the joint deflections. The member strains and stresses to be determined demand that for each element the strains and stress are to be recovered in the local coordinate system. For example, the axial stress $\sigma_{\bar{x}\bar{x}}$ is evaluated using

$$\sigma_{\bar{x}\bar{x}} = E[\mathbf{B}][\mathbf{L}]\{\mathbf{q}\}, \tag{2.88}$$

where $[\mathbf{B}] = \begin{bmatrix} \frac{-1}{\ell} & \frac{+1}{\ell} \end{bmatrix}$.

Note that a positive stress implies that the member is in tension and a negative stress implies compression.

Engineering students perform linear static analysis of plane/space trusses using the method of joints and the method of sections. These methods become tedious to apply for complex, statistically indeterminate, space trusses. The finite element method described here can handle statically determinate and indeterminate trusses alike. In principle, every member can be made out of a different material and have uniform or varying cross-section. Joint defections, member forces, support reactions, are all calculated at one stretch. Initial strains induced by forcing members into place that are either too long or too short can be routinely taken into account. Elastic supports, inclined supports, can all be routinely handled. Finally, pultruded fibre reinforced plastic tubular members widely used in spacecraft construction do not pose any additional complexity.

2.9 Test Problems

Students are advised to solve these problems using progressively refined meshes and to compare the converged solution with the known analytical solutions.

***Problem* 1** A prismatic bar subjected to an axial load $P = 60 \times 10^3$ N is shown in Fig. 2.9. Determine the displacement and stress distribution, and support reactions. Assume Young's modulus $E = 20 \times 10^3$ N/mm^2.

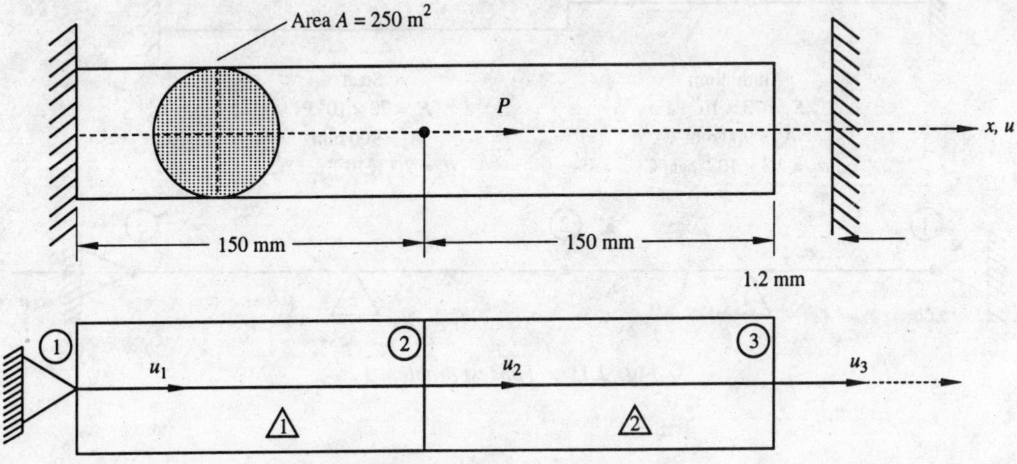

Fig. 2.9 *For test problem 1*

***Problem* 2** Figure 2.10 shows a prismatic bar rotating at a constant angular velocity $\omega = 30$ (rad/s). Determine the deformation, stress distribution and support reaction. Assume $L = 300$ mm, area = 250 mm^2, the material to be steel.

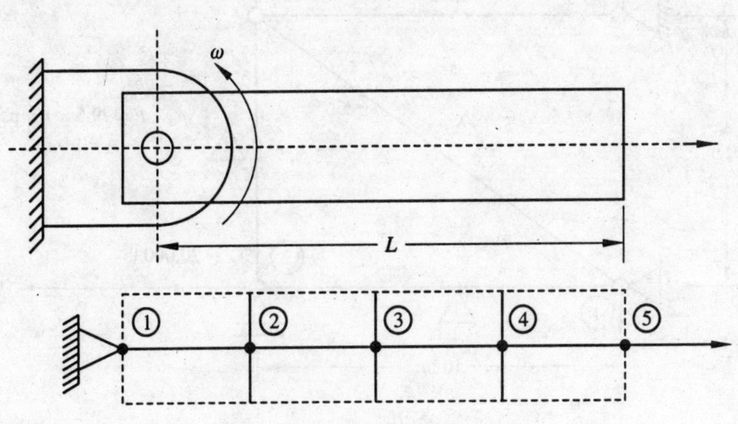

Fig. 2.10 *For test problem 2*

***Problem* 3** An axial load $P = 300 \times 10^3$ N is applied at 20°C to the rod as shown in Fig. 2.11. The temperature is then raised to 60°C. Determine the displacements, stresses and support reactions.

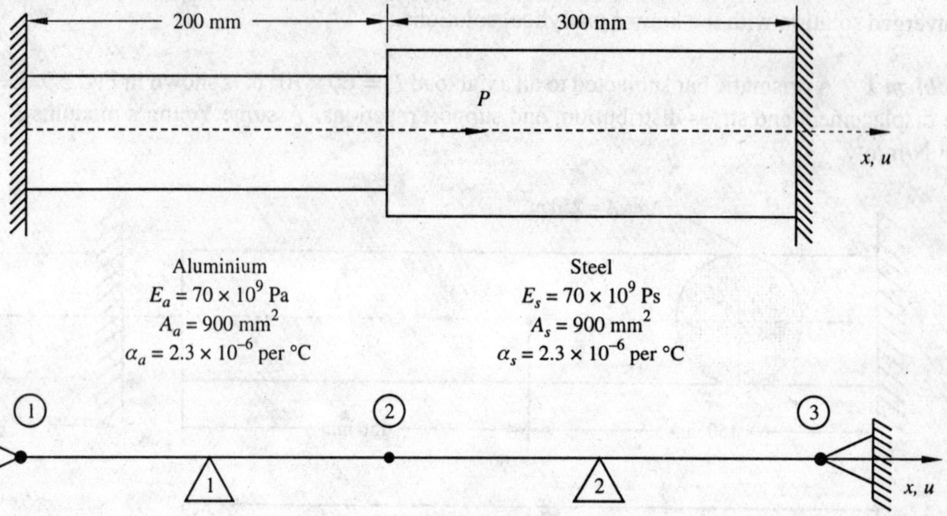

Fig. 2.11 *For test problem 3*

***Problem* 4** A four-member plane truss is shown in Fig. 2.12. Assuming $E = 29.5 \times 10^6$ psi and $A = 1.0$ in^2 for all members, determine joint displacements, member stresses, and support reactions.

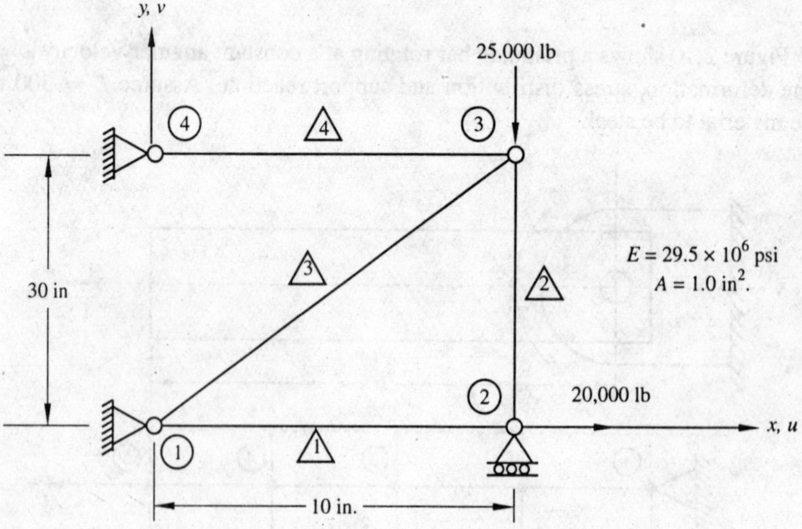

Fig. 2.12 *For test problem 4*

***Problem* 5** For the space truss shown in Fig. 2.7, all the members are made out of structural steel. Determine joint displacements, member stresses, and support reactions.

2.10 Applications

NASA has identified the need for large deployable plane and space trusses, whose members will be constructed of lightweight materials (pultruded carbon fibre reinforced plastic tubes) and will contain a large number of spherical joints. This need is evident for programs such as the space station. These deployable and retractable space structures may have precision shape requirements and need for vibration suppression during in-orbit operation.

One such space structure is the mast flight system, which is shown in Fig. 2.13. The basic element in the system is a 60.7 meter-long truss beam, which is attached to the shuttle orbiter. Included at the tip of the space truss are the primary actuators and collocated sensors. A deployment/retractor subsystem, which also secures the stowed beam package during launch and landing is provided. The system uses a large motor to move the structure. The need for application of finite element method in general and the 3-D bar elements in particular for thermal distortion and vibration analysis of such complex space trusses is therefore evident.

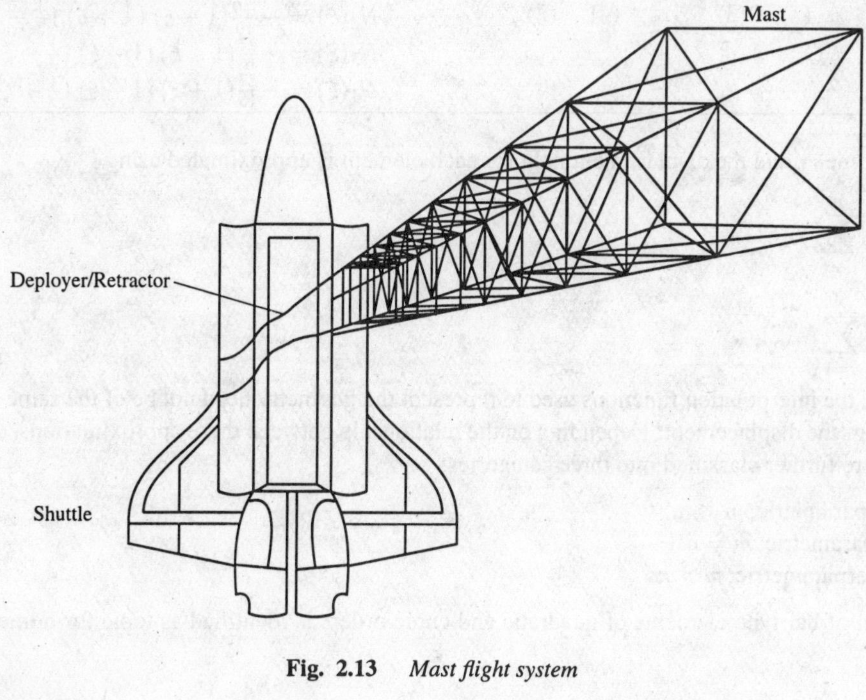

Fig. 2.13 *Mast flight system*

2.11 Higher Order Elements

The bar elements presented here are linear in order and are the simplest one-dimensional elements. This simplicity often demands a fine mesh to achieve a desired accuracy of the numerical solution. In order to achieve faster convergence with mesh refinement, improved accuracy for a given mesh and computational efficiency, the development of higher order elements is desirable.

A family of bar type elements of increasing orders can be developed on a pattern indicated in Table 2.1. The number of nodes in each element coincides with the number of polynomial terms employed in the approximation. A complete polynomial of linear, quadratic, and cubic order is used. Explicit shape functions for these elements are also given in Table 2.1. The reader is advised to create these functions, and produce plots of their variation over each element domain.

Table 2.1 *Shape functions for a family of 1-D finite elements*

Order	Parent element	Shape functions
Linear	−1 (1) —— 0 —— +1 (2) → ξ	$N_1(\xi) = \frac{1}{2}(1-\xi)$ $N_2(\xi) = \frac{1}{2}(1+\xi)$
Quadratic	−1 (1) —— 0 (3) —— +1 (2) → ξ	$N_1(\xi) = -\frac{1}{2}\xi(1-\xi)$ $N_2(\xi) = \frac{1}{2}\xi(1+\xi)$ $N_3(\xi) = (1+\xi)(1-\xi)$
Cubic	−1 (1) — (3) — 0 — (4) — +1 (2) → ξ	$N_1(\xi) = -\frac{-9}{16}(1-\xi)\left(\frac{1}{3}+\xi\right)\left(\frac{1}{3}-\xi\right)$ $N_2(\xi) = -\frac{-9}{16}(1+\xi)\left(\frac{1}{3}+\xi\right)\left(\frac{1}{3}-\xi\right)$ $N_3(\xi) = -\frac{27}{16}(1+\xi)(1-\xi)\left(\frac{1}{3}-\xi\right)$ $N_4(\xi) = -\frac{27}{16}(1+\xi)(1-\xi)\left(\frac{1}{3}+\xi\right)$

The geometry and the displacement field for each element is approximated using

$$\mathbf{X} = \sum_{i=1}^{m} N_i(\xi)X_i.$$

$$\mathbf{u} = \sum_{i=1}^{n} N_i(\xi)u_i.$$

In general, the interpolation functions used to represent the geometry need not be of the same order as that used for the displacement. Depending on the relationship between these approximations, the finite elements are further classified into three categories:

1. Subparametric; $m < n$.
2. Isoparametric; $m = n$.
3. Superparametric; $m > n$.

The design of bar type elements of quadratic and cubic orders is identified as topic for home assignment.

2.12 Closure

In this chapter, the design of one-dimensional elements and their application to solid mechanics problems has been presented. A step-by-step procedure for finite element modelling and analysis of bars and trusses has also been given. The test problems when solved using a commercial finite element analysis program, will help the reader in establishing a feeling of comfort with the finite element procedure for engineering analysis. The skills required to use a commercial FEM system can be acquired by the reader by solving these problems with progressively refining the finite element models. The design of one-dimensional elements for heat transfer analysis and fluid flow analysis is identified as a topic for advanced study. The reader is advised to refer Baker and Pepper [2.4] and Reddy [2.5] for details.

2.13 References

[2.1] Dym, C.L. and Shames, I.H. *Solid Mechanics: A Variational Approach*. McGraw-Hill, New York, 1973.

[2.2] Chandrupatla, T.R. and Belegundu, A.D. *Introduction to Finite Elements in Engineering*. Prentice Hall of India, New Delhi, 2nd ed., 2001.

[2.3] Bathe, K.J. *Finite Element Procedures*. Prentice Hall of India, New Delhi, 1996.

[2.4] Baker, A.J. and Pepper, D.W. *Finite Elements 1-2-3*. McGraw Hill, New York, 1991.

[2.5] Reddy, J.N. *An Introduction to the Finite element Method*. McGraw Hill, New York, 2nd ed., 1993.

3 Two-Dimensional Elements
Analysis of Plane Elasticity Problems

3.1 Overview

In many branches of engineering, the solution of stress and strain distributions in elastic continua is required. Special cases of such problems are two-dimensional plane stress or plane strain distributions. A state of plane stress exists in a thin plate under in-plane loading only. By definition, stresses normal to the plane are zero. In a plane strain state, the strain normal to the plane is assumed to be zero. Theory of elasticity [3.1] provides the basic equations for the analysis of plane stress and plane strain problems. These equations are presented and used to construct the potential energy functional.

A library of two-dimensional finite elements of different shapes (triangular and quadrilateral) and increasing orders (linear, quadratic, cubic etc.) are available. Among these only linear elements, namely, the three-noded triangular element (TRIA 3) and the four-noded quadrilateral element (QUAD 4) are presented in detail.

Users of a commercial FEM system have to validate the elements, meshes, constraints and solution procedures they wish to employ in practice using benchmarks. Keeping this in mind, a set of standard test problems along with available solutions for comparison is also presented.

Accurate finite element solutions to some practical problems are included to illustrate the scale and complexities. The chapter closes with a brief description of higher order elements and identification of topics for advanced study.

3.2 Three-Noded Triangular Element (TRIA 3)

Figure 3.1 shows the TRIA 3 element, with nodes 1, 2, 3, numbered in an anti-clockwise order. The nodal degrees of freedom are listed as a vector.

$$\{\mathbf{q}\}^T = (u_1, v_1, u_2, v_2, u_3, v_3). \tag{3.1}$$

The corresponding nodal forces are listed as

$$\{\mathbf{f}\}^T = (F_{x1}, F_{y1}, F_{x2}, F_{y2}, F_{x3}, F_{y3}). \tag{3.2}$$

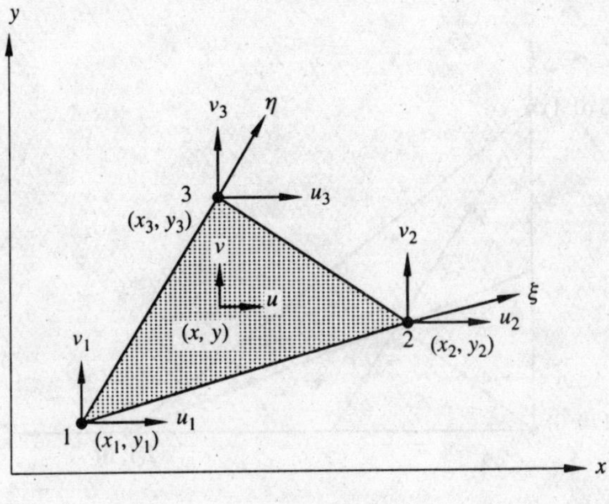

Fig. 3.1 *Three-noded triangular finite element (TRIA 3)*

3.2.1 Admissible displacement functions

The displacements u, v at any point (x, y) within the element have to be uniquely defined by nodal displacements. The simplest representation is given by

$$\{\mathbf{u}\} = [\mathbf{N}]\{\mathbf{q}\}, \tag{3.3}$$

where

$$\{\mathbf{u}\}^T = (u, v)$$

and

$$[\mathbf{N}] = \begin{bmatrix} N_1 & 0 & N_2 & 0 & N_3 & 0 \\ 0 & N_1 & 0 & N_2 & 0 & N_3 \end{bmatrix}. \tag{3.4}$$

In order to derive explicit expressions for the shape functions N_i $(i = 1, 2, 3)$ in the shape function matrix **N**, it is now a standard practice to work in the parametric space (ξ, η) and use the concept of a parent element as shown in Fig. 3.2. The parent element is a right-angled isosceles triangle and the natural coordinates (ξ, η) are non-dimensional. Any variable Φ within the parent element has to be uniquely defined by its nodal values Φ_1, Φ_2 and Φ_3. The simplest representation is given by the linear polynomial

$$\Phi = \alpha_1 + \alpha_2\xi + \alpha_3\eta. \tag{3.5}$$

Since $\Phi = \Phi_1$ at node 1 $(\xi = 0, \eta = 0)$, $\Phi = \Phi_2$ at node 2 $(\xi = 1, \eta = 0)$ and $\Phi = \Phi_3$ at node 3 $(\xi = 0, \eta = 1)$, (3.5) gives

$$\alpha_1 = \Phi_1,$$

$$\alpha_1 + \alpha_2 = \Phi_2,$$

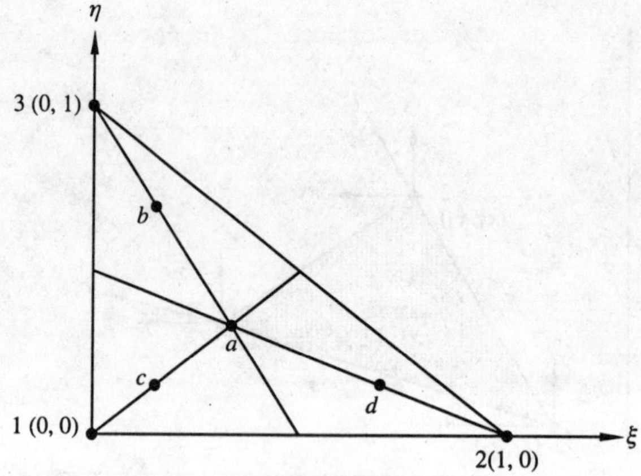

Fig. 3.2 *Parent element and natural coordinates for TRIA 3*

$$\alpha_1 + \alpha_3 = \Phi_3.$$

The polynomial coefficients are evaluated by solving the above equations as $\alpha_1 = \Phi_1$, $\alpha_2 = \Phi_2 - \Phi_1$ and $\alpha_3 = \Phi_3 - \Phi_1$. Substituting these values in (3.5), we can write

$$\Phi = N_1\Phi_1 + N_2\Phi_2 + N_3\Phi_3, \tag{3.6}$$

where

$$N_1 = 1 - \xi - \eta,$$

$$N_2 = \xi,$$

$$N_3 = \eta. \tag{3.7}$$

It can be verified that $N_i = 1$ at node i and zero at the other two nodes while $N_1 + N_2 + N_3 = 1$. Let us now map, uniquely, the parent element (Fig. 3.2) into the TRIA 3 element (Fig. 3.1). The required relationship between Cartesian and natural coordinates to accomplish this is

$$\begin{Bmatrix} X \\ Y \end{Bmatrix} = \sum_{i=1}^{3} N_i(\xi, \eta) \begin{Bmatrix} X_i \\ Y_i \end{Bmatrix}, \tag{3.8}$$

where (X_i, Y_i) denotes Cartesian coordinates of node i.

With the geometry of the TRIA 3 element now defined by the parametric equation (3.8), the admissible displacement functions are written as

$$\begin{Bmatrix} u \\ v \end{Bmatrix} = \sum_{i=0}^{3} N_i(\xi, \eta) \begin{Bmatrix} u_i \\ v_i \end{Bmatrix}. \tag{3.9}$$

Since the same order interpolation functions are used to define the geometry as well as to approximate the displacement field, the formulation will be called "isoparametric".

The chosen displacement functions guarantee inter-element displacement continuity (compatibility) with adjacent elements.

3.2.2 Element strain matrix

The state of strain at any point within the TRIA 3 element is completely characterized by the three-strain components $\varepsilon_{xx}, \varepsilon_{yy}$ and γ_{xy}. These are determined using the well-known strain displacement relations [3.1]. Assuming small deformations, and using matrix notation, the strain–displacement relations can be written as

$$\{\boldsymbol{\varepsilon}\} = \begin{Bmatrix} \varepsilon_{xx} \\ \varepsilon_{yy} \\ \gamma_{xy} \end{Bmatrix} = \begin{bmatrix} \dfrac{\partial}{\partial x} & 0 \\ 0 & \dfrac{\partial}{\partial y} \\ \dfrac{\partial}{\partial y} & \dfrac{\partial}{\partial x} \end{bmatrix} \begin{Bmatrix} u \\ v \end{Bmatrix}$$

or

$$\{\boldsymbol{\varepsilon}\} = [\mathbf{L}]\{\mathbf{u}\}. \tag{3.10}$$

Substituting (3.9) we have

$$\{\boldsymbol{\varepsilon}\} = [\mathbf{B}]\{\mathbf{q}\}, \tag{3.11}$$

where the element strain matrix is

$$[\mathbf{B}] = \begin{bmatrix} \dfrac{\partial N_1}{\partial x} & 0 & \dfrac{\partial N_2}{\partial x} & 0 & \dfrac{\partial N_3}{\partial x} & 0 \\ 0 & \dfrac{\partial N_1}{\partial y} & 0 & \dfrac{\partial N_2}{\partial y} & 0 & \dfrac{\partial N_3}{\partial y} \\ \dfrac{\partial N_1}{\partial y} & \dfrac{\partial N_1}{\partial x} & \dfrac{\partial N_2}{\partial y} & \dfrac{\partial N_2}{\partial x} & \dfrac{\partial N_3}{\partial y} & \dfrac{\partial N_3}{\partial x} \end{bmatrix}. \tag{3.12}$$

To evaluate matrix **B**, we need the Cartesian derivatives of the shape functions N_i, which are defined in terms of the natural coordinates (ξ, η). By the usual rules of partial differentiation, we can write

$$\begin{aligned} \frac{\partial N_i}{\partial \xi} &= \frac{\partial N_i}{\partial x}\frac{\partial x}{\partial \xi} + \frac{\partial N_i}{\partial y}\frac{\partial y}{\partial \xi}, \\ \frac{\partial N_i}{\partial \eta} &= \frac{\partial N_i}{\partial x}\frac{\partial x}{\partial \eta} + \frac{\partial N_i}{\partial y}\frac{\partial y}{\partial \eta}, \end{aligned} \tag{3.13}$$

which in matrix notation is written as

$$\begin{bmatrix} \dfrac{\partial N_i}{\partial \xi} \\ \dfrac{\partial N_i}{\partial \eta} \end{bmatrix} = \begin{bmatrix} \dfrac{\partial x}{\partial \xi} & \dfrac{\partial y}{\partial \xi} \\ \dfrac{\partial x}{\partial \eta} & \dfrac{\partial y}{\partial \eta} \end{bmatrix} \begin{bmatrix} \dfrac{\partial N_i}{\partial x} \\ \dfrac{\partial N_i}{\partial y} \end{bmatrix} = [\mathbf{J}] \begin{bmatrix} \dfrac{\partial N_i}{\partial x} \\ \dfrac{\partial N_i}{\partial y} \end{bmatrix}. \tag{3.14}$$

The matrix **J** known as the Jacobian matrix is evaluated using

$$[\mathbf{J}] = \sum_{j=1}^{3} \begin{bmatrix} \frac{\partial N_j}{\partial \xi} x_j & \frac{\partial N_j}{\partial \xi} y_j \\ \frac{\partial N_j}{\partial \eta} x_j & \frac{\partial N_j}{\partial \eta} y_j \end{bmatrix} = \begin{bmatrix} J_{11} & J_{12} \\ J_{21} & J_{22} \end{bmatrix}. \tag{3.15}$$

The Cartesian derivatives of shape functions are now obtained from Eqn. (3.14) as

$$\begin{bmatrix} \frac{\partial N_i}{\partial x} \\ \frac{\partial N_i}{\partial y} \end{bmatrix} = [\mathbf{J}]^{-1} \begin{bmatrix} \frac{\partial N_i}{\partial \xi} \\ \frac{\partial N_i}{\partial \eta} \end{bmatrix} \quad (i = 1, 2, 3), \tag{3.16}$$

where

$$[\mathbf{J}]^{-1} = \frac{1}{|\mathbf{J}|} \begin{bmatrix} J_{22} & -J_{12} \\ -J_{21} & J_{11} \end{bmatrix}. \tag{3.17}$$

It is essential that the determinant of the Jacobian matrix $|\mathbf{J}| = (J_{11}J_{12} - J_{21}J_{22})$ cannot be zero. An infinitesimal area element within the TRIA 3 element is

$$dxdy = |\mathbf{J}|\, d\xi d\eta. \tag{3.18}$$

Physically, $|\mathbf{J}|$ is twice the area of the TRIA 3 element. It will be noted that each nonzero element of the matrix **B** is a constant and hence independent of the position within the TRIA 3 element. Therefore this element is popularly known as the constant strain triangular (CST) element in literature. Obviously, the convergence criterion of recovering a state of constant strain is satisfied by the chosen displacement functions.

Initial strains, that is strains which are independent of stresses, most frequently due to temperature changes, will, in general, result in an initial strain vector

$$\left\{\boldsymbol{\varepsilon}^{\text{init}}\right\} = \begin{Bmatrix} \varepsilon_{xx}^{\text{init}} \\ \varepsilon_{yy}^{\text{init}} \\ \gamma_{xy}^{\text{init}} \end{Bmatrix}. \tag{3.19}$$

For the case of plane stress, and for a homogeneous isotropic linear elastic material model, the thermal strain vector is

$$\left\{\boldsymbol{\varepsilon}^{\text{init}}\right\} = \begin{Bmatrix} \alpha T \\ \alpha T \\ 0 \end{Bmatrix} = \begin{Bmatrix} \alpha \sum_{i=1}^{3} N_i T_i \\ \alpha \sum_{i=1}^{3} N_i T_i \\ 0 \end{Bmatrix}. \tag{3.20}$$

In Eqn. (3.20), T_i denotes temperature change at node i and α is the coefficient of thermal expansion, a material property.

In plane strain, the thermal strain vector is

$$\left\{\boldsymbol{\varepsilon}^{\text{ init}}\right\} = \begin{Bmatrix} (1+\nu)\,\alpha T \\ (1+\nu)\,\alpha T \\ 0 \end{Bmatrix} = \begin{Bmatrix} (1+\nu)\,\alpha \sum_{i=1}^{3} N_i T_i \\ (1+\nu)\,\alpha \sum_{i=1}^{3} N_i T_i \\ 0 \end{Bmatrix}, \tag{3.21}$$

where ν is the Poisson's ratio, a material property:

3.2.3 Element stress recovery

The state of stress at any point within the element can be defined by the three components $\{\sigma\}^T = (\sigma_{xx}, \sigma_{yy}, \tau_{xy})$. These stresses are related to the strains through the constitutive law:

$$\{\sigma\} = [\mathbf{D}](\{\boldsymbol{\varepsilon}\} - \{\boldsymbol{\varepsilon}^{\text{ init}}\}). \tag{3.22}$$

The matrix **D** can be explicitly defined for assumed material models. For a homogeneous, isotropic, linear, elastic material model in plane stress state, the derivation of matrix **D** is as follows. We start with the generalized Hooke's law.

$$\begin{aligned} \varepsilon_{xx} &= \frac{\sigma_{xx}}{E} - \frac{\nu}{E}\sigma_{yy} + \varepsilon_{xx}^{\text{init}}, \\ \varepsilon_{yy} &= \frac{\sigma_{yy}}{E} - \frac{\nu}{E}\sigma_{xx} + \varepsilon_{yy}^{\text{init}}, \\ \gamma_{xy} &= \frac{2\,(1+\nu)}{E}\tau_{xy}. \end{aligned} \tag{3.23}$$

Solving the above equations for stresses, we get the matrix **D** as

$$[\mathbf{D}] = \frac{E}{(1-\nu^2)} \begin{bmatrix} 1 & \nu & 0 \\ \nu & 1 & 0 \\ 0 & 0 & \dfrac{(1-\nu)}{2} \end{bmatrix}, \tag{3.24}$$

where E is the Young's modulus of elasticity and ν is the Poisson's ratio.

For a plane strain state, the normal stress σ_{zz} also exists. Therefore, we have the generalized Hooke's law.

$$\begin{aligned} \varepsilon_{xx} &= \frac{\sigma_{xx}}{E} - \frac{\nu}{E}(\sigma_{yy} + \sigma_{zz}) + \varepsilon_{xx}^{\text{init}}, \\ \varepsilon_{yy} &= \frac{\sigma_{yy}}{E} - \frac{\nu}{E}(\sigma_{xx} + \sigma_{zz}) + \varepsilon_{yy}^{\text{init}}, \\ \gamma_{xy} &= \frac{2\,(1+\nu)}{E}\tau_{xy}. \end{aligned} \tag{3.25}$$

In addition, for a plane strain state $\varepsilon_{zz} = 0$ and therefore

$$\sigma_{zz} = \nu(\sigma_{xx} + \sigma_{yy}) - E\alpha T.$$

On eliminating σ_{zz} and solving for the stresses, we get the matrix **D** as

$$[\mathbf{D}] = \frac{E(1-\nu)}{(1+\nu)(1-2\nu)} \begin{bmatrix} 1 & \frac{\nu}{1-\nu} & 0 \\ \frac{\nu}{1-\nu} & 1 & 0 \\ 0 & 0 & \frac{(1-2\nu)}{2(1-\nu)} \end{bmatrix}. \tag{3.26}$$

Substituting (3.11) for $\boldsymbol{\varepsilon}$ into (3.22), we have

$$\{\sigma\} = [\mathbf{D}][\mathbf{B}]\{\mathbf{q}\} - [\mathbf{D}]\{\boldsymbol{\varepsilon}^{\text{init}}\}. \tag{3.27}$$

It is a standard practice to evaluate the stresses either at the centroid of the element ($\xi = 1/3, \eta = 1/3$) or at the Gauss point locations (points a, b, c, and d in Fig. 3.2). However, if nodal point stresses are required, extrapolation of stresses computed at the Gauss point locations to the nodes and averaging of contributions from all the elements connected to a node are essential.

Once the stress components are known, it is possible to compute the principal stresses σ_1, σ_2 using

$$(\sigma_1, \sigma_2) = \frac{(\sigma_{xx} + \sigma_{yy})}{2} \pm \frac{1}{2}\sqrt{(\sigma_{xx} - \sigma_{yy})^2 + 4\tau_{xy}^2}\,. \tag{3.28}$$

Using the principal stresses, an equivalent stress based on any of the known failure criteria can then be calculated. For example, the Von-Mises equivalent stress, widely used in engineering design to predict the onset of yielding can be calculated as

$$\sigma_e = \left[\frac{\sigma_1^2 + \sigma_2^2 - \sigma_1\sigma_2}{2}\right]^{1/2}. \tag{3.29}$$

We can then define a failure index (FI) as

$$\text{FI} = \frac{\sigma_e}{S_{yt}}, \tag{3.30}$$

where S_{yt} is the yield strength intension of the material.

3.2.4 Element stiffness matrix

The formula to compute the TRIA 3 element stiffness matrix is defined from an expression for the strain energy (SE).

$$\text{SE} = \frac{1}{2}\int \{\boldsymbol{\varepsilon}\}^T[\mathbf{D}]\{\boldsymbol{\varepsilon}\}h\,dx\,dy, \tag{3.31}$$

where h denotes the thickness and the integration is taken over the area of the TRIA 3 element. If the element has varying thickness, denoting the thicknesses at nodes 1, 2, and 3 by h_1, h_2 and h_3 respectively, the thickness h at any point (ξ, η) within the element is evaluated using

$$h = \sum_{i=1}^{3} N_i(\xi, \eta) h_i. \tag{3.32}$$

Substituting (3.11), (3.18) and (3.32) into (3.31), we can write

$$\text{SE} = \frac{1}{2}\{\mathbf{q}\}^T[\mathbf{K}]\{\mathbf{q}\}, \tag{3.33}$$

where

$$[\mathbf{K}] = \int [\mathbf{B}]^T[\mathbf{D}][\mathbf{B}] \left(\sum_{i=1}^{3} N_i(\xi, \eta) h_i \right) |\mathbf{J}| d\xi d\eta \tag{3.34}$$

and the integration is now taken over the area of the parent element.

The symmetric matrix $\mathbf{K}$ is the stiffness matrix for the TRIA 3 element. Equation (3.34) is sufficiently explicit for computation with the matrix operations left to the computer. In practice, numerical integration using Gauss–Legendre quadrature formulae [3.3] is employed to compute TRIA 3 element matrices. Table 3.1 gives the sampling point locations and weighting coefficients for integration over triangular domain. Choosing n Gauss points, one can arrive at a summation expression for the computation of $\mathbf{K}$ as

$$[\mathbf{K}] = \frac{1}{2} \sum_{i=1}^{n} W_i[\Phi(\xi_i, \eta_i)], \tag{3.35}$$

where

$$[\Phi(\xi, \eta)] = [\mathbf{B}]^T[\mathbf{D}][\mathbf{B}] \left(\sum_{i=1}^{n} N_i(\xi, \eta) h_i \right) |\mathbf{J}|.$$

The factor $\frac{1}{2}$ appears in (3.35) because $|\mathbf{J}|$ is equal to twice the area of the TRIA 3 element. The number of Gauss points n depends upon the nature of Φ, the functions to be integrated. For the TRIA 3 element, a four-point formula ($n = 4$) is recommended. If the thickness h is constant, even a one-point formula ($n = 1$) is sufficient [2.2]. The required weighting factors W_i ($i = 1, 4$) and Gauss point locations (ξ_i, η_i) ($i = 1, 4$) are given in Table 3.1.

3.2.5 Consistent nodal force vector: Body force

For both plane stress and plane strain analysis, the body force per unit volume is defined by

$$\{\mathbf{b}\} = \begin{Bmatrix} b_x \\ b_y \end{Bmatrix}, \tag{3.36}$$

where b_x and b_y denote body forces in the x and y directions.

The potential energy (PE) of this external load is

$$\mathrm{PE} = -\int \{\mathbf{u}\}^T\{\mathbf{b}\} h\,dx\,dy, \tag{3.37}$$

where h denotes thickness and integration is over the area of the TRIA 3 element.

Substituting (3.2), (3.18) and (3.32) into (3.37), we can write

$$\mathrm{PE} = -\{\mathbf{q}\}^T\{\mathbf{f}^b\}, \tag{3.38}$$

where

$$\{\mathbf{f}^b\} = \int [\mathbf{N}]^T\{\mathbf{b}\}\left(\sum_{i=1}^{3} N_i(\xi,\eta)h_i\right)|\mathbf{J}|d\xi\,d\eta. \tag{3.39}$$

Table 3.1 *Numerical integration formulae for triangles*

Order	Figure	Sampling points	Natural ξ	Coordinates η	Weighting factors W
Linear	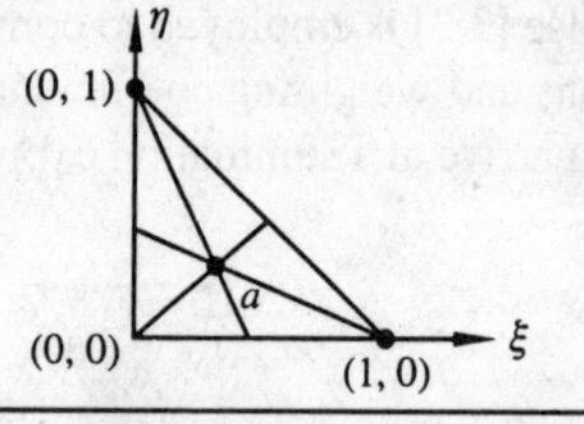	*a*	1/3	1/3	1
Quadratic	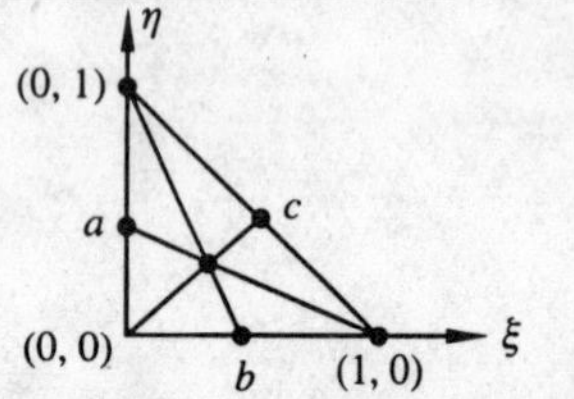	*a*	0	1/2	1/3
		b	1/2	0	1/3
		c	1/2	1/2	1/3
Cubic	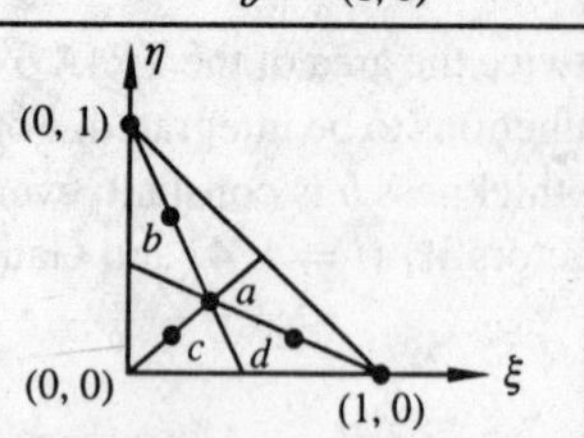	*a*	1/3	1/3	27/48
		b	0.2	0.6	25/48
		c	0.2	0.2	25/48
		d	0.6	0.2	25/48
Quintic	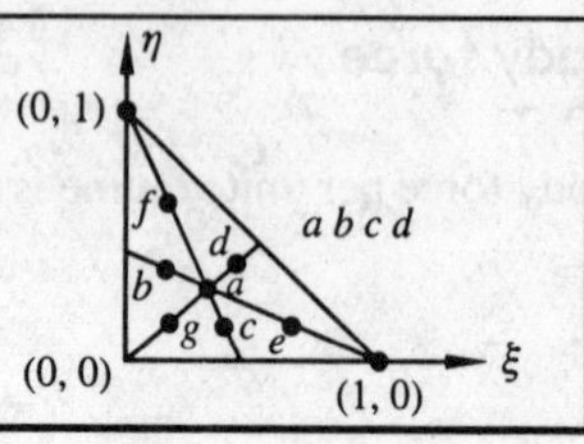	*a*	1/3	1/3	0.2250000000
		b	0.0597158717	0.4701420641	0.1323941527
		c	0.4701420641	0.0597158717	0.1323941527
		d	0.4701420641	0.4701420641	0.1323941527
		e	0.7974269853	0.1012865073	0.1259391805
		f	0.1012865073	0.7974269853	0.1259391805
		g	0.1012865073	0.1012865073	0.1259391805

and the integration is now taken over the area of the parent element. The vector $\mathbf{f}^b$ is called the consistent nodal force vector due to body force.

Specifically, for the case of centrifugal loading due to rotation about the z-axis at an angular velocity ω (rad/sec), the body forces are

$$\{\mathbf{b}\} = \begin{Bmatrix} b_x \\ b_y \end{Bmatrix} = \begin{Bmatrix} \rho\omega^2 x \\ \rho\omega^2 y \end{Bmatrix}, \tag{3.40}$$

where ρ is the density of the material.

Using a four-point Gauss–Legendre quadrature formula, one can arrive at a summation expression for the computation of $\mathbf{f}^b$ as

$$\mathbf{f}^b = \frac{1}{2}\sum_{i=1}^{4} W_i \{\psi(\xi_i, \eta_i)\}, \tag{3.41}$$

where

$$\{\psi(\xi, \eta)\} = [\mathbf{N}]^T \begin{Bmatrix} \rho\omega^2 \sum_{i=1}^{3} N_i(\xi, \eta)\, x_i \\ \rho\omega^2 \sum_{i=1}^{3} N_i(\xi, \eta)\, y_i \end{Bmatrix} \left(\sum_{i=1}^{3} N_i(\xi, \eta)\, h_i\right) |\mathbf{J}|. \tag{3.42}$$

3.2.6 Consistent nodal force vector: Thermal strain

Nodal forces to be applied to simulate thermal expansion can be evaluated using

$$\{\mathbf{f}^{\text{init}}\} = \int [\mathbf{B}]^T [\mathbf{D}] \{\varepsilon^{\text{init}}\} \left(\sum_{i=1}^{3} N_i(\xi, \eta) h_i\right) |\mathbf{J}| d\xi\, d\eta, \tag{3.43}$$

where the integration is now over the area of the parent element. The initial strains $\boldsymbol{\varepsilon}^{\text{init}}$ are calculated using (3.20) for plane stress state and (3.21) for plane strain state. Using a four-point Gauss-Legendre quadrature formula, we arrive at the following summation expression for the computation of $\mathbf{f}^{\text{init}}$.

$$\{\mathbf{f}^{\text{init}}\} = \frac{1}{2}\sum_{i=0}^{4} W_i \{\chi(\xi_i, \eta_i)\}, \tag{3.44}$$

where

$$\{\chi(\xi, \eta)\} = [\mathbf{B}]^T [\mathbf{D}] \{\boldsymbol{\varepsilon}^{\text{init}}\} \left(\sum_{i=0}^{3} N_i(\xi, \eta) h_i\right) |\mathbf{J}|. \tag{3.45}$$

3.2.7 Element equations

The potential energy functional for the TRIA 3 element is evaluated as

$$\pi_p(\mathbf{q}) = \frac{1}{2}\{\mathbf{q}\}^T [\mathbf{K}] \{\mathbf{q}\} - \{\mathbf{q}\}^T \{\mathbf{f}\}, \tag{3.46}$$

where

$$\{\mathbf{f}\} = \{\mathbf{f}^c\} + \{\mathbf{f}^b\} + \{\mathbf{f}^{\text{init}}\} + \{\mathbf{f}^p\}$$

and $\mathbf{f}^c$ is a vector of concentrated forces applied at the nodes; $\mathbf{f}^b$ is a vector of nodal forces due to body force; $\mathbf{f}^{\text{init}}$ is a vector of nodal forces to simulate thermal expansion, and $\mathbf{f}^p$ is a vector of nodal forces due to surface tractions, if any.

For equilibrium to be ensured, the potential energy functional π_p must be stationary for variation of admissible displacements. In elastic situations, π_p is not only stationary but is a minimum. We can therefore write the equilibrium conditions as

$$\frac{\partial \pi_p}{\partial\{\mathbf{q}\}} = 0$$

or

$$[\mathbf{K}]\{\mathbf{q}\} = \{\mathbf{f}\}. \tag{3.47}$$

Equations (3.47) are known as finite element equations.

3.3 Four-Noded Quadrilateral Element (QUAD 4)

Figure 3.3 shows a two-dimensional finite element, quadrilateral in shape and linear in order. The four nodes 1, 2, 3, and 4 are numbered in an anti-clockwise order. The displacement of a point within the element has two components u, v and the nodal degrees of freedom vector is

$$\{\mathbf{q}\}^T = (u_1, v_1, u_2, v_2, u_3, v_3, u_4, v_4). \tag{3.48}$$

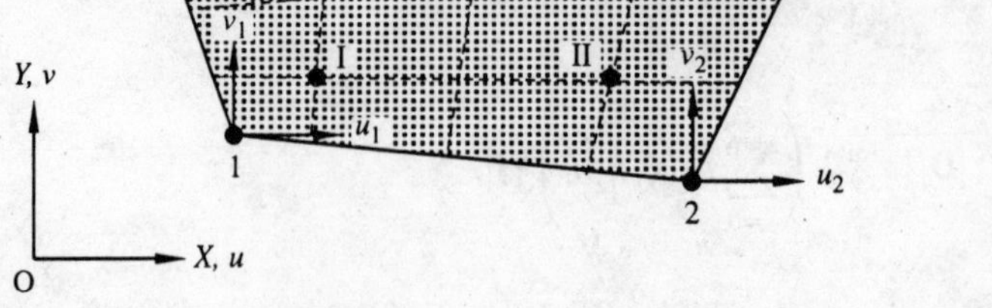

Fig. 3.3 *Four-noded quadrilateral finite element (QUAD 4)*

The forces acting at a node i also have two components F_{xi}, F_{yi} and the vector of nodal forces is

$$\{\mathbf{f}\}^T = (F_{x1}, F_{y1}, F_{x2}, F_{y2}, F_{x3}, F_{y3}, F_{x4}, F_{y4}). \tag{3.49}$$

The derivation of element equations is now described in a step-by-step manner.

3.3.1 Admissible displacement functions

The displacements u, v at any point (x, y) within the element have to be uniquely defined by their nodal values. The simplest representation is given by

$$\{\mathbf{u}\} = [\mathbf{N}]\{\mathbf{q}\}, \tag{3.50}$$

where

$$\{\mathbf{u}\} = \begin{Bmatrix} u \\ v \end{Bmatrix} \tag{3.51}$$

and

$$[\mathbf{N}] = \begin{bmatrix} N_1 & 0 & N_2 & 0 & N_3 & 0 & N_4 & 0 \\ 0 & N_1 & 0 & N_2 & 0 & N_3 & 0 & N_4 \end{bmatrix}. \tag{3.52}$$

In order to create the shape functions N_i $(i = 1, 4)$ so as to a priori obey the convergence criteria, it is now a standard practice to use the parent element and the natural coordinates (ξ, η) shown in Fig. 3.4. The parent element is a bi-unit square and the natural coordinates are non-dimensional. Any variable Φ over the parent element domain has to be uniquely defined by its nodal values Φ_1, Φ_2, Φ_3, and Φ_4. The simplest representation is given by the bi-linear polynomial

$$\Phi = \alpha_1 + \alpha_2\xi + \alpha_3\eta + \alpha_4\xi\eta. \tag{3.53}$$

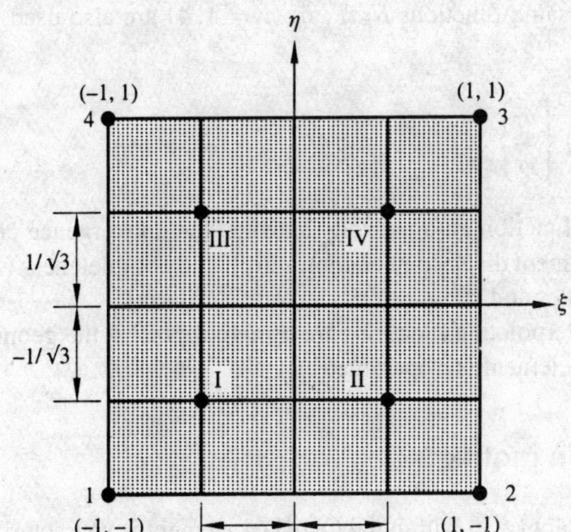

Fig. 3.4 *Parent element and natural coordinates for QUAD 4*

It is then possible to evaluate the polynomial coefficients α_i $(i = 1, 4)$ in terms of Φ_i $(i = 1, 4)$. We can thus write

$$\Phi = \sum_{i=1}^{4} N_i(\xi, \eta)\Phi_i, \tag{3.54}$$

with

$$\begin{aligned}
N_1(\xi, \eta) &= \frac{1}{4}(1 - \xi)(1 - \eta), \\
N_2(\xi, \eta) &= \frac{1}{4}(1 + \xi)(1 - \eta), \\
N_3(\xi, \eta) &= \frac{1}{4}(1 + \xi)(1 + \eta), \\
N_4(\xi, \eta) &= \frac{1}{4}(1 - \xi)(1 + \eta).
\end{aligned} \tag{3.55}$$

It is to be noted that the above functions are such that $N_i = l$ at node i and $N_i = 0$ at other nodes, and $N_1 + N_2 + N_3 + N_4 = 1.0$.

We now map, uniquely, the parent element on to the QUAD 4 element. The required relationship between the Cartesian and natural coordinates to accomplish this is

$$\begin{Bmatrix} x \\ y \end{Bmatrix} = \sum_{i=1}^{4} N_i(\xi, \eta) \begin{Bmatrix} x_i \\ y_i \end{Bmatrix}, \tag{3.56}$$

where (x_i, y_i) denote Cartesian coordinates of node i of the QUAD 4 element. However, uniqueness of such a mapping should be ensured. With the geometry of the QUAD 4 element now defined by the parametric (3.56), the same functions N_i (ξ, η) $(i = 1, 4)$ are also used to define the admissible displacement function as

$$\begin{Bmatrix} u \\ v \end{Bmatrix} = \sum_{i=1}^{4} N_i(\xi, \eta) \begin{Bmatrix} u_i \\ v_i \end{Bmatrix}. \tag{3.57}$$

The chosen displacement function automatically satisfies the convergence criteria, namely, continuity, compatibility (inter-element displacement continuity), and completeness (zero strain energy corresponding to rigid body modes and constant first derivative criteria).

Since the same order interpolation functions are used to represent the geometry as well as to define the displacement field, the element is isoparametric.

3.3.2 Element strain matrix

The state of strain at any point (x, y) within the QUAD 4 element is characterized by the three components ε_{xx}, ε_{yy}, and γ_{xy}. These strains are determined using (3.10) along with (3.57) as

$$\{\boldsymbol{\varepsilon}\} = [\mathbf{B}]\{\mathbf{q}\}, \tag{3.58}$$

where

$$[\mathbf{B}] = \begin{bmatrix} \frac{\partial N_1}{\partial x} & 0 & \frac{\partial N_2}{\partial x} & 0 & \frac{\partial N_3}{\partial x} & 0 & \frac{\partial N_4}{\partial x} & 0 \\ 0 & \frac{\partial N_1}{\partial y} & 0 & \frac{\partial N_2}{\partial y} & 0 & \frac{\partial N_3}{\partial y} & 0 & \frac{\partial N_4}{\partial y} \\ \frac{\partial N_1}{\partial y} & \frac{\partial N_1}{\partial x} & \frac{\partial N_2}{\partial y} & \frac{\partial N_2}{\partial x} & \frac{\partial N_3}{\partial y} & \frac{\partial N_3}{\partial x} & \frac{\partial N_4}{\partial y} & \frac{\partial N_4}{\partial x} \end{bmatrix}. \tag{3.59}$$

To compute the element strain matrix **B**, we therefore need to evaluate the Cartesian derivates of the shape functions N_i using

$$\begin{bmatrix} \frac{\partial N_i}{\partial x} \\ \frac{\partial N_i}{\partial y} \end{bmatrix} = [\mathbf{J}]^{-1} \begin{bmatrix} \frac{\partial N_i}{\partial \xi} \\ \frac{\partial N_i}{\partial \eta} \end{bmatrix} \quad (i = 1, 2, 3, 4). \tag{3.60}$$

The Jacobian matrix **J** is evaluated using

$$[\mathbf{J}] = \sum_{j=1}^{4} \begin{bmatrix} \frac{\partial N_j}{\partial \xi} x_j & \frac{\partial N_j}{\partial \xi} y_j \\ \frac{\partial N_j}{\partial \eta} x_j & \frac{\partial N_j}{\partial \eta} y_j \end{bmatrix} = \begin{bmatrix} J_{11} & J_{12} \\ J_{21} & J_{22} \end{bmatrix} \tag{3.61}$$

and the inverse of the matrix **J** is obtained as

$$[\mathbf{J}]^{-1} = \frac{1}{|\mathbf{J}|} \begin{bmatrix} J_{22} & -J_{12} \\ -J_{21} & J_{11} \end{bmatrix}, \tag{3.62}$$

where

$$|\mathbf{J}| = (J_{11}J_{22} - J_{12}J_{21}).$$

It should be checked that $|\mathbf{J}|$ should not become zero and that the sign of $|\mathbf{J}|$ should remain unchanged at all points within the QUAD 4 element. This will ensure that the mapping is unique.

It is interesting to note that $B_{ij} = B_{ij}(\xi, \eta)$. Initial strains, i.e., strains that are independent of stress, frequently result in an initial strain vector $\boldsymbol{\varepsilon}^{\text{init}}$ due to change in temperature. For the case of plane stress, the thermal strains are evaluated using

$$\left\{\boldsymbol{\varepsilon}^{\text{init}}\right\} = \begin{Bmatrix} \alpha T \\ \alpha T \\ 0 \end{Bmatrix} = \begin{Bmatrix} \alpha \sum_{i=1}^{4} N_i(\xi, \eta) T_i \\ \alpha \sum_{i=1}^{4} N_i(\xi, \eta) T_i \\ 0 \end{Bmatrix}, \tag{3.63}$$

while for the case of plane strain,

$$\left\{\boldsymbol{\varepsilon}^{\text{init}}\right\} = \begin{Bmatrix} (1+\nu)\,\alpha \sum_{i=1}^{4} N_i(\xi, \eta) T_i \\ (1+\nu)\,\alpha \sum_{i=1}^{4} N_i(\xi, \eta) T_i \\ 0 \end{Bmatrix} \tag{3.64}$$

In (3.63) and (3.64), T_i denotes temperature change at node i, α denotes the coefficient of thermal expansion, and ν the Poisson's ratio.

3.3.3 Element stress recovery

The state of stress at any point (x, y) within the QUAD 4 element is characterized by the stress vector $\{\sigma\}^T = (\sigma_{xx}, \sigma_{yy}, \tau_{xy})$ and these are determined using

$$\{\sigma\} = [\mathbf{D}][\mathbf{B}]\{\mathbf{q}\} - [\mathbf{D}]\{\boldsymbol{\varepsilon}^{\text{init}}\}. \tag{3.65}$$

For plane stress problems, the matrices $\mathbf{D}$ and $\boldsymbol{\varepsilon}^{\text{init}}$ are evaluated using (3.24) and (3.63). For plane strain problems, (3.26) and (3.64) are used to evaluate $\mathbf{D}$ and $\boldsymbol{\varepsilon}^{\text{init}}$.

By default, the stresses are recovered at the centroid of the element ($\xi = 0$, $\eta = 0$). Gauss point locations (points I, II, III, IV in Fig. 3.4) are optimal for accurate prediction of stresses. However, if nodal point stresses are required, extrapolation of the Gauss point stresses to the nodes and averaging of contributions from all elements connected to a node are essential.

3.3.4 Element stiffness matrix

The formula to compute the QUAD 4 element stiffness matrix is

$$[\mathbf{K}] = \int [\mathbf{B}]^T[\mathbf{D}][\mathbf{B}]h\,dx\,dy, \tag{3.66}$$

where h denotes the thickness and integration is now taken over the area of the QUAD 4 element.

For varying thickness, denoting the thickness at node i as h_i, h at any point within the QUAD 4 element is evaluated using

$$h = \sum_{i=0}^{4} N_i(\xi, \eta)h_i, \tag{3.67}$$

with $dxdy = |\mathbf{J}|\, d\xi d\eta$. Equation (3.66) becomes

$$[\mathbf{K}] = \int_{-1}^{+1}\int_{-1}^{+1} [\mathbf{B}]^T[\mathbf{D}][\mathbf{B}] \left(\sum_{i=0}^{4} N_i(\xi, \eta)h_i\right)|\mathbf{J}|d\xi\, d\eta \tag{3.68}$$

and the integration is now taken over the area of the parent element.

In (3.68), the matrix operations being left to the computer, we note that both $\mathbf{B}$ and $\mathbf{J}$ being involved functions of ξ and η, the evaluation of the integral has to be performed numerically. The Gaussian quadrature formula is ideally suited and hence exclusively used.

To find numerically the integral of a function $f(\xi)$ of one variable ξ, the Gauss quadrature formula is [3.2]

$$I = \int_{-1}^{+1} f(\xi)d\xi = \sum_{i=1}^{n} W_i f(\xi_i). \tag{3.69}$$

In (3.69), n denotes number of Gauss points, W_i are the weighting coefficients, and ξ_i denotes the location of sampling points. Table 3.2 gives the sampling point locations and weighting coefficients for Gauss quadrature formulae. By choosing n sampling points, a polynomial of degree $(2n - 1)$ could be constructed and exactly integrated.

Table 3.2 *Sampling point locations and weight coefficients of the Gauss quadrature formula*

$$I = \int_{-1}^{+1} f(\xi)d\xi = \sum_{i=1}^{n} W_i f(\xi_i)$$

	Location ξ_i	Weight W_i
$n = 1$	0.0	2.0
$n = 2$	+0.57735 02691 89626	1.0
	−0.577350269189626	1.0
$n = 3$	+0.774596669241483	0.5555555551
	0.0	0.8888888888
	−0.774596669241483	0.5555555551
$n = 4$	+0.861136311594055	0.34785 48451 37454
	+0.339981043584886	0.652145154862546
	−0.339981043584886	0.652145154862546
	−0.861136311594055	0.347854845137454

The most obvious way of evaluating the integral of a function $f(\xi, \eta)$ of two variables ξ, η is

$$I = \int_{-1}^{+1}\int_{-1}^{+1} f(\xi, \eta)d\xi d\eta = \sum_{i=1}^{n}\sum_{j=1}^{n} W_i W_j f(\xi_i, \eta_j). \tag{3.70}$$

Choosing n Gauss points and using Eqn. (3.70), we can arrive at a summation expression for the computation of $\mathbf{K}$ for the QUAD 4 element as

$$[\mathbf{K}] = \sum_{i=1}^{n}\sum_{j=1}^{n} W_i W_j [\mathbf{\Phi}(\xi_i, \eta_j)], \tag{3.71}$$

where

$$[\mathbf{\Phi}(\xi, \eta)] = [\mathbf{B}]^T[\mathbf{D}][\mathbf{B}]\left(\sum_{i=1}^{n} N_i(\xi, \eta)h_i\right)|\mathbf{J}|. \tag{3.72}$$

The number of Gauss points n has to be chosen depending upon the nature of Φ, the integrand. For the QUAD 4 element, a two-point formula ($n = 2$) is recommended and the location of the four Gauss points is shown in Fig. 3.4.

3.3.5 Consistent nodal force vector: Body forces

Work equivalent nodal forces due to body force **b** are evaluated using

$$\left\{\mathbf{f}^{b}\right\} = \int [\mathbf{N}]^{T} \{\mathbf{b}\}\, h dx\, dy, \tag{3.73}$$

where h denotes thickness and the integration is taken over the area of the QUAD 4 element.

For the specific case of centrifugal loading due to rotation about the z-axis at an angular velocity ω (rad./s), the body force vector is evaluated using

$$\{\mathbf{b}\} = \begin{Bmatrix} b_x \\ b_y \end{Bmatrix} = \begin{Bmatrix} \rho\omega^2 x \\ \rho\omega^2 y \end{Bmatrix} = \begin{Bmatrix} \rho\omega^2 \sum_{i=1}^{4} N_i(\xi,\eta)\, x_i \\ \rho\omega^2 \sum_{i=1}^{4} N_i(\xi,\eta)\, y_i \end{Bmatrix}, \tag{3.74}$$

where ρ denotes the density of the material

Choosing four Gauss points, one can arrive at the following summation expression for the computation of $\mathbf{f}^{b}$

$$\left\{\mathbf{f}^{b}\right\} = \sum_{i=1}^{2}\sum_{j=1}^{2} W_i W_j \left\{\psi\left(\xi_i, \eta_j\right)\right\}, \tag{3.75}$$

where

$$\{\psi(\xi,\eta)\} = [\mathbf{N}]^{T} \begin{bmatrix} \rho\omega^2 \sum_{i=1}^{4} N_i(\xi,\eta)\, x_i \\ \rho\omega^2 \sum_{i=1}^{4} N_i(\xi,\eta)\, y_i \end{bmatrix} \left(\sum_{i=1}^{4} N_i(\xi,\eta)\, h_i\right) |\mathbf{J}|. \tag{3.76}$$

3.3.6 Consistent nodal force vector: Initial strain

Nodal forces corresponding to initial strain can be evaluated using

$$\{\mathbf{f}^{\text{init}}\} = \int [\mathbf{B}]^{T}[\mathbf{D}]\{\boldsymbol{\varepsilon}^{\text{init}}\}\, h dx\, dy, \tag{3.77}$$

where the integration is taken over the area of the QUAD 4 element. For the important case of thermal stress analysis, the initial strains $\boldsymbol{\varepsilon}^{\text{init}}$ are evaluated using (3.63) for plane stress and (3.64) for plane strain problems.

Again choosing four Gauss points, one can arrive at a summation expression as

$$\{\mathbf{f}^{\text{init}}\} = \sum_{i=1}^{2}\sum_{j=1}^{2} W_i W_j \{\chi(\xi,\eta)\}, \tag{3.78}$$

where

$$\{\chi(\xi,\eta)\} = [\mathbf{B}]^T[\mathbf{D}]\{\boldsymbol{\varepsilon}^{\text{init}}\}\left(\sum_{i=1}^{4} N_i(\xi,\eta)h_i\right)|\mathbf{J}|. \tag{3.79}$$

3.3.7 Consistent nodal force vector: Surface traction

In general, on each of the four surfaces of the QUAD 4 element, one can specify traction force. Let us consider face 2-3 of the QUAD 4 element on which $\xi = +1$ and $-1 \leq \eta \leq 1$. Let **p** denote the uniform traction vector on this face. Then the consistent nodal force vector due to surface traction is

$$\{\mathbf{f}^p\} = \frac{h_{2-3}l_{2-3}}{2}\{0,0,p_x,p_y,p_x,p_y,0,0\}, \tag{3.80}$$

where h_{2-3} and l_{2-3} denote the thickness and length of edge 2-3. For varying pressure, we may express P_x and P_y in terms of their values at the nodes 2 and 3 using linear interpolation functions. Numerical integration can then be employed to compute the corresponding $\mathbf{f}^p$.

3.4 Higher Order Elements

The linear triangular and quadrilateral elements developed here are the simplest of the available two-dimensional elements. The simplicity of TRIA 3 and QUAD 4 generally demands a fine mesh to model the geometry as well as to approximate the variables over the problem domain. Therefore, development of higher order elements is desirable for practical applications.

A family of triangular elements can be developed on a pattern indicated in Fig. 3.5. The number of nodes in each member coincides exactly with the number of polynomial terms employed. This follows by comparison with Pascal's triangle also shown in Fig. 3.5. A complete polynomial of the order needed to ensure inter-element displacement continuity, is used. Interpolation functions derived from them are termed as Lagrange functions. The derivation of explicit shape functions and use of Gauss–Legendre quadrature formulae will be preferred when element matrices for the quadratic and cubic order elements have to be computed.

Figure 3.6 shows linear, quadratic, and cubic order two-dimensional elements of the so-called "serendipity" family. In each element, a progressively increasing and equal number of nodes is placed on the element boundary. The number of polynomial terms used to approximate any field variable over each element domain coincides with the number of nodes. This follows by comparison with the Pascal's triangle also shown in Fig. 3.6, although all the polynomials may not be complete. Explicit shape functions for these elements have been originally derived by inspection, and progression to yet higher orders is quite difficult and requires some ingenuity. It was therefore appropriate to name this family "serendipity" after the famous princes of Serendipity, noted for their chance discoveries [3.2].

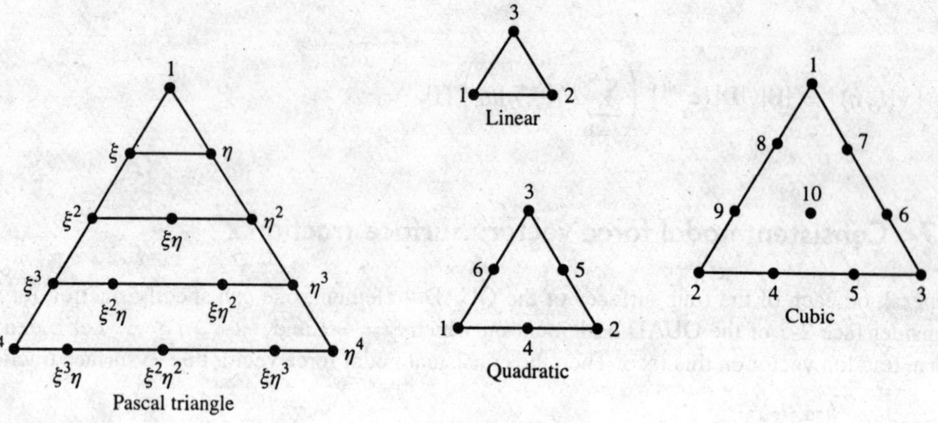

Fig. 3.5 *A family of triangular finite elements*

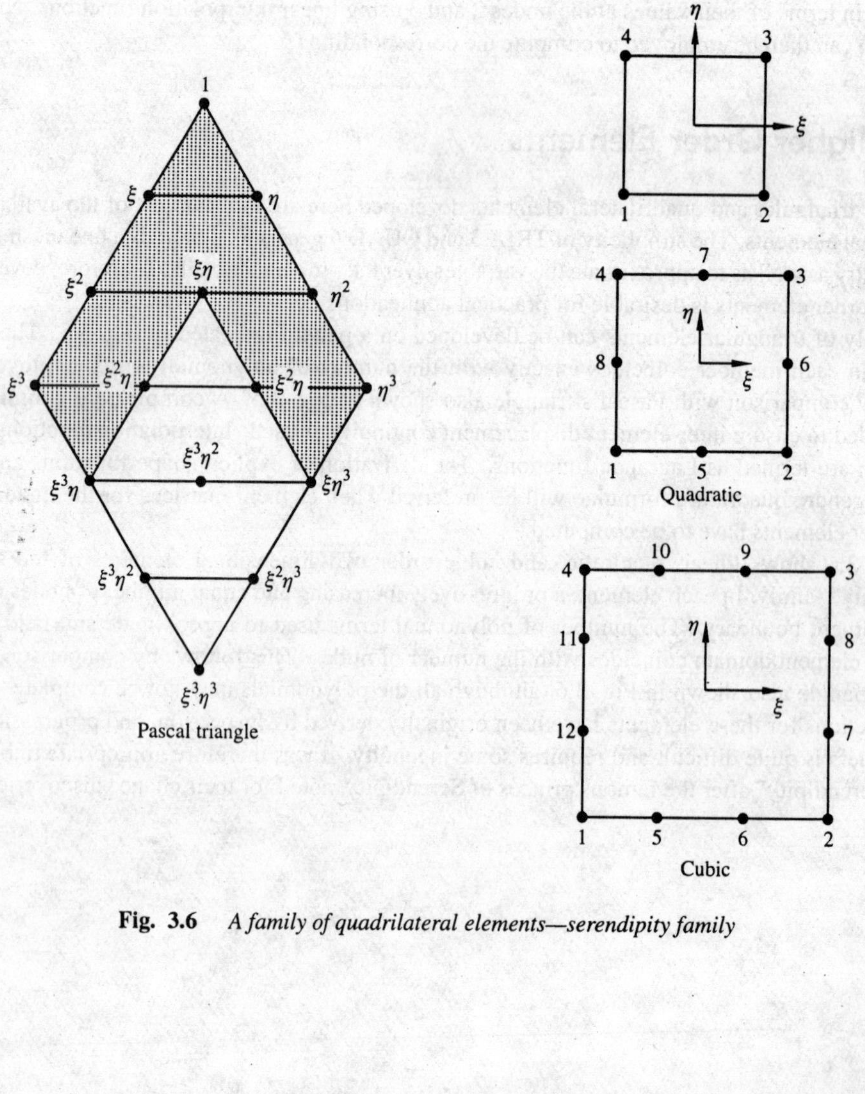

Fig. 3.6 *A family of quadrilateral elements—serendipity family*

Figure 3.7 shows two-dimensional elements of the Lagrange family, which are linear, quadratic and cubic in order. Explicit shape functions for these elements are easy to generate as products of Lagrange polynomials in the two natural coordinates ξ and η. The polynomials associated with these elements follow by comparison with Pascal's triangle, also shown in Fig. 3.7. Explicit shape functions for these elements are available [3.2] and progression to yet higher orders is quite feasible. However, these elements are difficult to use in practice due to the presence of the internal nodes and are not implemented in commercial FEM systems.

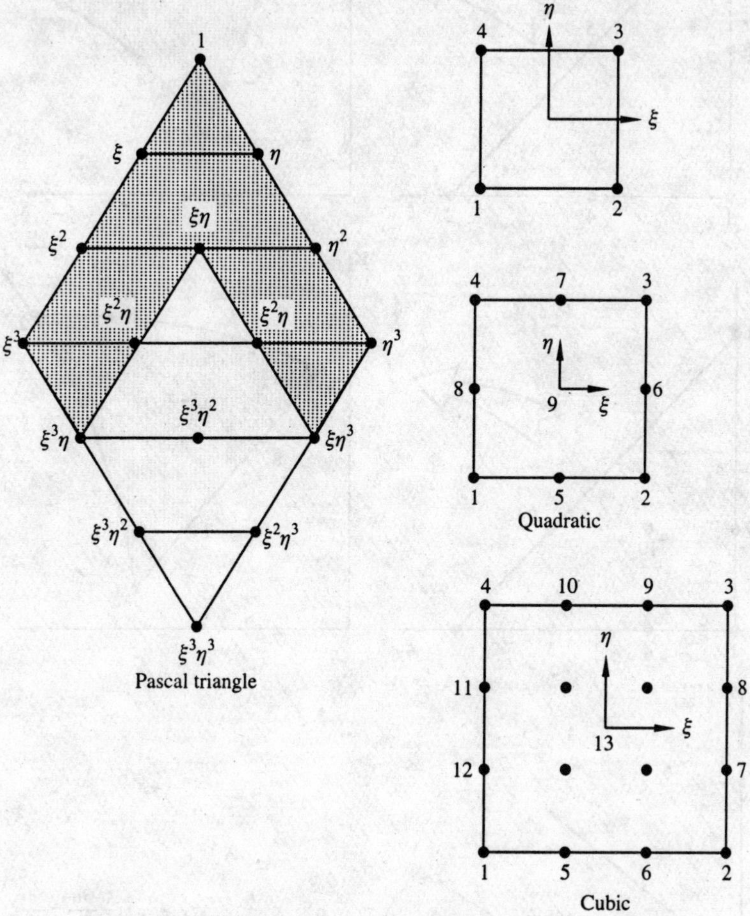

Fig. 3.7 *A family of quadrilateral elements—Lagrange family*

The quadratic and cubic order elements can also be distorted to have curved edges to fit realistic boundaries and methods for doing this are illustrated in Figs. 3.8 and 3.9. Shape functions for TRIA 6 (six-noded triangular), QUAD 8 (eight-noded quadrilateral) and QUAD 12 (twelve-noded quadrilateral) elements are presented in Tables 3.3, 3.4, and 3.5 respectively.

Triangular and quadrilateral elements with drilling degrees of freedom are also included in Figs. 3.8 and 3.9. A drilling degree of freedom is a rotational degree of freedom whose vector is normal to the

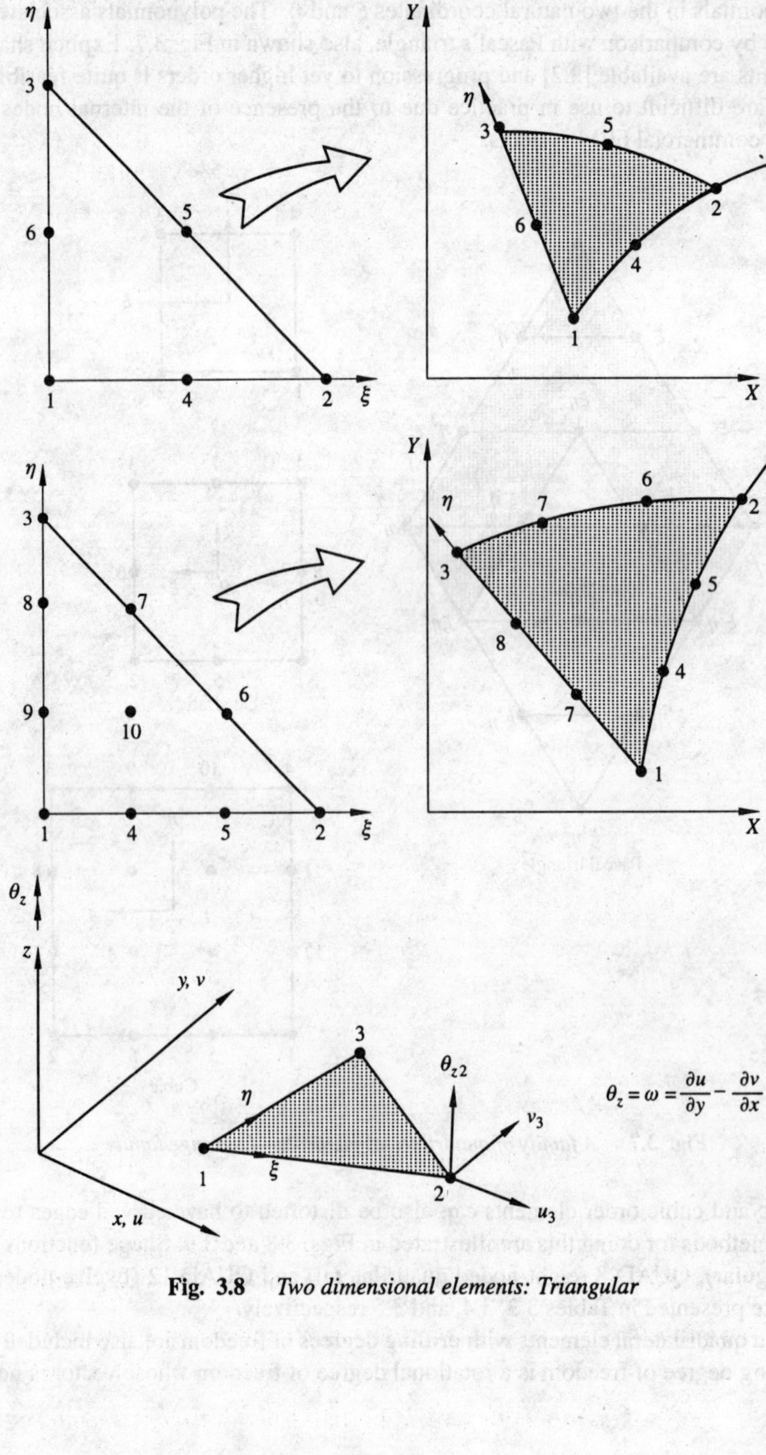

Fig. 3.8 *Two dimensional elements: Triangular*

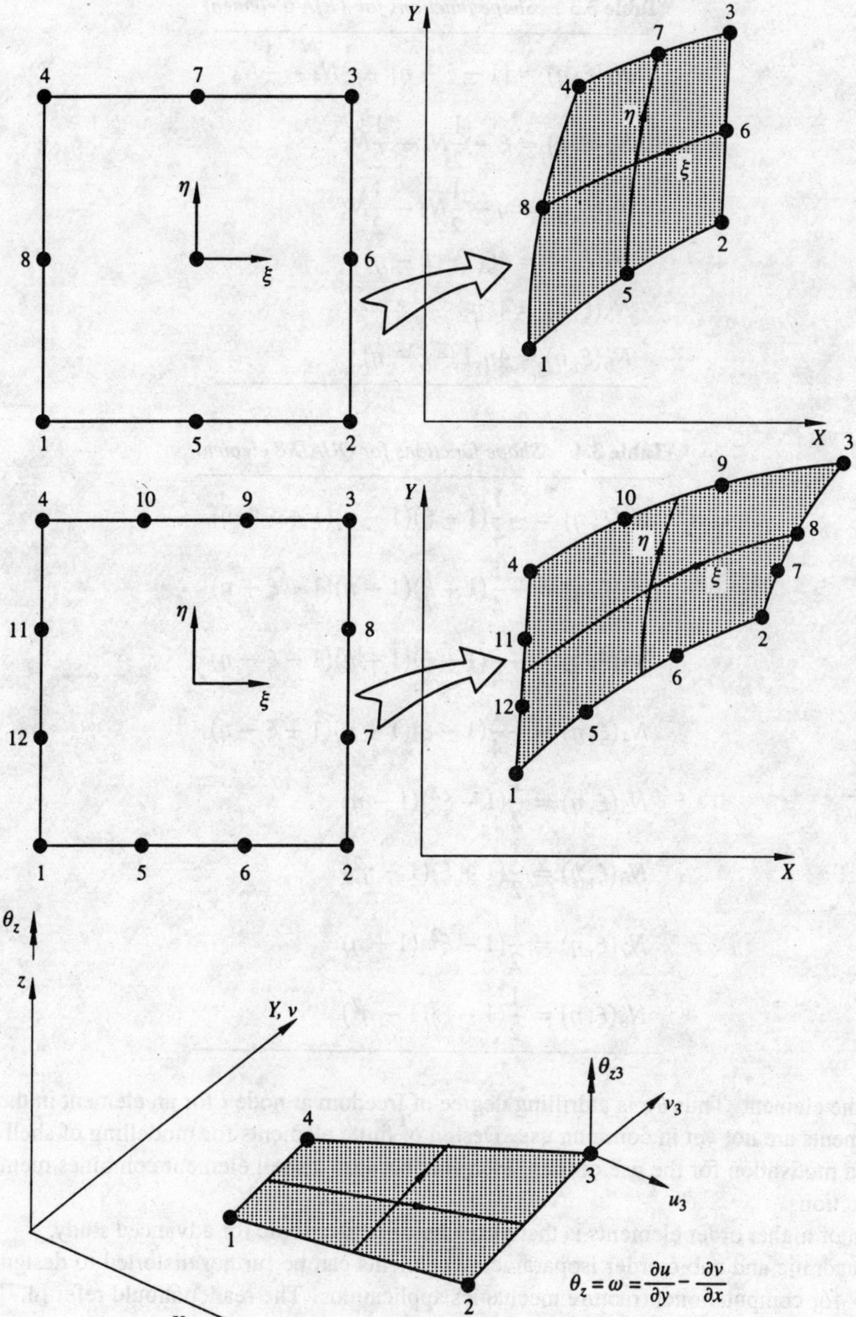

Fig. 3.9 *Two dimensional elements: Quadrilateral*

Table 3.3 *Shape functions for TRIA 6 element*

$$N_1(\xi,\eta) = (1-\xi-\eta) - \frac{1}{2}N_4 - \frac{1}{2}N_6$$
$$N_2(\xi,\eta) = \xi - \frac{1}{2}N_4 - \frac{1}{2}N_5$$
$$N_3(\xi,\eta) = \eta - \frac{1}{2}N_5 - \frac{1}{2}N_6$$
$$N_4(\xi,\eta) = 4\xi(1-\xi-\eta)$$
$$N_5(\xi,\eta) = 4\xi\eta$$
$$N_6(\xi,\eta) = 4\eta(1-\xi-\eta)$$

Table 3.4 *Shape functions for QUAD 8 element*

$$N_1(\xi,\eta) = -\frac{1}{4}(1-\xi)(1-\eta)(1+\xi+\eta)$$
$$N_2(\xi,\eta) = -\frac{1}{4}(1+\xi)(1-\eta)(1-\xi+\eta)$$
$$N_3(\xi,\eta) = -\frac{1}{4}(1+\xi)(1+\eta)(1-\xi-\eta)$$
$$N_4(\xi,\eta) = -\frac{1}{4}(1-\xi)(1+\eta)(1+\xi-\eta)$$
$$N_5(\xi,\eta) = \frac{1}{2}(1-\xi^2)(1-\eta)$$
$$N_6(\xi,\eta) = \frac{1}{2}(1+\xi)(1-\eta^2)$$
$$N_7(\xi,\eta) = \frac{1}{2}(1-\xi^2)(1+\eta)$$
$$N_8(\xi,\eta) = \frac{1}{2}(1-\xi)(1-\eta^2)$$

plane of the element. Thus θ_{zi} is a drilling degree of freedom at node i for an element in the xy plane. Such elements are not yet in common use. Design of finite elements for modelling of shell structures provides a motivation for the use of these elements because a shell element combines membrane and bending actions.

Design of higher order elements is therefore identified as a topic for advanced study.

The quadratic and cubic order isoparametric elements can be further distorted to design "singular elements" for computational fracture mechanics applications. The reader should refer [3.7] and [3.8] for details.

Table 3.5 *Shape functions for QUAD 12 element*

$$N_1(\xi,\eta) = \frac{1}{32}(1-\xi)(1-\eta)[-10+9(\xi^2+\eta^2)]$$
$$N_2(\xi,\eta) = \frac{1}{32}(1+\xi)(1-\eta)[-10+9(\xi^2+\eta^2)]$$
$$N_3(\xi,\eta) = \frac{1}{32}(1+\xi)(1+\eta)[-10+9(\xi^2+\eta^2)]$$
$$N_4(\xi,\eta) = \frac{1}{32}(1-\xi)(1+\eta)[-10+9(\xi^2+\eta^2)]$$
$$N_5(\xi,\eta) = \frac{9}{32}(1-\eta)(1-\xi^2)(1-3\xi)$$
$$N_6(\xi,\eta) = \frac{9}{32}(1-\eta)(1-\xi^2)(1+3\xi)$$
$$N_7(\xi,\eta) = \frac{9}{32}(1+\xi)(1-\eta^2)(1-3\eta)$$
$$N_8(\xi,\eta) = \frac{9}{32}(1+\xi)(1-\eta^2)(1+3\eta)$$
$$N_9(\xi,\eta) = \frac{9}{32}(1+\eta)(1-\xi^2)(1+3\xi)$$
$$N_{10}(\xi,\eta) = \frac{9}{32}(1+\eta)(1-\xi^2)(1-3\xi)$$
$$N_{11}(\xi,\eta) = \frac{9}{32}(1-\xi)(1-\eta^2)(1+3\eta)$$
$$N_{12}(\xi,\eta) = \frac{9}{32}(1-\xi)(1-\eta^2)(1-3\eta)$$

3.5 Test Problems–Assessment of Accuracy

There is no doubt that the finite element solution to plane stress and plane strain problems is, in the limit of subdivision, an exact solution. What is important for the engineer to know is the level of accuracy achievable in specific problems with a certain fineness of element subdivision. In any particular problem, the error can be assessed by comparison with the exact solution (if known) or by a convergence study, using progressive model refinement. In this section, attention will be focused on some test problems for which exact solutions are available.

3.5.1 Bending of a cantilever beam

Figure 3.10(a) shows the problem of a cantilever beam subjected to a tip load, mathematically modelled as a plane stress problem and discretized using a 4×10 mesh of QUAD 4 elements. The material is modelled to be homogeneous and isotropic and its behaviour is assumed to be linear and elastic. The nodes along the Y axis are constrained to simulate clamped boundary conditions. The load P is applied at node C. A linear static analysis is carried out. The predicted deflection at node C agrees with the Euler–Bernoulli beam theory solution. The predicted bending stress (σ_{xx}) and transverse shear stress (τ_{xy}) distributions are presented in Figs. 3.10(b) and 3.10(c) respectively. These straddle the beam theory solutions and in fact, if the stress values are recovered at the Gauss points of the elements and plotted, the best fit represents the exact solution.

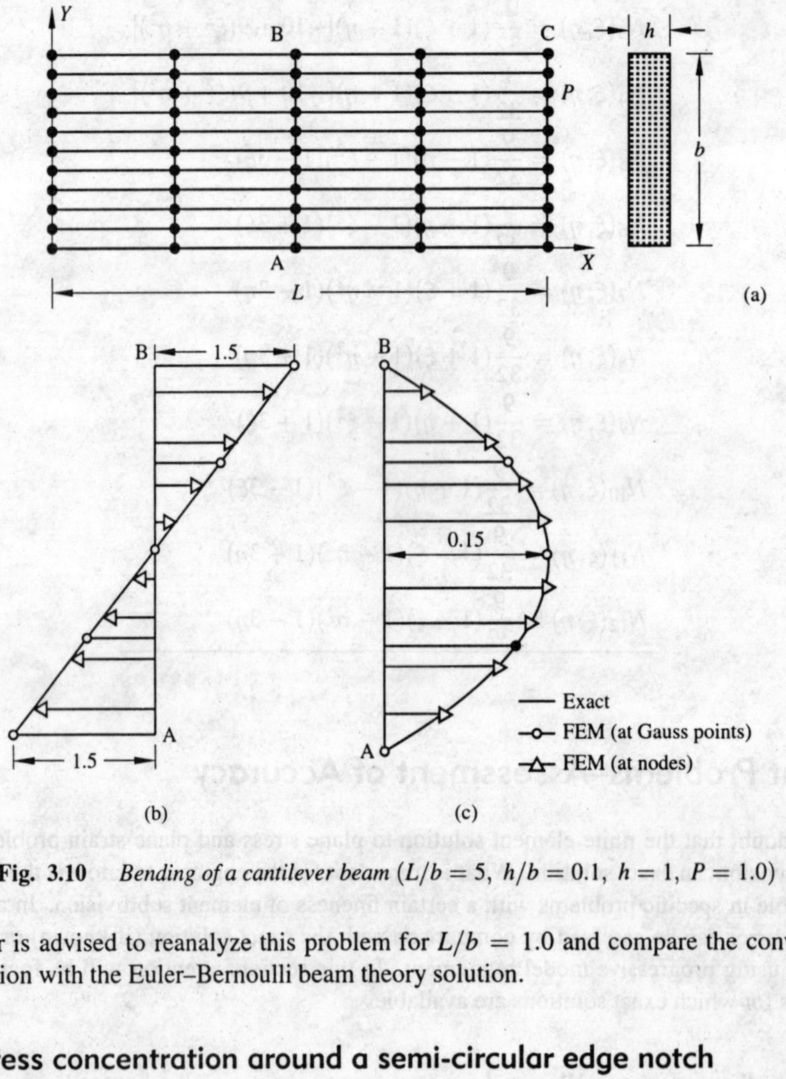

Fig. 3.10 *Bending of a cantilever beam* $(L/b = 5,\ h/b = 0.1,\ h = 1,\ P = 1.0)$

The reader is advised to reanalyze this problem for $L/b = 1.0$ and compare the converged finite element solution with the Euler–Bernoulli beam theory solution.

3.5.2 Stress concentration around a semi-circular edge notch

A more realistic stress analysis problem is shown in Fig. 3.11(a). Here the study of stress concentration due to a semi-circular edge notch of radius b in a rectangular plate of length $2L$, width W and thickness h subjected to uniform uniaxial tension is considered. A graded mesh of QUAD 4 elements as shown in Fig. 3.10(b) is designed to capture the steep stress gradients in the vicinity of the notch. The degree of accuracy achievable can be assessed from Fig. 3.11(c) where the predicted stress σ_{xx} is compared against a reference solution [3.4].

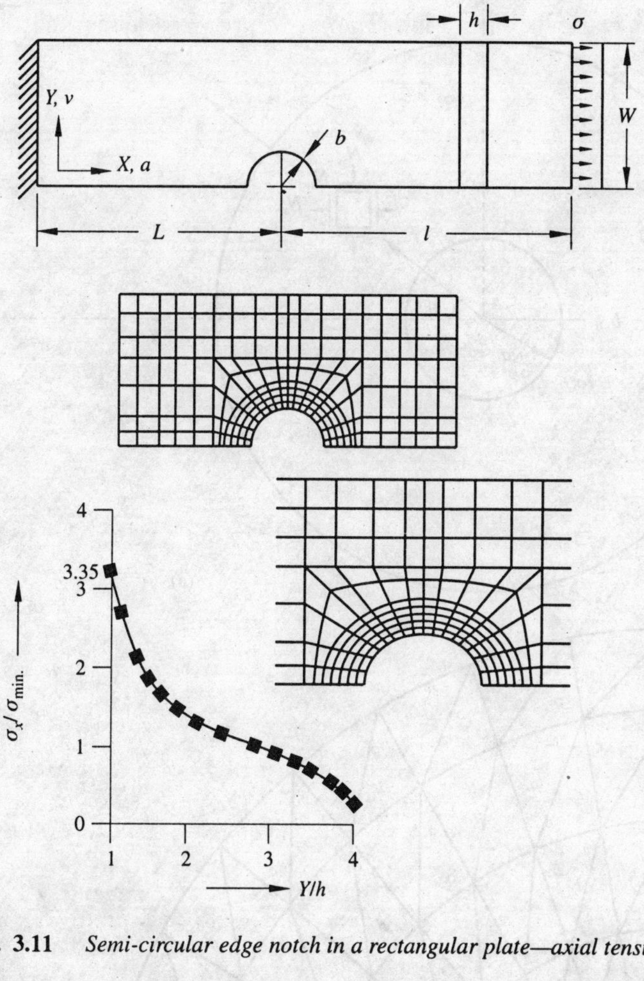

Fig. 3.11 *Semi-circular edge notch in a rectangular plate—axial tension*

It is to be noted that converged accurate solutions can also be obtained using a coarse mesh of quadratic or cubic order elements, which can also be distorted to fit the curved boundaries more accurately. The reader is advised to experiment with this idea and report results for loading by a moment.

3.5.3 Rotating annular disk

This test problem shown in Fig. 3.12(a) is used to assess the accuracy of the consistent nodal force evaluation due to body forces, such as centrifugal loading in turbomachinery parts. The body force b_r varies proportional to r as

$$b_r = \rho\omega^2 r,$$

where ω is the angular velocity and ρ is the density for the material.

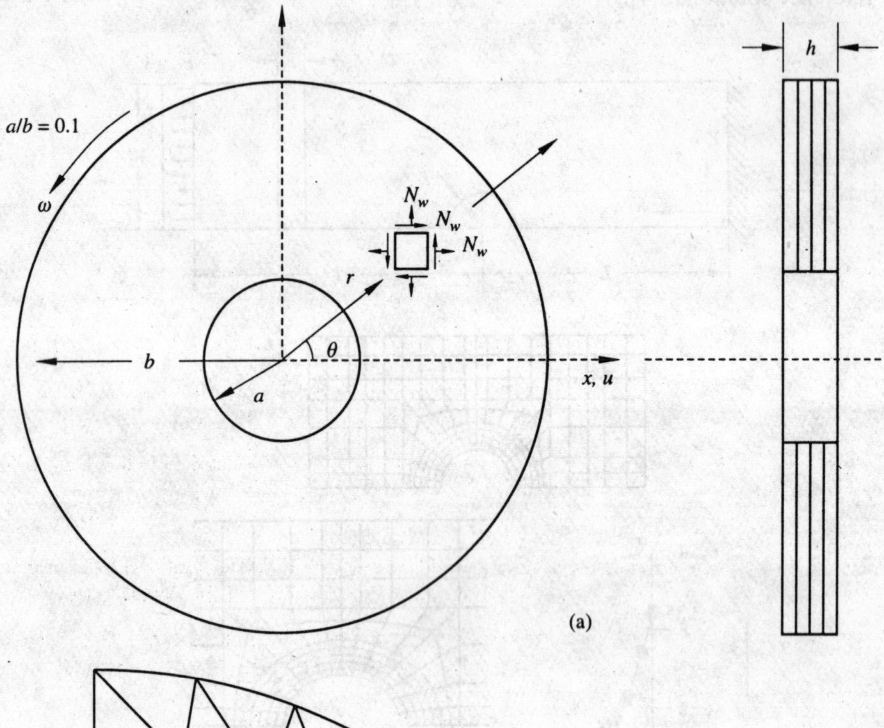

Fig. 3.12 *Rotating annular disk (continues)*

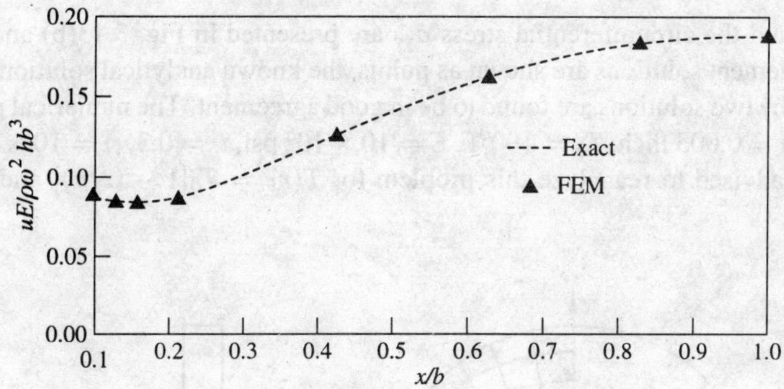

(c)

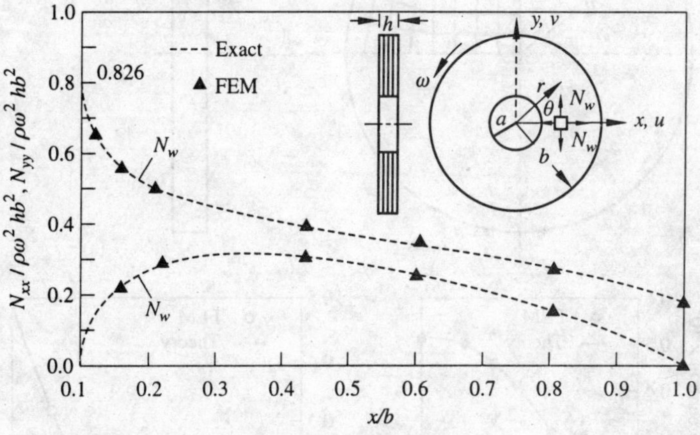

(d)

Fig. 3.12 *Rotating annular disk (concluded)*

Exploiting the symmetry, only one quadrant of the disk is discretized using a 7×9 mesh of TRIA 3 elements as shown in Fig. 3.12(b). The predicted variation along the X-axis of the displacement u and stress resultants $N_{xx} = \sigma_{xx}.\ h$ and $N_{yy} = \sigma_{yy}, h$ is presented in Fig. 3.12(c) and (d) respectively. These are seen to be in excellent agreement with the known two-dimensional elasticity theory solutions [3.1].

In fact, converged accurate solutions can also be obtained using a coarse mesh of quadrilateral elements of quadratic or cubic orders, which can be distorted to fit the circular boundaries more accurately. The reader is advised to perform numerical experiments with this idea for different values of a/b.

3.5.4 Thermal stress analysis of a solid circular disk

A solid circular disk of radius b and thickness h subjected to radially varying thermal loading is considered. This test problem is designed to assess the accuracy of the consistent nodal force evaluation due to thermal strain. The temperature varies with the radius r as $T = T_0[1 - (r/b)^2]$ where T_0 is the temperature change at the center of the disk.

Exploiting the symmetry, only one quadrant of the disk is discretized using the mesh of QUAD 4 elements as shown in Fig. 3.13(a). The predicted variation along the x-axis of the displacement u, the radial stress σ_{xx} and the circumferential stress σ_{yy} are presented in Fig. 3.13(b) and (c) respectively. While the finite element solutions are shown as points, the known analytical solution is plotted as continuous curves. The two solutions are found to be in good agreement. The numerical results correspond to $b = 0.1$ inch, $h = 0.005$ inch, $T_0 = 100°F$, $E = 10 \times 10^6$ psi, $\nu = 0.3$, $\alpha = 10 \times 10^{-6}$ (/°F).

The reader is advised to reanalyze this problem for $T(r) = T_0[1 - (r/b)]$ and comment on the results.

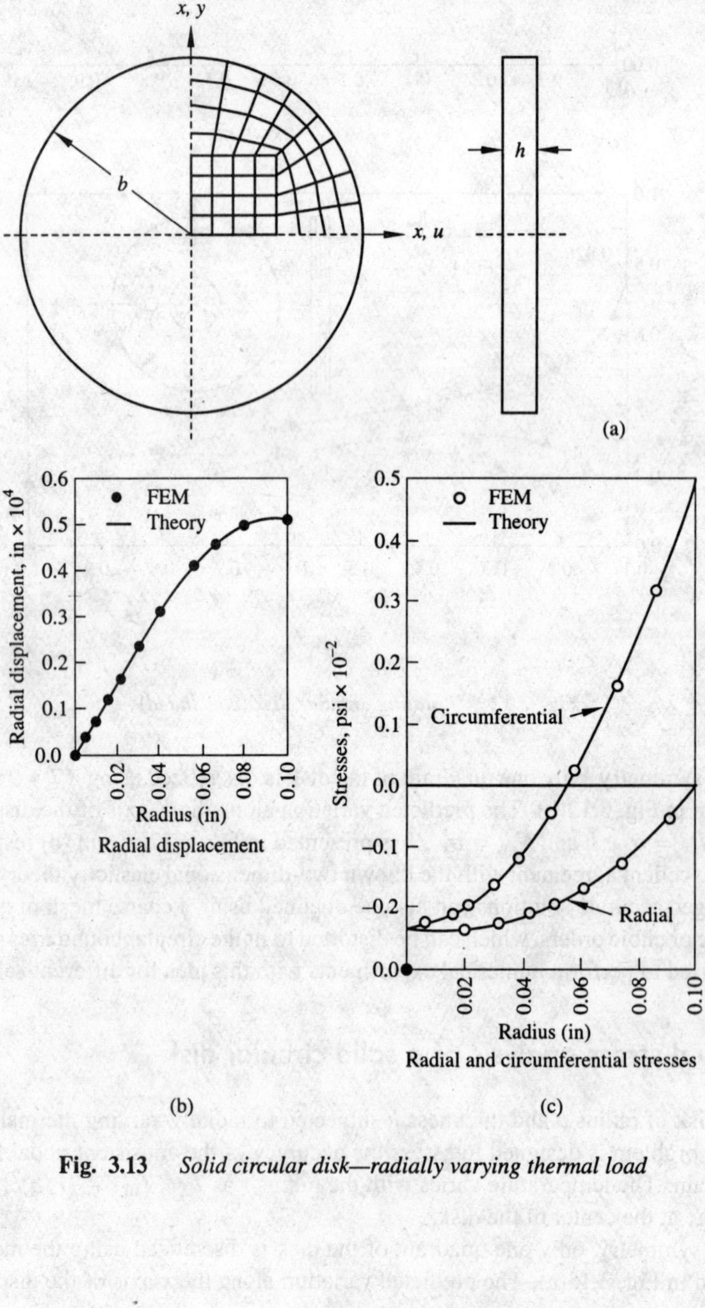

Fig. 3.13 *Solid circular disk—radially varying thermal load*

3.6 Some Practical Applications

Obviously, the practical applications of two-dimensional finite elements are limitless and finite element analysis has superseded experimental techniques for plane elasticity problems because of its high accuracy, low cost, and versatality. The ease of modelling complex geometries, heterogeneous anisotropic materials, body forces and thermal loads add to its advantage. A few examples of the application of two-dimensional elements in engineering practice will now be given.

3.6.1 Stress analysis of a tapered skew cantilever plate

This problem, whose details are given in Fig. 3.14(a) is modelled as a plane stress problem and discretized using a 5 × 4 mesh of two dimensional isoparametric elements, quadrilateral in shape and of quadratic order. The nodes along the clamped edge are all constrained and the load P is applied at all the nodes along the right edge. A linear static analysis is carried out. The predicted displacement at point C in the Y direction is 23.718 while the target value is 23.91. Fig. 3.14(b) shows the variation of the bending stress σ_{xx} along the line AB. The bending stress distribution along another line EF is shown in Fig. 3.14(c). It is highly nonlinear and clearly brings out the effect of the re-entrant corner which is a singular point. Obviously, a local mesh refinement around the point S is required to capture the singular stress state at the point S. The reader is welcome to suggest an improved finite element model for reliable and accurate stress analysis of the problem.

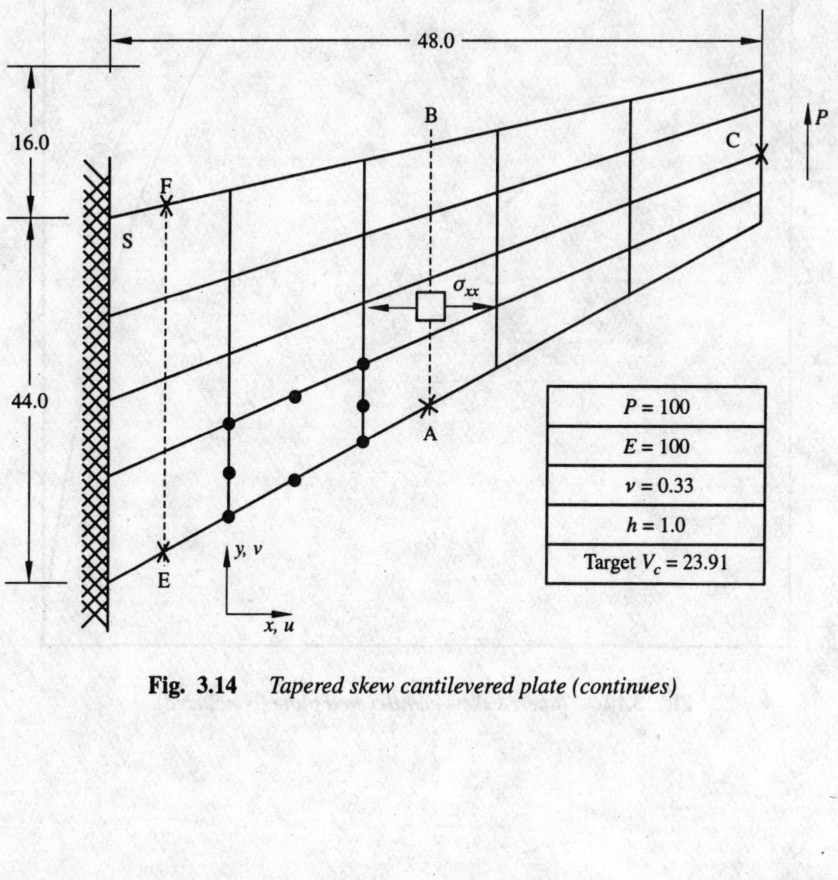

Fig. 3.14 *Tapered skew cantilevered plate (continues)*

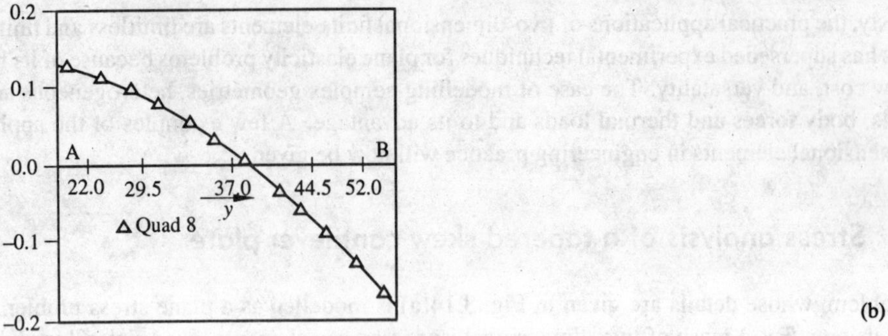

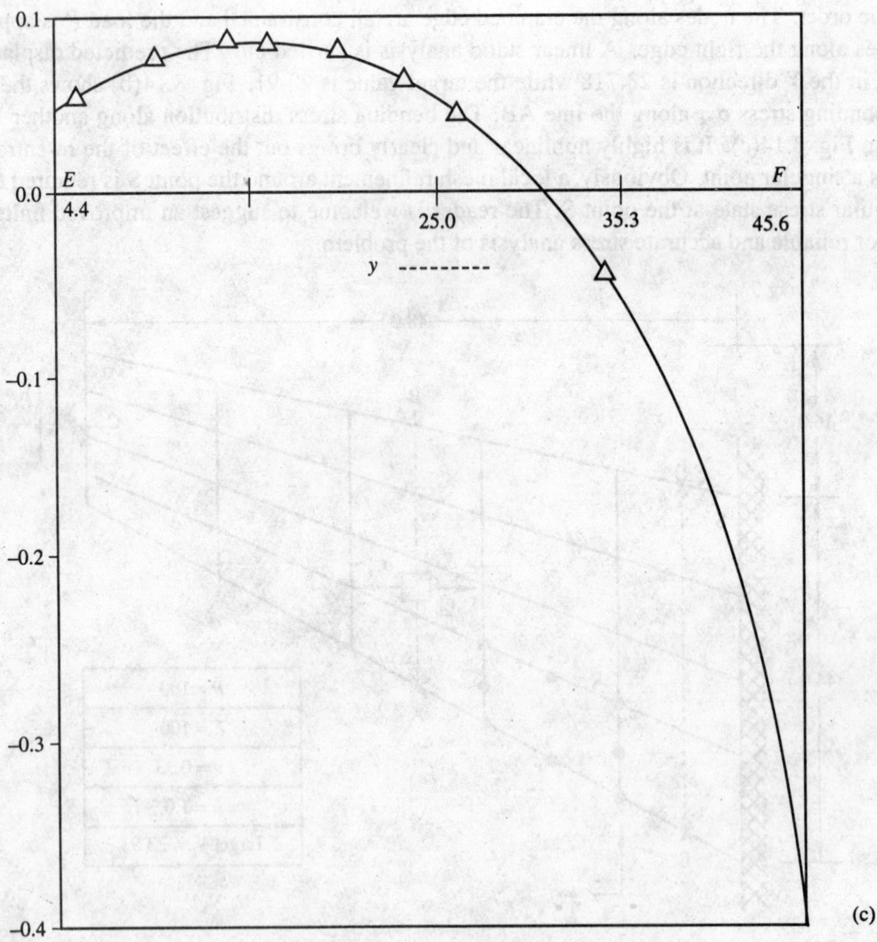

Fig. 3.14 *Tapered skew cantilevered plate (concluded)*

3.6.2 Stress concentration around a semi-elliptical edge notch

The notch geometry, loading, and support conditions are shown in Fig. 3.15(a). Assuming plane stress state, the domain is discretized using four-nodded quadrilateral elements and a locally refined mesh around the notch as shown in Fig. 3.15(b). The predicted variation of the stress σ_{xx} at the critical section is presented in Fig. 3.15(c). The computed stress concentration factor being 4.46 clearly shows the severity of this semi-elliptical ($b/a = 2$) notch in comparison with the semi-circular ($b/a = 1$) notch. The reader is advised to perform similar analysis with $b/a = 3$, 4, etc., and comment on the results. Accurate stress analysis is an important pre-requisite for strength prediction, which employs a variety of failure criteria. Some of these criteria are based on not only the maximum stress at the root of the notch but also on their gradients. The finite element method is ideally suited for providing this information.

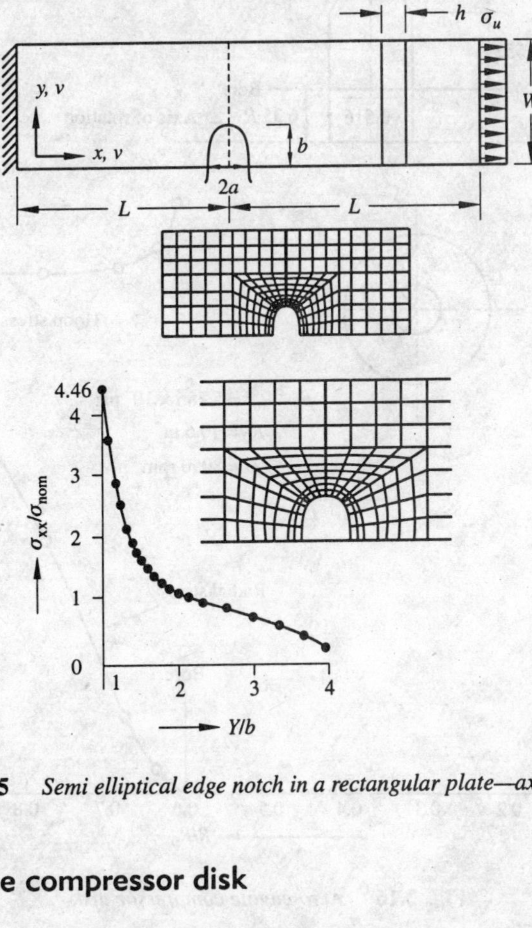

Fig. 3.15 *Semi elliptical edge notch in a rectangular plate—axial tension*

3.6.3 Aero-engine compressor disk

The cross-section of the disk between the bore and the neck is shown in Fig. 3.16(a). The loading considered is that due to rotation of the disk at 8050 rpm. The chosen finite element mesh of eight-noded quadrilateral elements (QUAD 8) is similar to that shown in Fig. 3.12(b). Symmetry conditions are enforced at nodes along X and Y axes. The inner surface is free and a uniform radial pressure simulating the action of blade loads is applied at the outer surface. Kinematically consistent nodal forces due to the centrifugal load and the pressure load are generated within the program used. The material of the disk is steel. Variations of radial and hoop stresses along a radial line are given in Fig. 3.16(b).

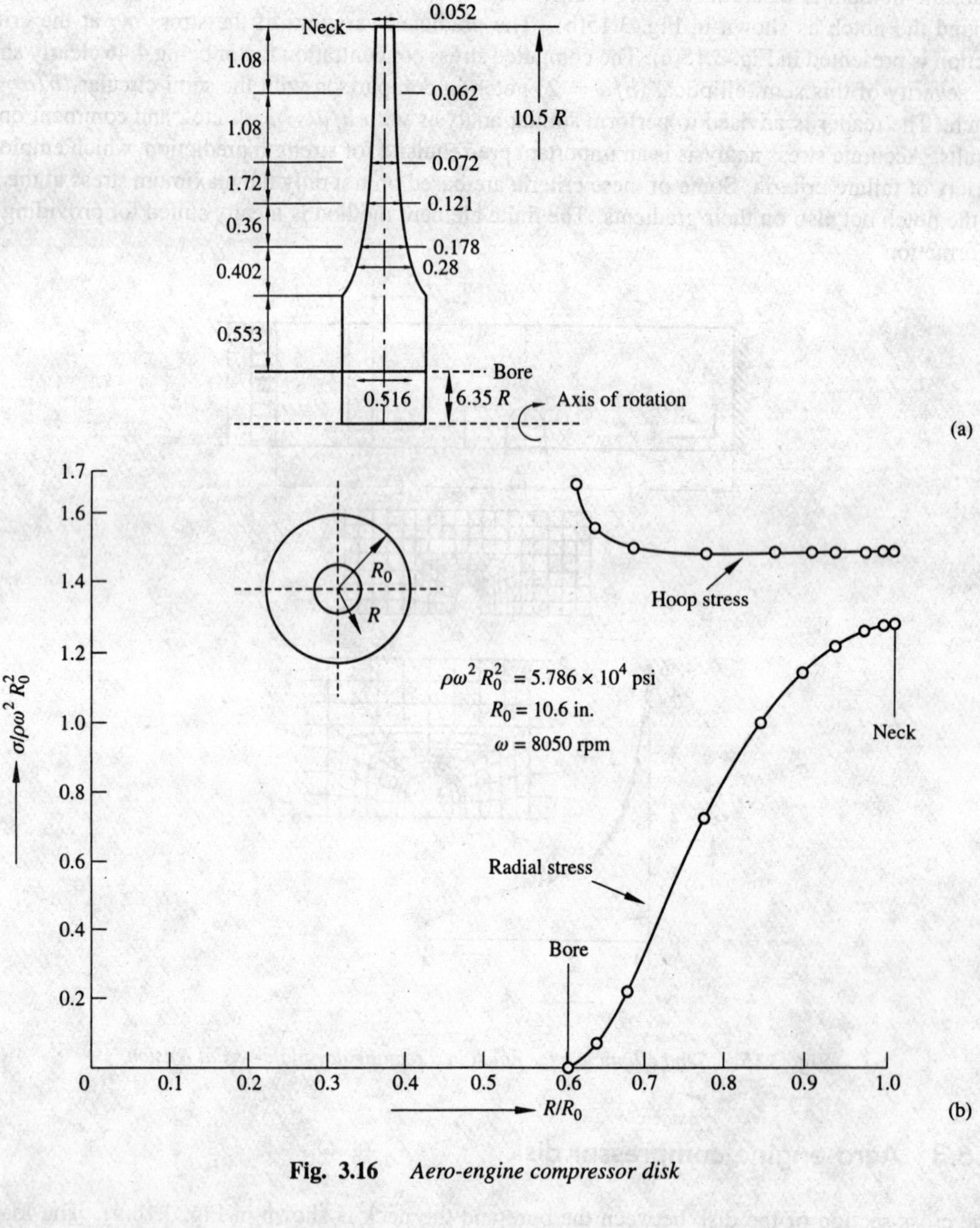

Fig. 3.16 *Aero-engine compressor disk*

In practice, such a disk is assembled on to a shaft using interference. The reader is advised to simulate the interference between this disk and a rigid shaft and plot the contact pressure distribution for various values of the radial interference.

Quite often the disk is an integral part of a shaft and its stress analysis is left to the reader.

3.7 Closure

A library of two-dimensional elements of different shapes (triangular and quadrilateral) and increasing orders (linear, quadratic, and cubic) has been presented to the reader. Some of these elements are implemented in commercial FEM systems and widely used in practice.

The ability of these elements to model complex domains with irregular geometric shapes makes the FEM a valuable tool for the solution of not only boundary value problems but also eigen value and initial value problems arising in various branches of engineering. The interested reader should refer Reddy [3.5] for details.

The combination of explicit shape functions and Gaussian quadrature formulae presented in this chapter are the key ingredients of element design. They are applicapable in other fields namely structural mechanics (analysis of beams, plates, and shells), heat transfer (prediction of temperature distribution), fluid dynamics, etc., in addition to the solid mechanics applications considered so far. The interested reader should refer Bathe [3.6] and Silvester [3.9] for details.

3.8 References

[3.1] Timoshenko, S.P. and Goodier J.N. *Theory of Elasticity*. McGraw Hill, 1970, 3rd ed., New York.

[3.2] Zienkiewicz, O.C. *The Finite Element Method*. Tata McGraw Hill, New Delhi, 3rd ed., 1979.

[3.3] Cowper, G.R. Gaussian quadrature formulae for triangles. *Int. Journal for Numerical Methods in Engineering*, 7:405–408, 1973.

[3.4] Lakshminarayana, H.V. Stress distribution around a semi-circular edge-notch in a finite size laminated composite plate under uniaxial tension. *Journal of Composite Materials*, 17:357–367, 1983.

[3.5] Reddy, J.N. *An Introduction to the Finite element Method*. McGraw Hill, New York, 2nd ed., 1993.

[3.6] Bathe, K.J. *Finite Element Procedures*. Prentice Hall of India, New Delhi, 1996.

[3.7] Barsoum, R.S. On the use of isoparametric finite elements in linear elastic fracture mechanics. *Int. J. Num. Meth. Engng*, 10:25–37, 1976.

[3.8] Pu, S.L., Hussain M.A., and Lorensen W.E. The collapsed cubic isotroparmetric element as singular element for crack problems. *Int. J. Num, Meth. Engng*, 12:1727–1742, 1978.

[3.9] Silventer, P.P., and Ferrari, R.L. *Finite ELements for Electrical Engineers*. Cambridge University Press, 1990, 2nd ed.

3.9 Computational Problems

In the following plane problems, report significant results. Exploit symmetry if possible. Choose convenient numbers and consistent units for material properties, dimensions, and loads. Unless otherwise stated, assume plane stress state, constant thickness, and homogeneous isotropic material model, and linear elastic behaviour.

3.1 An annular disk is supported and loaded as shown. Compute and plot stresses along AB and CD.

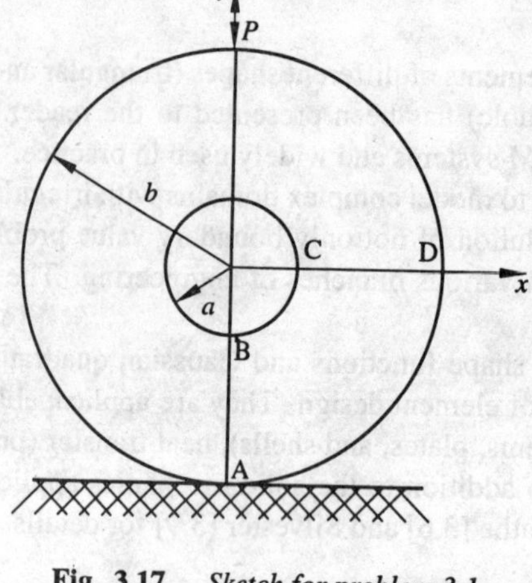

Fig. 3.17 *Sketch for problem 3.1*

3.2 An annular disk is clamped and subjected to a parabolic temperature distribution. Compute and plot the thermal stress distribution.

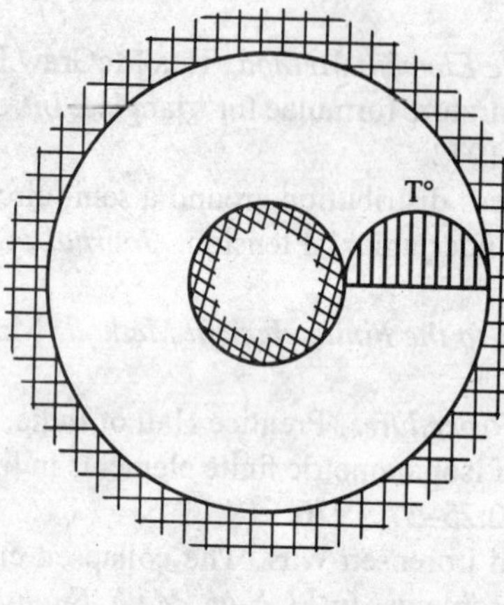

Fig. 3.18 *Sketch for problem 3.2*

3.3 The rotating compressor disk shown in Fig. 3.16 contains eight holes located on a pitch circle of radius 9.0 inches. The hole diameter is 1.0. An accurate stress analysis is required to locate the points where maximum stresses are induced and hence to compute the stress concentration factor.

3.4 A flat plate with central hole is loaded by internal pressure. Plane stress state prevails. Magnitudes and locations of maximum principal stress are desired.

The sketch shows a central crack of length $2a$ in a long rectangular plate whose width is $2c$. The stresses are singular at the crack tips and vary as $1/\sqrt{r}$, where r is the radial distance measured from a crack tip. The mode I stress intensity factor is $\mathbf{K}_I$. A handbook gives a formula for $\mathbf{K}_I$ as

$$\frac{\mathbf{K}_I}{\sigma_0\sqrt{\pi a}} = Y\left(\frac{a}{c}\right) \equiv \frac{1 - 0.5(a/c) + 0.326(a/c)^2}{[1 - (a/c)]^{0.5}}$$

Assign convenient dimensions, create finite element models, do computations, use appropriate methods for evaluation of $\mathbf{K}_I$ and compare the computed and formula values of $\mathbf{K}_I$.

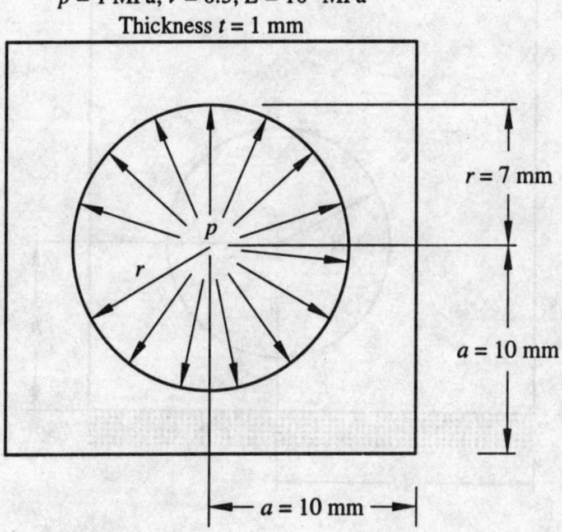

Fig. 3.19 *Sketch for problem 3.4*

3.5 The sketch shows a circular hole of radius R in a rectangular plate whose width is H. It is securely bonded to a rigid base on the bottom and to a rigid bar on the top. Peak values of Von-Mises equivalent stress around the hole are required. Load the model by translating the rigid bar by an amount (a) u_0 in the x direction or (b) v_0 in the y direction. Assign convenient dimensions, material properties and numbers.

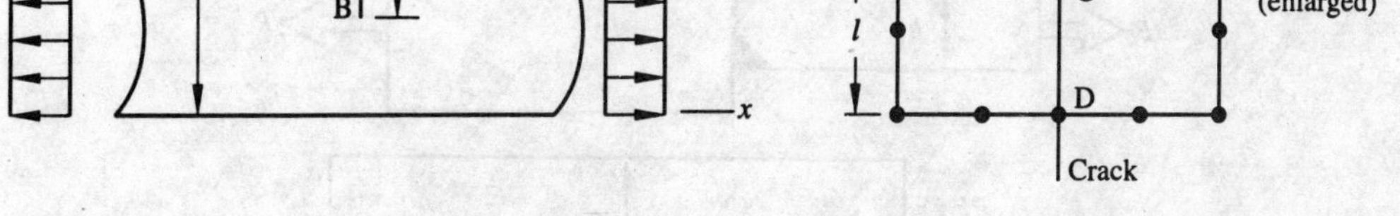

Fig. 3.20 *Sketch for problem 3.5*

3.6 As a variant of Problem 3.6, locally reinforce the hole by doubling the z direction thickness of material from the edge of the hole out to a radius $1.3R$. Predict the alleviation of stress concentration due to the local reinforcement.

3.7 The geometry, dimensions, supports and loading for the Izod and Charpy impact tests on materials is well-known. The stress concentration around the notch is of special interest. A finite

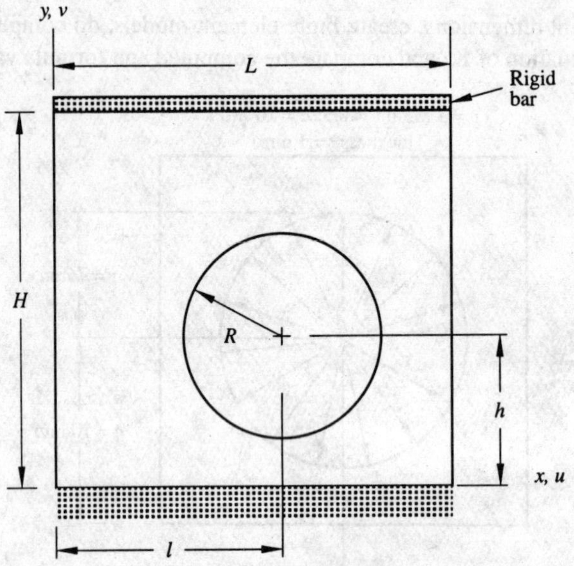

Fig. 3.21 *Sketch for problem 3.6*

element analysis can be undertaken, using progressive mesh refinement near the point of peak stress, until error is reduced to less than (say) 5%.

3.8 Welding is often used to join two plates of different thicknesses and different materials using both lap and butt joints. A finite element analysis can be undertaken in support of the design of such joints.

3.9 Any of the test problems considered in Section 3.5 can be modified by modelling the material as orthotropic. With L and T as orthogonal material axes, E_L, E_T, G_{LT}, and ν_{LT} as basic elastic constants, the axis L may be oriented arbitrarily with respect to the global x-axis. Computations can be performed using the known properties of glass/epoxy, Kevlar/epoxy, and graphite/epoxy composites and for various orientation of the fibre direction with the x-axis. The material anisotropy will have a significant influence on the results.

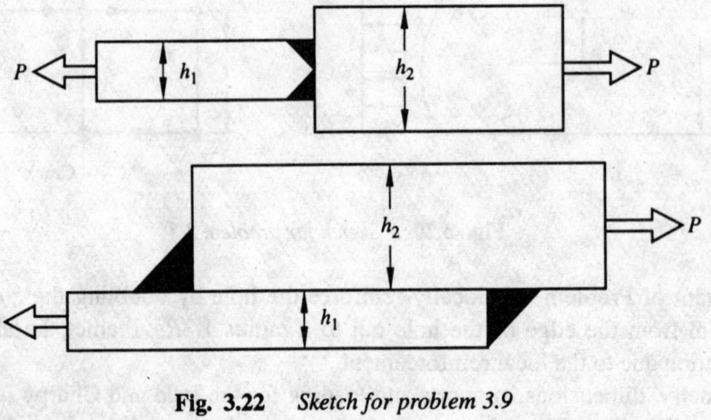

Fig. 3.22 *Sketch for problem 3.9*

4 Axisymmetric Solid Elements
Analysis of Bodies of Revolution

4.1 Overview

A body of revolution, also called an axisymmetric solid, is generated by revolving a plane figure about an axis in the plane. Common examples include a cast flywheel (Fig. 4.1), and a rocket motor casing (Fig. 4.2). We consider the case where geometry, loads, and supports are all axisymmetric. Consequently, nothing varies with the circumferential coordinate θ and material points displace only radially (r) and axially (z). Thus the analysis of the problem is mathematically two-dimensional.

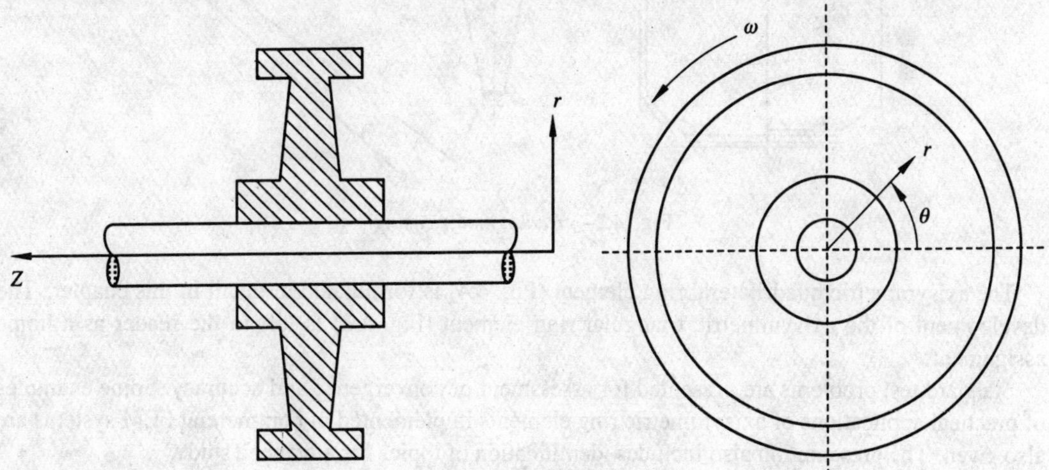

Fig. 4.1 *Rotating flywheel*

Finite elements for axisymmetric solids are pictured as triangular (Fig. 4.3) or quadrilateral (Fig. 4.4), but these shapes are actually cross-sections of ring elements and what appear to be nodal points are actually nodal circles. It can be readily seen that the same interpolation functions as those used in Chapter 3 can be used to represent the geometry as well as to define the displacement functions.

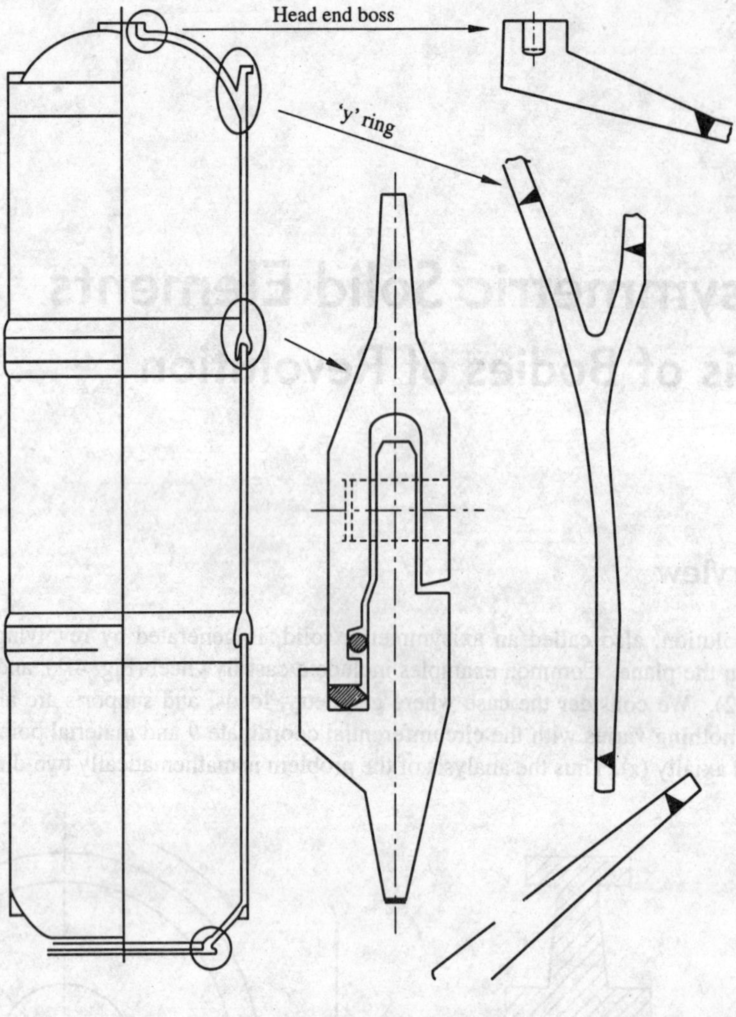

Fig. 4.2 *Rocket motor casing*

The axisymmetric quadrilateral ring element (Fig. 4.4) is formulated in detail in this chapter. The development of the axisymmetric triangular ring element (Fig. 4.3) is left to the reader as a home assignment.

Standard test problems are presented for assessment of convergence and accuracy. Some examples of practical applications of axisymmetric ring elements implemented in commercial FEM systems are also given. The presentation also includes identification of topics for advanced study.

4.2 Axisymmetric Quadrilateral Ring Element

The cross-section of a typical body of revolution can be discretized using a four-noded quadrilateral. In reality, each element is a ring obtained by revolving the quadrilateral about the z-axes. It has four nodal circles, which are identified as nodes in the r, z plane. Each node i has two displacement degrees

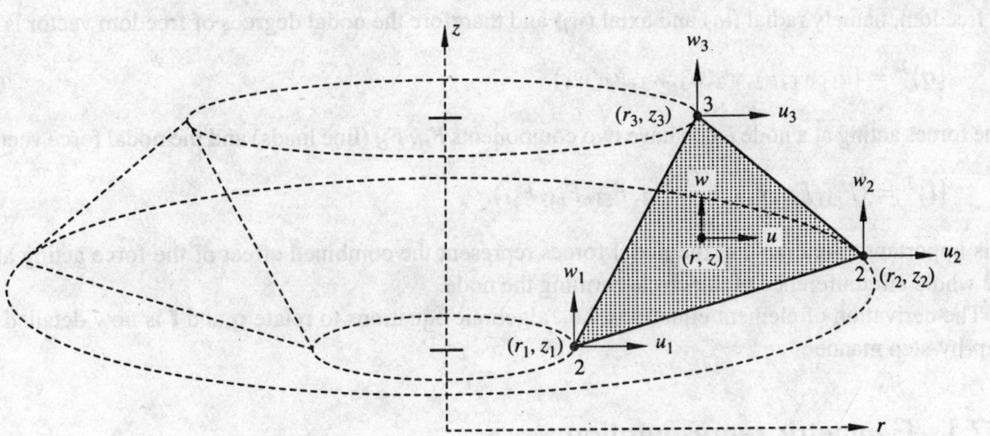

Fig. 4.3 *Three-node triangular ring element*

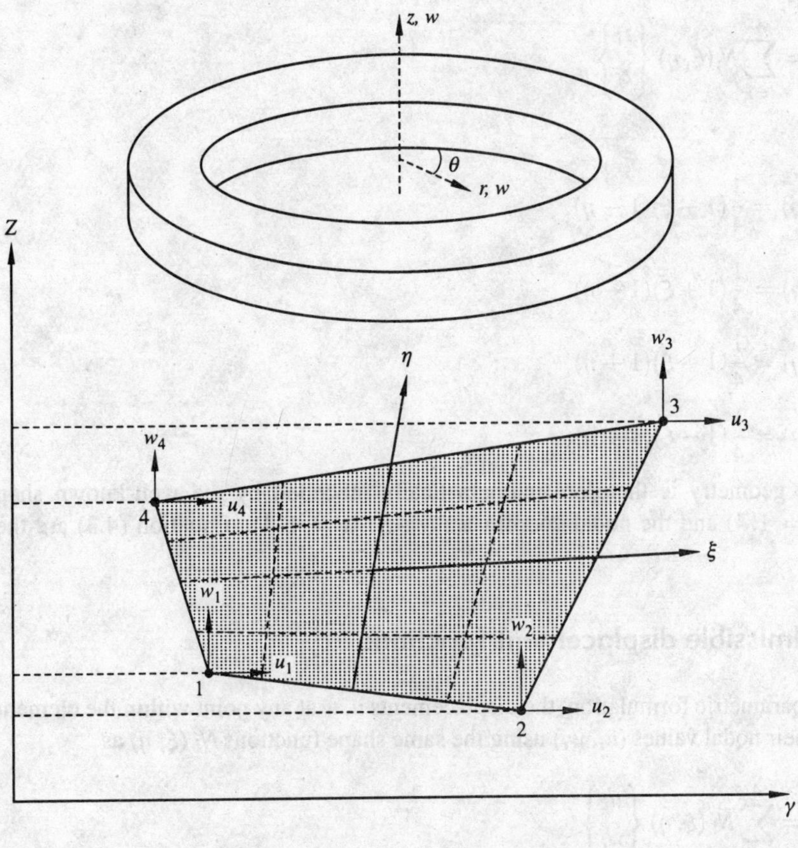

Fig. 4.4 *Four-node quadrilateral ring element*

of freedom, namely radial (u_i) and axial (w_i) and therefore the nodal degrees of freedom vector is

$$\{\mathbf{q}\}^T = (u_1, w_1, u_2, w_2, u_3, w_3, u_4, w_4). \tag{4.1}$$

The forces acting at a node i also have two components F_{ri}, F_{zi} (line loads) and the nodal force vector is

$$\{\mathbf{f}\}^T = (F_{r1}, F_{z1}, F_{r2}, F_{z2}, F_{r3}, F_{z3}, F_{r4}, F_{z4}). \tag{4.2}$$

It is important to realize that the nodal forces represent the combined effect of the force acting along the whole circumference of the circle forming the node.

The derivation of element equations, i.e., algebraic equations to relate **q** and **f** is now detailed in a step-by-step manner.

4.2.1 Geometric representation

The (r, z) coordinates of any point within an element have to be uniquely defined by their nodal values. The simplest representation is clearly given by

$$\begin{Bmatrix} r \\ z \end{Bmatrix} = \sum_{i=1}^{4} N_i(\xi, \eta) \begin{Bmatrix} r_i \\ z_i \end{Bmatrix} \tag{4.3}$$

where

$$\begin{aligned} N_1(\xi, \eta) &= \frac{1}{4}(1-\xi)(1-\eta), \\ N_2(\xi, \eta) &= \frac{1}{4}(1+\xi)(1-\eta), \\ N_3(\xi, \eta) &= \frac{1}{4}(1+\xi)(1+\eta), \\ N_4(\xi, \eta) &= \frac{1}{4}(1-\xi)(1+\eta). \end{aligned} \tag{4.4}$$

The element geometry is therefore mathematically modelled by the well-known shape functions N_i (ξ, η) $(i = 1, 4)$ and the nodal coordinates (r_i, z_i) $(i = 1, 4)$. Equation (4.3) are the parametric equations.

4.2.2 Admissible displacement functions

By using isoparametric formulation, the displacements u, w at any point within the element are defined in terms of their nodal values (u_i, w_i) using the same shape functions N_i (ξ, η) as

$$\begin{Bmatrix} u \\ w \end{Bmatrix} = \sum_{i=1}^{4} N_i (\xi, \eta) \begin{Bmatrix} u_i \\ w_i \end{Bmatrix}. \tag{4.5}$$

It can be verified that this displacement field satisfies the convergence criteria.

4.2.3 Element strain matrix

In the axisymmetric elasticity problem, any radial displacement u automatically induces a strain $\varepsilon_{\theta\theta}$ in the circumferential (θ) direction and the stress in this direction $\sigma_{\theta\theta}$ (the hoop stress) is certainly nonzero. The strain components possible in an axisymmetric deformation are the radial strain ε_{rr}, circumferential strain $\varepsilon_{\theta\theta}$, the axial strain ε_{zz}, and the shear strain γ_{rz}. These are evaluated using the strain displacement relations [4.1] as

$$\{\boldsymbol{\varepsilon}\} = \begin{Bmatrix} \varepsilon_{rr} \\ \varepsilon_{\theta\theta} \\ \varepsilon_{zz} \\ \gamma_{rz} \end{Bmatrix} = \begin{bmatrix} \frac{\partial}{\partial r} & 0 \\ \frac{1}{r} & 0 \\ 0 & \frac{\partial}{\partial z} \\ \frac{\partial}{\partial z} & \frac{\partial}{\partial r} \end{bmatrix} \begin{Bmatrix} u \\ w \end{Bmatrix}. \tag{4.6}$$

Using Eqn. (4.5) we can write

$$\{\boldsymbol{\varepsilon}\} = [\mathbf{B}]\{\mathbf{q}\}, \tag{4.7}$$

where

$$[\mathbf{B}] = \begin{bmatrix} \frac{\partial N_1}{\partial r} & 0 & \frac{\partial N_2}{\partial r} & 0 & \frac{\partial N_3}{\partial r} & 0 & \frac{\partial N_4}{\partial r} & 0 \\ \frac{N_1}{r} & 0 & \frac{N_2}{r} & 0 & \frac{N_3}{r} & 0 & \frac{N_4}{r} & 0 \\ 0 & \frac{\partial N_1}{\partial z} & 0 & \frac{\partial N_2}{\partial z} & 0 & \frac{\partial N_3}{\partial z} & 0 & \frac{\partial N_4}{\partial z} \\ \frac{\partial N_1}{\partial z} & \frac{\partial N_1}{\partial r} & \frac{\partial N_2}{\partial z} & \frac{\partial N_2}{\partial r} & \frac{\partial N_3}{\partial z} & \frac{\partial N_3}{\partial r} & \frac{\partial N_4}{\partial z} & \frac{\partial N_4}{\partial r} \end{bmatrix}. \tag{4.8}$$

The required derivatives of the shape functions $N_i\ (\xi, \eta)$ with respect to r and z are evaluated in the usual manner as

$$\begin{bmatrix} \frac{\partial N_i}{\partial r} \\ \frac{\partial N_i}{\partial z} \end{bmatrix} = [\mathbf{J}]^{-1} \begin{bmatrix} \frac{\partial N_i}{\partial \xi} \\ \frac{\partial N_i}{\partial \eta} \end{bmatrix} \quad (i = 1, 2, 3, 4). \tag{4.9}$$

The Jacobian matrix $\mathbf{J}$ is evaluated as

$$[\mathbf{J}] = \sum_{j=1}^{4} \begin{bmatrix} \frac{\partial N_j}{\partial \xi} r_j & \frac{\partial N_j}{\partial \xi} z_j \\ \frac{\partial N_j}{\partial \eta} r_j & \frac{\partial N_j}{\partial \eta} z_j \end{bmatrix} = \begin{bmatrix} J_{11} & J_{12} \\ J_{21} & J_{22} \end{bmatrix}, \tag{4.10}$$

where (r_j, z_j) denote the (r, z) coordinates of node j.

The inverse of $\mathbf{J}$ is calculated as

$$[\mathbf{J}]^{-1} = \frac{1}{|J|} \begin{bmatrix} J_{22} & -J_{12} \\ -J_{21} & J_{11} \end{bmatrix} \tag{4.11}$$

and $|\mathbf{J}| = (J_{11} * J_{22} - J_{12} * J_{21})$.

Obviously, the determinant of the Jacobian matrix cannot become zero. Furthermore, for the distorted shape of an arbitrary quadrilateral to be reasonable, the sign of $|\mathbf{J}|$ should not change from point to point within the quadrilateral element.

Equation (4.7) thereby enables evaluation of the four strain components at any point within the element once $\mathbf{q}$ is known.

The most frequently encountered case of initial strains will be that due to a thermal expansion. Assuming a homogeneous isotropic material model, the initial strain vector is

$$\left\{\boldsymbol{\varepsilon}^{\text{init}}\right\} = \begin{Bmatrix} \alpha \sum_{i=1}^{4} N_i(\xi,\eta)\, T_i \\ \alpha \sum_{i=1}^{4} N_i(\xi,\eta)\, T_i \\ \alpha \sum_{i=1}^{4} N_i(\xi,\eta)\, T_i \\ 0 \end{Bmatrix}, \tag{4.12}$$

where T_i denotes temperature change at node i and α is the coefficient of thermal expansion.

4.2.4 Element stress recovery

The state of stress in a body of revolution undergoing axisymmetric deformation is characterized by the radial stress σ_{rr}, the circumferential (hoop) stress $\sigma_{\theta\theta}$, the axial stress σ_{zz} and the shear stress τ_{rz} as shown in Fig. 4.5. These stresses are related to the strains through the constitutive law of the form

$$\{\sigma\} = [\mathbf{D}](\{\boldsymbol{\varepsilon}\} - \{\boldsymbol{\varepsilon}^{\text{init}}\}), \tag{4.13}$$

where

$$\{\sigma\}^T = (\sigma_{rr}\sigma_{\theta\theta}\sigma_{zz}\tau_{rz}).$$

The matrix $\mathbf{D}$ can be explicitly defined for assumed material models. For a homogeneous isotropic material and linear and elastic behaviour, the generalized Hooke's law is

$$\begin{aligned}
\varepsilon_{rr} &= \frac{\sigma_{rr}}{E} - \frac{\nu}{E}(\sigma_{\theta\theta} + \sigma_{zz}) + \varepsilon_{rr}^{\text{init}}, \\
\varepsilon_{\theta\theta} &= \frac{\sigma_{\theta\theta}}{E} - \frac{\nu}{E}(\sigma_{rr} + \sigma_{zz}) + \varepsilon_{\theta\theta}^{\text{init}}, \\
\varepsilon_{zz} &= \frac{\sigma_{zz}}{E} - \frac{\nu}{E}(\sigma_{rr} + \sigma_{\theta\theta}) + \varepsilon_{zz}^{\text{init}}, \\
\gamma_{rz} &= \frac{\tau_{rz}}{G} = \frac{2(1+\nu)}{E}\tau_{rz} + 0.
\end{aligned} \tag{4.14}$$

On solving for stresses, we define

$$[\mathbf{D}] = \frac{\mathrm{E}}{(1+\nu)(1-2\nu)} \begin{bmatrix} (1-\nu) & \nu & \nu & 0 \\ \nu & (1-\nu) & \nu & 0 \\ \nu & \nu & (1-\nu) & 0 \\ 0 & 0 & 0 & \frac{(1-2\nu)}{2} \end{bmatrix}. \tag{4.15}$$

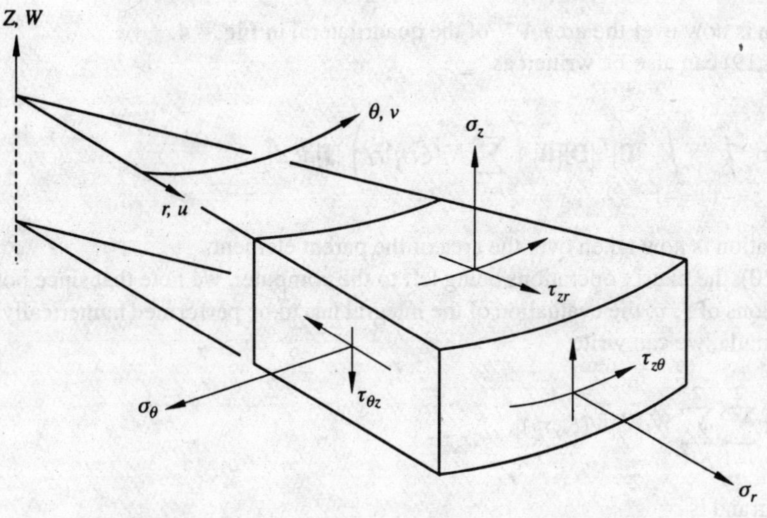

Fig. 4.5 *Stresses and displacements in a solid of revolution*

Using Eqn. (4.7) in Eqn. (4.13) we get

$$\{\sigma\} = [\mathbf{D}][\mathbf{B}]\{\mathbf{q}\} - [\mathbf{D}]\{\boldsymbol{\epsilon}^{\text{init}}\}. \tag{4.16}$$

The stress state is three-dimensional and varies from point to point within the element. The Gauss point locations ($\xi = \pm 1/\sqrt{3}, \eta = \pm 1/\sqrt{3}$) are optimal for accurate stress recovery. However, if nodal point stresses are needed, extrapolation of the Gauss point stresses and averaging of contributions to a node from adjacent elements is essential. With the stresses σ known, the three principal stresses $\sigma_1 > \sigma_2 > \sigma_3$ can be determined and used to compute an equivalent stress. For example, the Von-Mises equivalent stress

$$\sigma_e^{\text{Von-Mises}} = \left[\frac{(\sigma_1 - \sigma_2)^2 + (\sigma_2 - \sigma_3)^2 + (\sigma_3 - \sigma_1)^2}{2}\right]^{1/2} \tag{4.17}$$

can be calculated to predict the onset of failure by yielding.

4.2.5 Element stiffness matrix

The formula for stiffness matrix is

$$[\mathbf{k}] = \int_{v^{(e)}} [\mathbf{B}]^T[\mathbf{D}][\mathbf{B}]dv, \tag{4.18}$$

where $v^{(e)}$ denotes that the integral has to be taken over the volume of the element. Equation (4.18) can be simplified as

$$[\mathbf{k}] = \int_{A^{(e)}} [\mathbf{B}]^T[\mathbf{D}][\mathbf{B}]2\pi r dr dz. \tag{4.19}$$

The integration is now over the area $A^{(e)}$ of the quadrilateral in Fig. 4.4.

Equation (4.19) can also be written as

$$[\mathbf{k}] = 2\pi \int_{\xi=-1}^{\xi=+1} \int_{\eta=-1}^{\eta=+1} [\mathbf{B}]^T[\mathbf{D}][\mathbf{B}] \left(\sum_{i=1}^{4} N_i(\xi,\eta) r_i \right) |\mathbf{J}| d\xi d\eta. \tag{4.20}$$

And the integration is now taken over the area of the parent element.

In Eqn. (4.20), the matrix operations being left to the computer, we note that since both **B** and **J** are involved functions of ξ, η, the evaluation of the integral has to be performed numerically. Using Gauss quadrature formula, we can write

$$[\mathbf{k}] = 2\pi \sum_{i=1}^{2} \sum_{j=1}^{2} W_i W_j [\Phi(\xi_i, \eta_j)], \tag{4.21}$$

where the integrand is

$$[\Phi(\xi,\eta)] = [\mathbf{B}]^T[\mathbf{D}][\mathbf{B}] \left(\sum_{i=1}^{4} N_i(\xi,\eta) r_i \right) |\mathbf{J}|. \tag{4.22}$$

In Eqn. (4.21), W_i and W_j are the weighting factors and (ξ_i, η_j) denotes the Gauss point locations given in Table 3.2.

A comment on the choice of number of Gauss points is in order. It can be shown that if numerical integration is of such an order that the volume of the element is exactly determined by it, then in the limit of subdivision, the finite element solution will converge to the exact solution. The two-point formula chosen here is of such a type, as it is well-known that the volume of a body of revolution is given exactly by the product of the area and the path swept around by its centroid.

4.2.6 Consistent nodal force vector: Body force

Body forces such as those due to gravity (if acting along the axis of revolution), and centrifugal force (acting along the radial direction), often occur in axisymmetric elasticity problems. Let such forces be denoted by

$$\{\mathbf{b}\} = \begin{Bmatrix} b_r \\ b_z \end{Bmatrix},$$

where b_r and b_z are body forces per unit volume of material in direction of r and z respectively. For example, in rotating machinery, we have the centrifugal force

$$b_r = \rho\omega^2 r,$$

where ω (rad/s) is the angular velocity and ρ is the density of the material.

The body force in the axial direction due to gravity is

$$b_z = -\rho g.$$

The consistent nodal force vector $\mathbf{f}^b$ due to body force is

$$\{\mathbf{f}^b\} = \int_{V^{(e)}} [\mathbf{N}]^T \{\mathbf{b}\} dv \tag{4.23}$$

and the integration is over the volume $v^{(e)}$ of the element. Equation (4.23) can also be written as

$$\{\mathbf{f}^b\} = \int_{A^{(e)}} [\mathbf{N}]^T \{\mathbf{b}\} 2\pi r dr dz \tag{4.24}$$

and the integration is now over the area $A^{(e)}$ of the quadrilateral (see Fig. 4.4).

Equation (4.24) can also be written as

$$\{\mathbf{f}^b\} = 2\pi \int_{\xi=-1}^{\xi=+1} \int_{\eta=-1}^{\eta=+1} [\mathbf{N}]^T \left\{ \begin{matrix} \rho\omega^2 \sum_{i=1}^{4} N_i(\xi,\eta)\, r_i \\ -\rho g \end{matrix} \right\} \left(\sum_{i=1}^{4} N_i(\xi,\eta)\, r_i \right) |\mathbf{J}|\, d\xi\, d\eta \tag{4.25}$$

and the integration is now over the parent element domain.

In Eqn. (4.25), the matrix operations being left to the computer, we note that the integrand is a very complicated function of ξ and η. Therefore, the integrals have to be evaluated numerically. Using a two-point Gauss quadrature formula, we get

$$\{\mathbf{f}^b\} = 2\pi \sum_{i=1}^{2} \sum_{j=1}^{2} W_i W_j \{\psi(\xi_i, \eta_j)\}, \tag{4.26}$$

where the integrand is

$$\{\psi(\xi,\eta)\} = [N]^T \left\{ \begin{matrix} \rho\omega^2 \sum_{i=1}^{4} N_i(\xi,\eta)\, r_i \\ -\rho g \end{matrix} \right\} \left(\sum_{i=1}^{4} N_i(\xi,\eta)\, r_i \right) |\mathbf{J}|\,. \tag{4.27}$$

In Eqn. (4.26) W_i, W_j are the weighting factors and (ξ_i, η_j) denote Gauss point locations given in Table 3.2.

4.2.7 Consistent nodal force vector: Initial strain

Consistent nodal force vector due to initial strain $\mathbf{f}^{\text{init}}$ is evaluated using

$$\{\mathbf{f}^{\text{init}}\} = \int_{V^{(e)}} [\mathbf{B}]^T [\mathbf{D}] \{\boldsymbol{\varepsilon}^{\text{init}}\} dv, \tag{4.28}$$

where the initial strains $\boldsymbol{\varepsilon}^{\text{init}}$ are evaluated using Eqn. (4.12) for the thermal load case.

Using Gauss quadrature formula to evaluate numerically the integral on the right-hand side of Eqn. (4.28) leads us to the following formula

$$\{\mathbf{f}^{\text{init}}\} = 2\pi \sum_{i=1}^{2}\sum_{j=1}^{2} W_i W_j \{\chi(\xi_i, \eta_j)\}, \tag{4.29}$$

where the integrand is

$$\{\chi(\xi,\eta)\} = [\mathbf{B}]^T[\mathbf{D}] \begin{Bmatrix} \alpha \sum_{i=1}^{4} N_i(\xi,\eta)T_i \\ \alpha \sum_{i=1}^{4} N_i(\xi,\eta)T_i \\ \alpha \sum_{i=1}^{4} N_i(\xi,\eta)T_i \\ 0 \end{Bmatrix} \left(\sum_{i=1}^{4} N_i(\xi,\eta)\right) |\mathbf{J}|. \tag{4.30}$$

As before, W_i, W_j are the weighting factors and (ξ_i, η_j) denote Gauss point locations.

4.2.8 Consistent nodal force vector: Surface traction

The axisymmetric quadrilateral ring element (see Fig. 4.4) is bounded by four faces. These are identified as face 1 ($\xi = +1$), face 2 ($\eta = +1$), face 3 ($\xi = -1$), and face 4 ($\eta = -1$). On each one of the faces a pressure load can be applied. For example, let us consider a uniform pressure of intensity **p** applied on face 1. This pressure multiplied by the surface area of face 1 is the surface force. Since this surface undergoes displacements, work is done. Equating this work to the work done by the nodal forces at nodes 2 and 3 associated with face 1, we can arrive at a formula to compute the consistent nodal force vector $\mathbf{f}^p$ due to surface traction $\{p\}$ as

$$\{\mathbf{f}^p\} = \int_{S_\sigma} [\bar{\mathbf{N}}]^T \{\mathbf{p}\} ds. \tag{4.31}$$

In Eqn. (4.31), S_σ denotes the surface area on which the pressure **p** is applied; $[\bar{\mathbf{N}}]$denotes the shape functions evaluated now on the face 1 ($\xi = 1, -1 \leq \eta \leq 1$). The pressure **p** acting normal to the face can be resolved into components p_x and p_y. In practice, the integral on the right-hand side of Eqn. (4.31) is evaluated numerically using a two-point Gauss quadrature formula.

4.2.9 Element equations

The potential energy functional for the axisymmetric quadrilateral ring element is evaluated as

$$\pi_p(\mathbf{q}) = \frac{1}{2}\{\mathbf{q}\}^T[\mathbf{K}]\{\mathbf{q}\} - \{\mathbf{q}\}^T\{\mathbf{f}\}, \tag{4.32}$$

where

$$\{\mathbf{f}\} = \{\mathbf{f}^c\} + \{\mathbf{f}^p\} + \{\mathbf{f}^b\} + \{\mathbf{f}^{\text{init}}\} \tag{4.33}$$

Invoking the variational principle of minimum potential energy to satisfy the equilibrium conditions, we get

$$\frac{\partial \pi_p}{\partial \{\mathbf{q}\}} = 0$$

or

$$[\mathbf{K}]\{\mathbf{q}\} = \{\mathbf{f}\}. \tag{4.34}$$

Equation (4.34) are known as element equations.

Assembly of such equations for all the elements in a finite element model gives

$$[\mathbf{K}]\{\mathbf{Q}\} = \{\mathbf{F}\}. \tag{4.35}$$

Equation (4.35) is the governing matrix equation for linear static analysis of a body of revolution subjected to axisymmetric loads by the finite element method. Solution of these linear algebraic equations gives the vector of nodal degrees of freedom $\mathbf{Q}$, using which the displacements $\mathbf{u}$, strains $\boldsymbol{\varepsilon}$, and stresses σ can be calculated at any point within every element in the mesh, thus providing a whole field solution.

The design of axisymmetric triangular ring element shown in Fig. 4.1 follows the same step-by-step procedure described above. The basic ingredients namely explicit shape functions and Gauss quadrature formulae are well documented in Chapter 3.

4.3 Topics for Advanced Study

4.3.1 Non-axisymmetric loading

The finite element modelling and analysis procedure described in the previous section can be extended to deal with a body of revolution subjected to non-axisymmetric loading. If the geometry is axisymmetric, but the applied loads and/or enforced boundary conditions are not (for example, a cooling tower in a power plant subjected to wind load), the problem becomes truly three-dimensional in the sense that all three displacement components are in general nonzero. Thus the displacement components u (radial), v (circumferential), and w (axial) are all functions of r, θ and Z. Similarly, all six strain components (ε_{rr}, $\varepsilon_{\theta\theta}$, ε_{zz}, $\gamma_{r\theta}$, $\gamma_{\theta z}$, γ_{rz}) and corresponding stress components (σ_{rr}, $\sigma_{\theta\theta}$, σ_{zz}, $\tau_{r\theta}$, $\tau_{\theta z}$, τ_{rz}) are in general nonzero and all are functions of r, θ and Z.

If the circumferential (θ) variation of the applied load is expressed in circular harmonics, then it is still possible to discretize the cross-section of a body of revolution using the triangular and quadrilateral ring elements. However, each node i will now carry three degrees of freedom, namely, u_i, v_i, w_i. One obvious advantage of this approach over the use of three-dimensional elements is that no discretization in the circumferential direction is required.

Extension of the axisymmetric triangular and quadrilateral ring elements to handle non-axisymmetric loads and their application to assess the structural integrity of practical cooling towers used in thermal and nuclear power plants subject to wind loads is therefore identified as a topic for advanced study. The reader can refer [4.2] for details.

4.3.2 Higher order elements

The axisymmetric ring elements, triangular or quadrilateral in shape studied so far are linear in order. To improve geometric modeling, achieve faster convergence with mesh refinement, ensure higher accuracy for a chosen mesh, and guarantee computational efficiency, there is a real need to develop higher order elements.

A family of axisymmetric ring type elements of different shapes (triangular and quadrilateral) and increasing orders (quadratic and cubic) can therefore be designed on a pattern indicated in Figs. 3.8 and 3.9. The availability of explicit shape functions and Gauss quadrature formulae (see Chapter three) are the pre-requisites for the development of higher order elements.

4.3.3 Heat transfer and fluid flow analysis

The axisymmetric ring type elements can also be applied in other disciplines such as heat transfer and fluid flow in addition to the solid mechanics problems considered here. These are topics for a next level course on the finite element method. The reader is advised to refer Zienkiewicz [4.3], Patankar [4.6], Baker and Pepper [4.4] for details.

4.4 Test Problems—Assessment of Accuracy

4.4.1 Thick-walled sphere subjected to internal pressure

A hollow sphere with inner radius a and outer radius b, under internal pressure p is a classic case of a body of revolution under axisymmetric loading. Fig. 4.6(a) shows one half of such a body discretized using an 8×16 mesh of axisymmetric quadrilateral ring elements. Symmetry conditions are enforced at nodes along the r- and z-axes. A consistent evaluation of nodal forces due to surface pressure is included in the FEM system used.

The deformed shape of the body superimposed over the undeformed configuration is shown in Fig. 4.6(b). The predicted radial displacements at the inner and outer surfaces are in excellent agreement with the exact solution given in Ref. [4.5]. The distribution of the radial stress σ_{rr}, the hoop stress $\sigma_{\theta\theta}$ and meridional stress $\sigma_{\phi\phi}$ along a radial line is in good agreement with the exact solution given in [4.5].

The reader is advised to compute and plot these results for various values of (a/b) and verify their accuracy. It is informative to compute and plot the Von-Mises equivalent stress contours also.

4.4.2 Thick walled sphere subjected to thermal load

A hollow sphere with inner radius a, and outer radius b is subjected to a steady state temperature variation (100°F at inner surface, 0°F at outer surface, linear gradient). Exploiting symmetry, only one half of the body of revolution is discretized using an 8×16 mesh of axisymmetric quadrilateral ring elements (see Fig. 4.6(a)). Symmetry conditions are enforced at nodes along the r- and z-axes. The temperature change at each node is input to the program. A consistent evaluation of nodal forces due to thermal strain is included in the analysis.

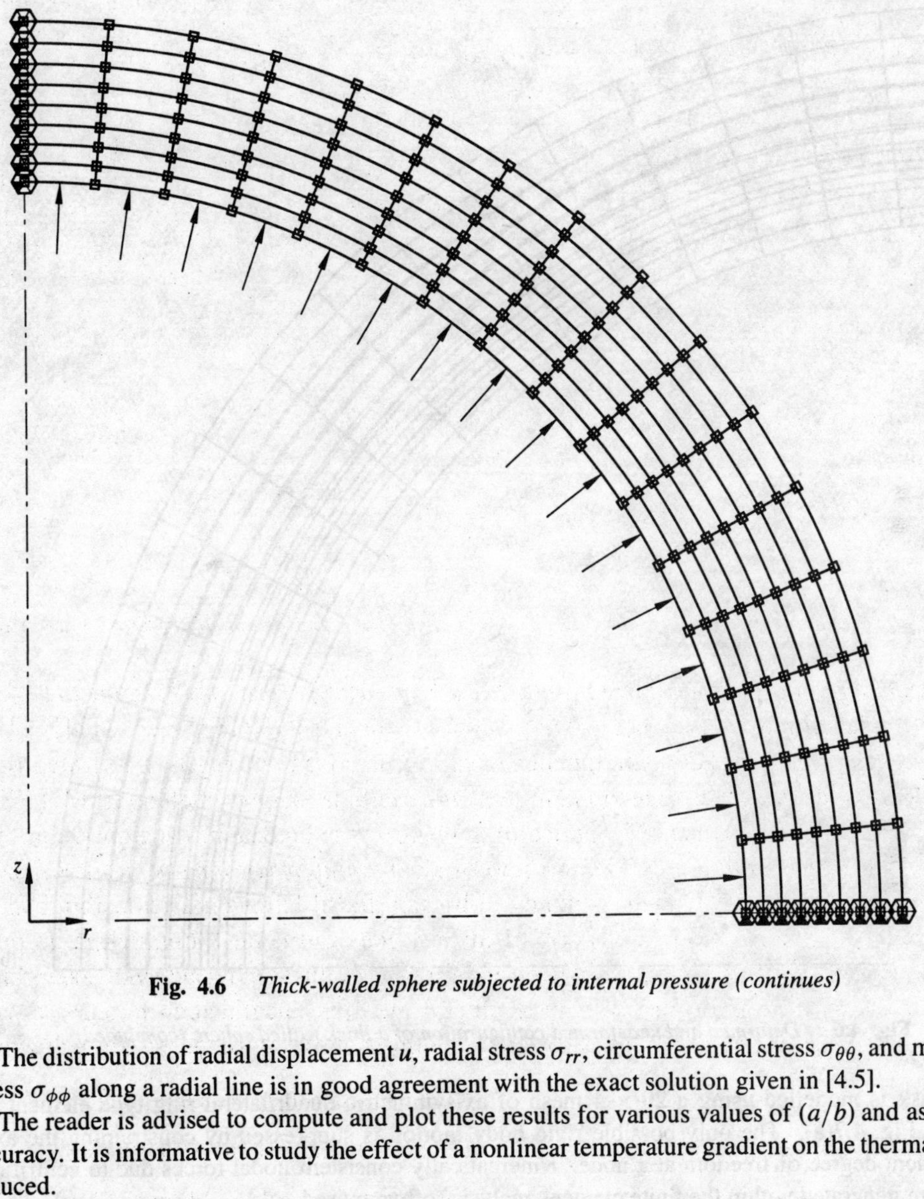

Fig. 4.6 *Thick-walled sphere subjected to internal pressure (continues)*

The distribution of radial displacement u, radial stress σ_{rr}, circumferential stress $\sigma_{\theta\theta}$, and meridional stress $\sigma_{\phi\phi}$ along a radial line is in good agreement with the exact solution given in [4.5].

The reader is advised to compute and plot these results for various values of (a/b) and assess their accuracy. It is informative to study the effect of a nonlinear temperature gradient on the thermal stresses induced.

Since the material is in a three-dimensional stresses state, it is recommended to compute and plot the Von-Mises equivalent stress contours and hence locate critical regions.

4.4.3 Rotating annular disk

An annular disk with inner radius a, outer radius b, and thickness h is made out of steel. The loading considered is that due to rotation of the disk at 8000 rpm (see insert in Fig. 4.7(d)).

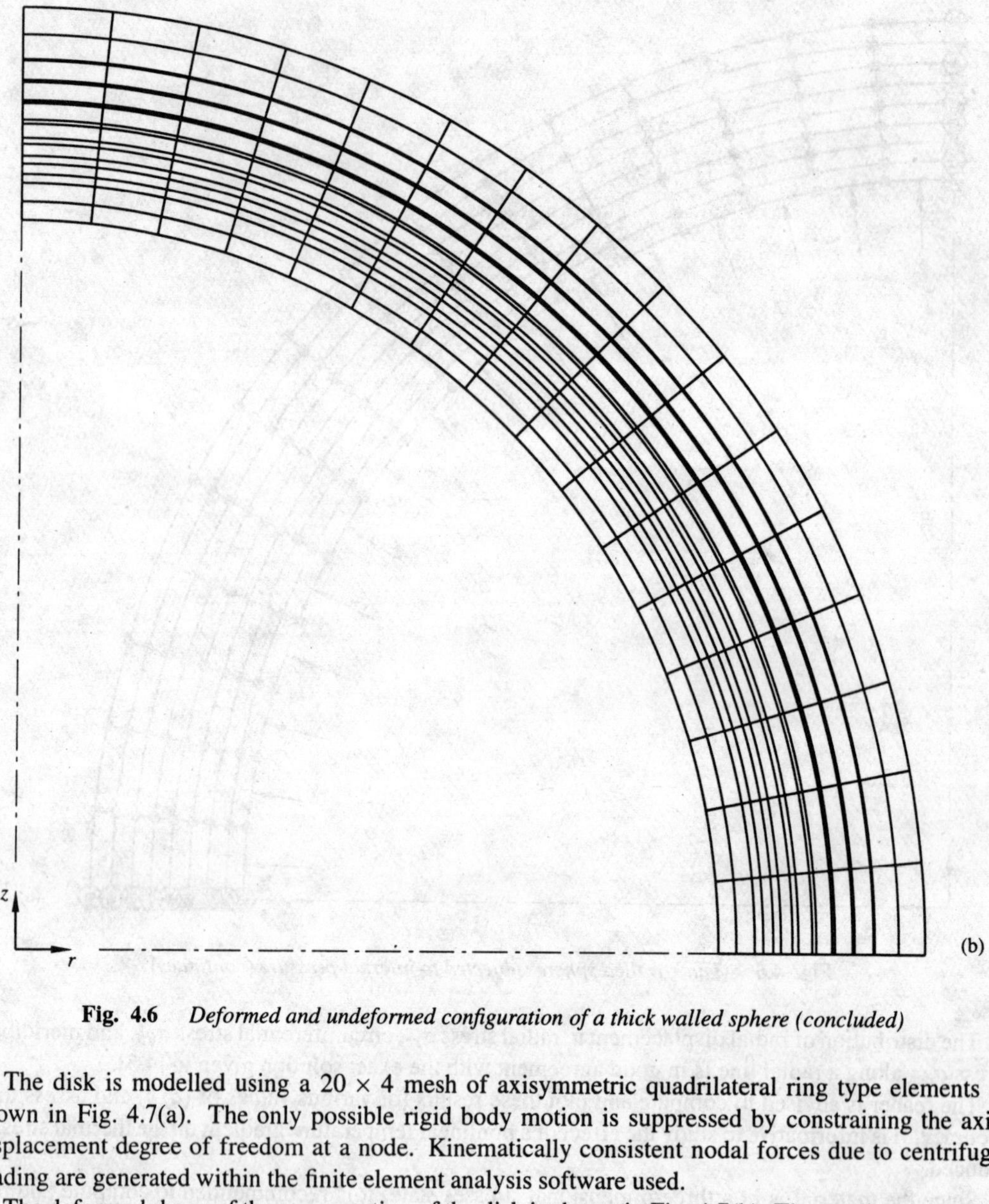

Fig. 4.6 *Deformed and undeformed configuration of a thick walled sphere (concluded)*

The disk is modelled using a 20 × 4 mesh of axisymmetric quadrilateral ring type elements as shown in Fig. 4.7(a). The only possible rigid body motion is suppressed by constraining the axial displacement degree of freedom at a node. Kinematically consistent nodal forces due to centrifugal loading are generated within the finite element analysis software used.

The deformed shape of the cross-section of the disk is shown in Fig. 4.7(b). The predicted variation of the radial displacement u is shown in Fig. 4.7(c). Computed radial (σ_{rr}) and circumferential ($\sigma_{\theta\theta}$) stress distribution is presented in Fig. 4.7(d). These results are in very good agreement with the plane stress solution given in [4.1].

If the disk has an integral solid shaft of outer radius a, the finite element model should include the shaft also. The reader is advised to generate finite element solutions and assess its accuracy.

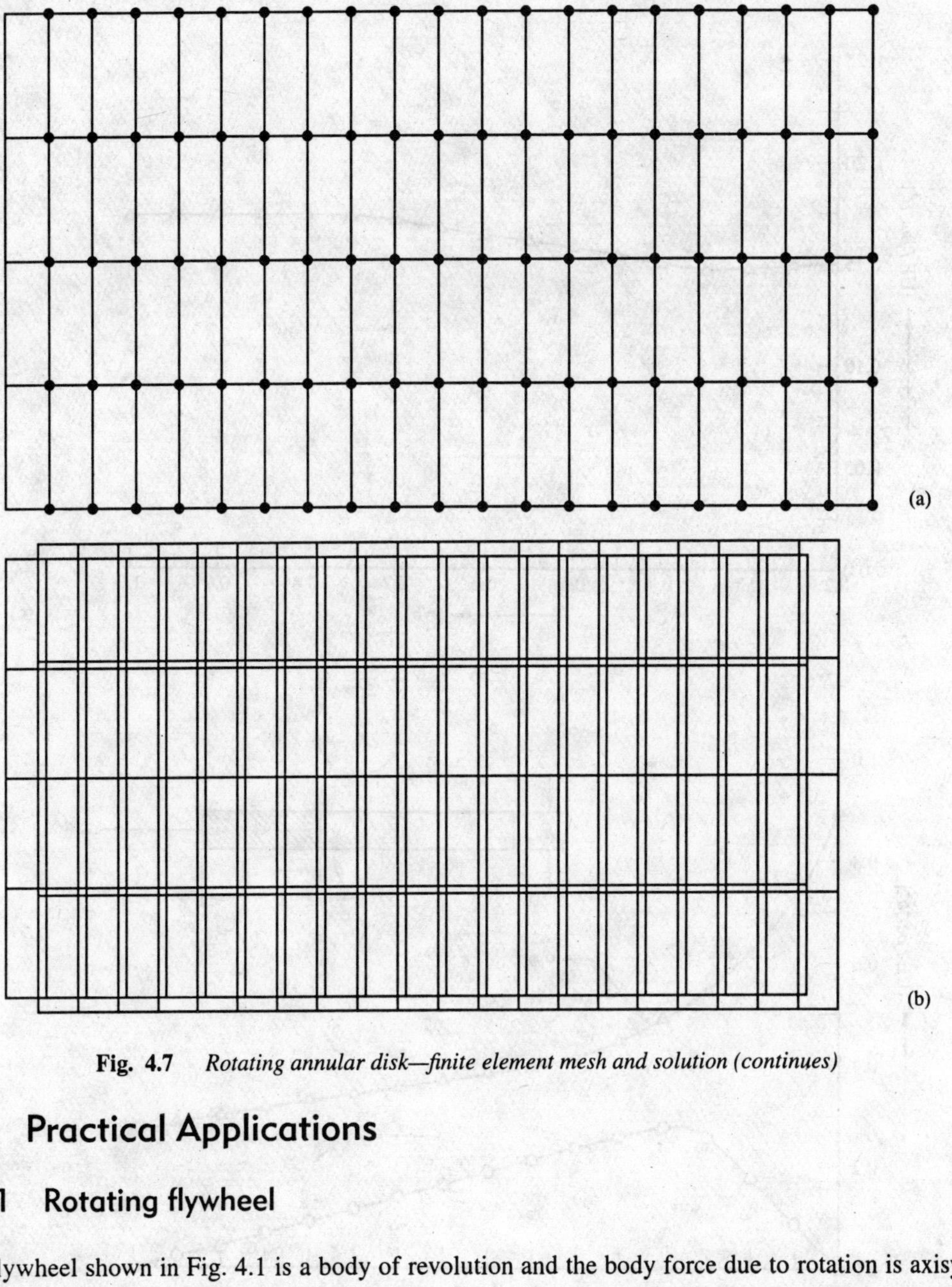

Fig. 4.7 *Rotating annular disk—finite element mesh and solution (continues)*

4.5 Practical Applications

4.5.1 Rotating flywheel

The flywheel shown in Fig. 4.1 is a body of revolution and the body force due to rotation is axisymmetric.

The cross-section of this cast iron wheel is discretized using axisymmetric triangular ring type elements as shown in Fig. 4.8(a). The only possible rigid body motion is suppressed by constraining the axial displacement degree of freedom at a node. Work-equivalent nodal forces due to centrifugal loads are generated within the finite element analysis program used.

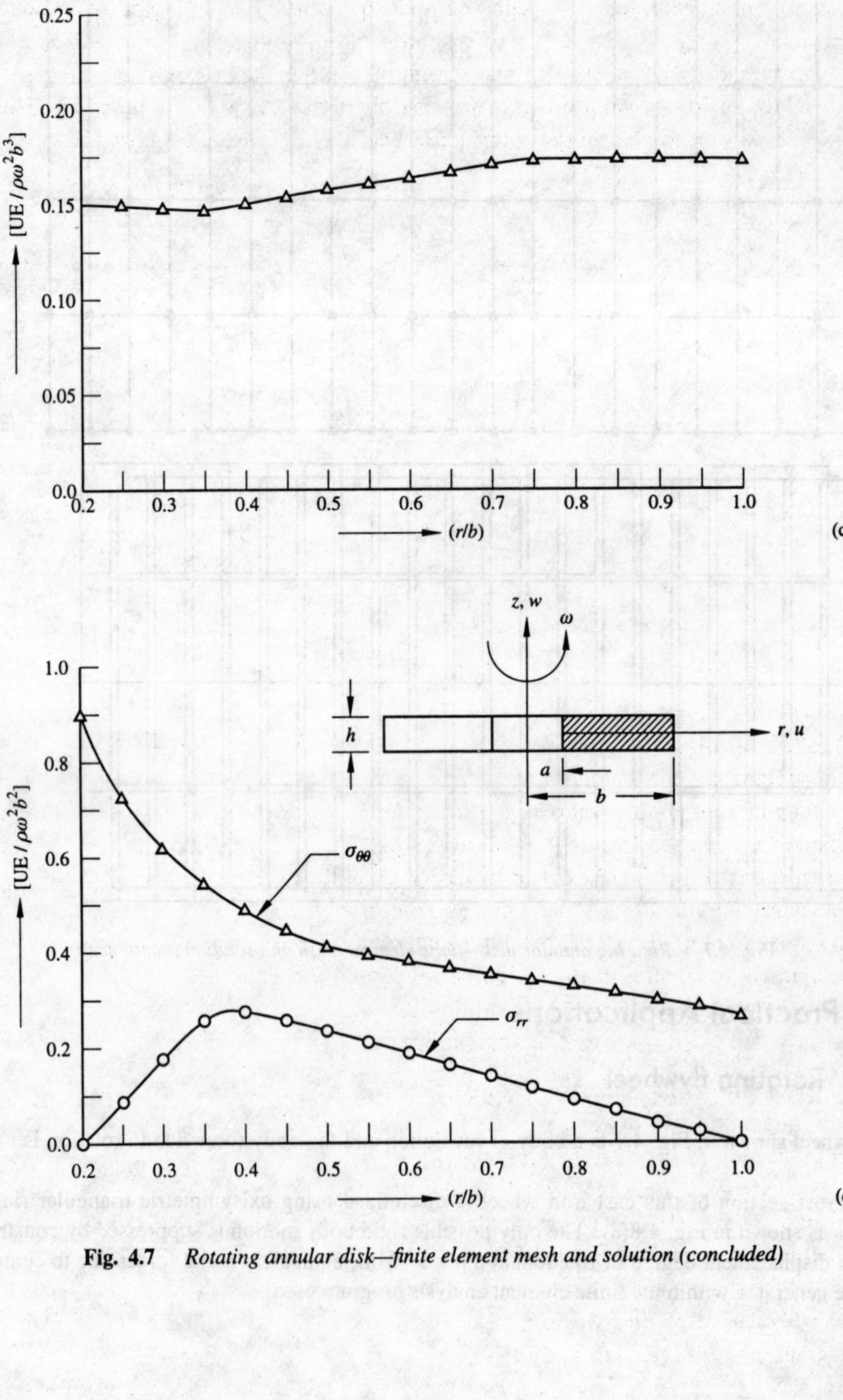

Fig. 4.7 *Rotating annular disk—finite element mesh and solution (concluded)*

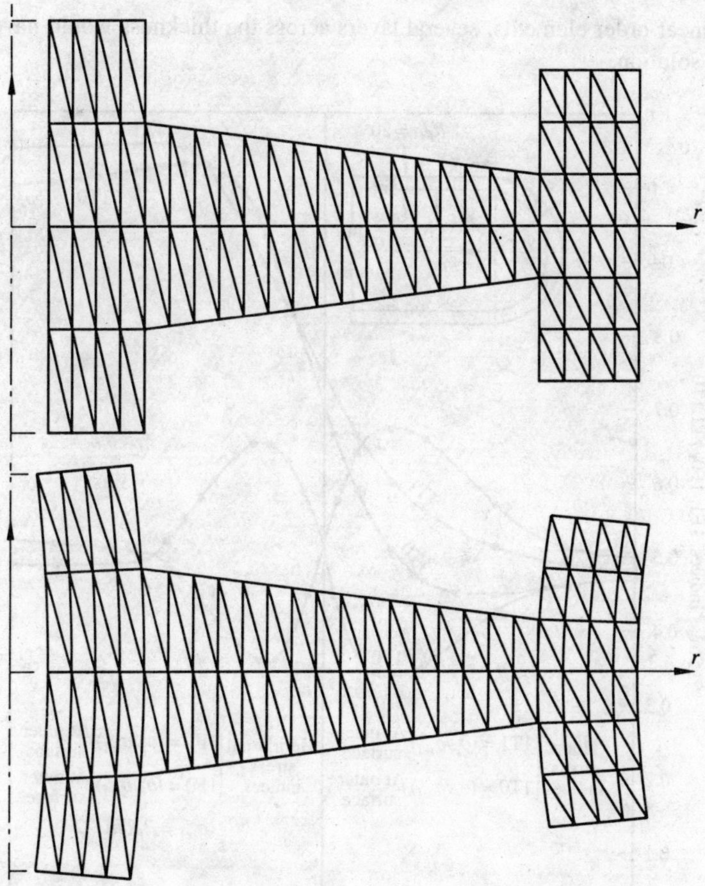

Fig. 4.8 *Finite element mesh and deformed shape of a rotating flywheel*

The deformed shape of the cross-section of the flywheel is shown in Fig. 4.8(b) and indicates possible loss of contact between the shaft and the hub.

The reader is advised to compute and plot the stress distribution in the wheel and comment on the results.

It is now a standard practice to use a welded steel flywheel. Here the hub, the web and flange are joined by welding. The design of welded joints demands a detailed finite element modelling of the hub, the web, the flange, and the welds, using axisymmetric ring elements.

4.5.2 Cylinder with hemispherical head under internal pressure

The problem shown as insert in Fig. 4.9 is recognized as a body of revolution under axisymmetric loading. It is descretized using a coarse mesh of axisymmetric ring type elements, quadrilateral in shape and quadratic in order. It is worth noting that only a single element across the thickness is adequate. With linear order elements, several layers across the thickness would have been needed to get an acceptable solution.

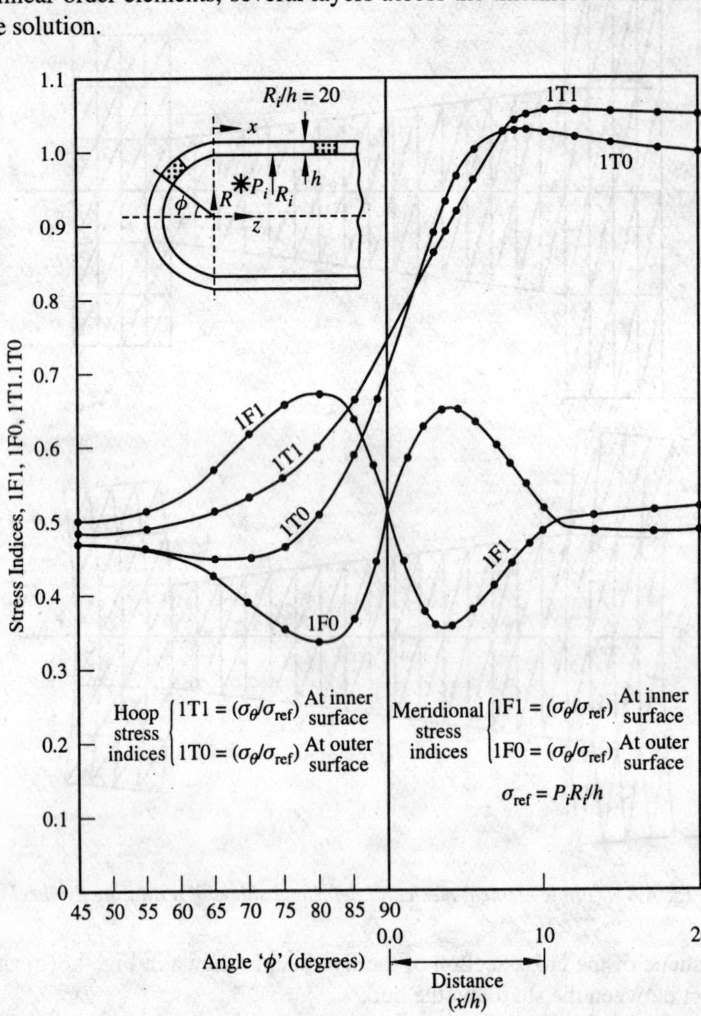

Fig. 4.9 *Stress distribution of a cylindrical pressure vessel with hemispherical head*

Significant results are summarized in Fig. 4.9. The meridional and hoop stress indices at the inner and outer surfaces are defined and their variation is presented. In particular, the oscillations of the meridional stress around the dome–cylinder junction is noteworthy. These oscillations can become more and more violent for smaller values of (R_i/h) and for different dome shapes.

The reader is advised to experiment with $(R_i/h) = 10$ (thick), 100 (thin) and report converged solutions along with mesh details to substantiate the above statement.

The reader can also experiment with semi-ellipsoidal and tori-spherical domes with different (R_i/h) ratios noted above and report converged solutions.

4.5.3 A rocket motor casing

Figure 4.2 shows the design details of a typical rocket motor casing used in space launch vehicles. The casing is fabricated in three segments, assembled together using segment joints and held together using a number of screws. In operation, it is subjected to axial thrust load on the skirts, internal pressure and thermal loads due to combustion of the solid propellant. The material is Chromium Vanadium steel and features extensive use in welded joints. The casing is about a metre in diameter and the minimum wall thickness is around 3.5 mm.

The rocket motor casing is recognized as a body of revolution and the applied loads are considered to be axisymmetric. It was discretized using a fine enough mesh of axisymmetric ring type elements, quadrilateral in shape and quadratic in order, and analyzed using a general purpose program. The Y-ring region and the segment joint regions are critical from a structural integrity point of view.

Among the voluminous results generated, variation of hoop stress index at the outer surface in the segment joint region is shown in Fig. 4.10.

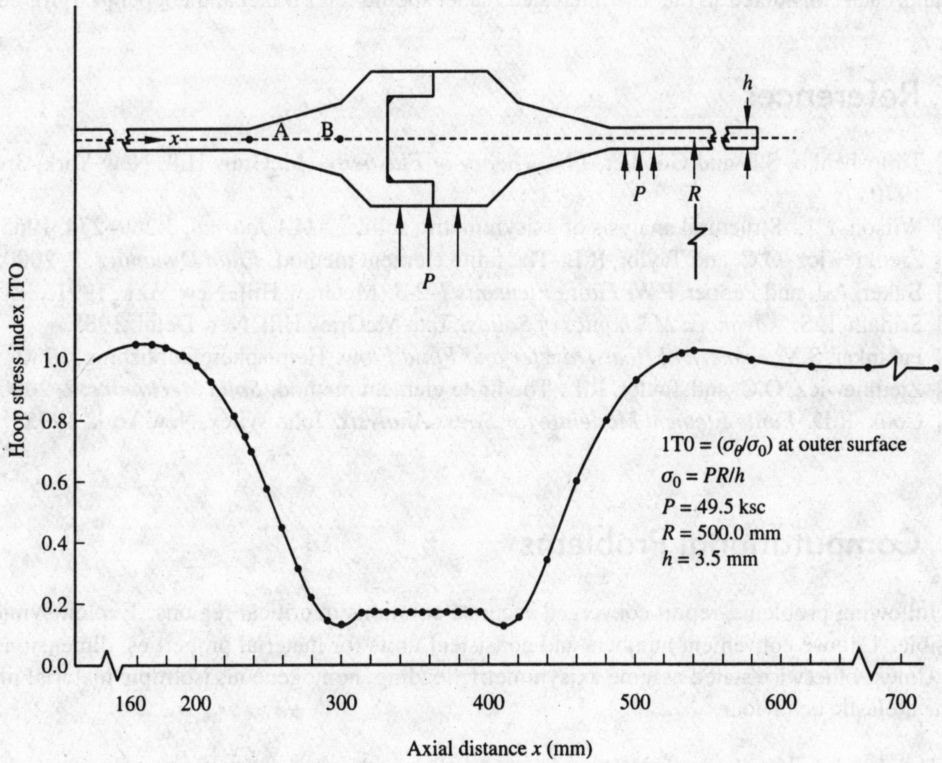

Fig. 4.10 *Stress distribution in the joint region of a rocket motor casing*

To certify the rocket motor as safe for flight, such results are correlated with test data collected during quality assurance testing.

4.6 Closure

A library of axisymmetric ring type elements of different shapes (triangular and quadrilateral) and increasing orders (linear, quadratic and cubic) has been presented in this chapter. These elements are implemented in commercial FEM systems and widely used in practice.

The ability of these elements to model bodies of revolution with irregular geometric shapes such as pressure vessels and piping makes FEM a valuable tool not only for linear static analysis but also for buckling, vibration, and transient response analyses also. The reader should refer Zienkiewicz and Taylor [4.7] for details.

Detailed local modelling of bolted, welded and adhesive bonded joints in practical pressure vessels and piping using the axisymmetric ring elements significantly improves the design by analysis of such joints.

Derivation of finite element equations for the analysis of field problems namely, heat transfer, fluid flow, and electromagnetic fields demands the use of weighted residual approach instead of the variational approach considered so far. The interested reader should refer Baker and Pepper [4.4] for details.

4.7 References

[4.1] Timoshenko, S.P. and Goodier, J.N. *Theory of Elasticity*. McGraw Hill, New York, 3rd ed., 1970.
[4.2] Wilson, E.L. Structural analysis of axisymmetric solids. *AIAA Journal*, 3:269–274, 1965.
[4.3] Zienkiewicz, O.C. and Taylor, R.L. The finite element method. *Fluid Dynamics*, 3, 2000.
[4.4] Baker, A.J. and Pepper, P.W. *Finite Elements 1-2-3*. McGraw Hill, New York, 1991.
[4.5] Srinath, L.S. *Advanced Mechanics of Solids*. Tata McGraw Hill, New Delhi, 1985.
[4.6] Patankar, S.V. *Numerical Heat Transfer and Fluid Flow*. Hemisphere Publishing, 1980.
[4.7] Zienkiewicz, O.C. and Taylor, R.L. The finite element method. *Solid Mechanics*, 2, 2000.
[4.8] Cook, R.D. *Finite Element Modeling for Stress Analysis*. John Wiley, New York, 1995.

4.8 Computational Problems

In the following problems, report converged accurate solutions in critical regions. Exploit symmetry if possible. Choose convenient numbers and consistent units for material properties, dimensions and loads. Unless otherwise stated assume axisymmetric loading, homogeneous isotropic material model, and linear elastic behaviour.

4.1 The sketch below shows the cross-section of a pipe with a supporting flange. For internal pressure loading, predict and plot the stress distribution.

4.2 The pipe considered in Problem 4.1 is filled with superheated liquid sodium at 644°K. Predict and plot the steady state temperature distribution assuming the outer surface to be in air at room temperature.

4.3 For the pipe considered in Problem 4.2, predict and plot the thermal stress distribution.

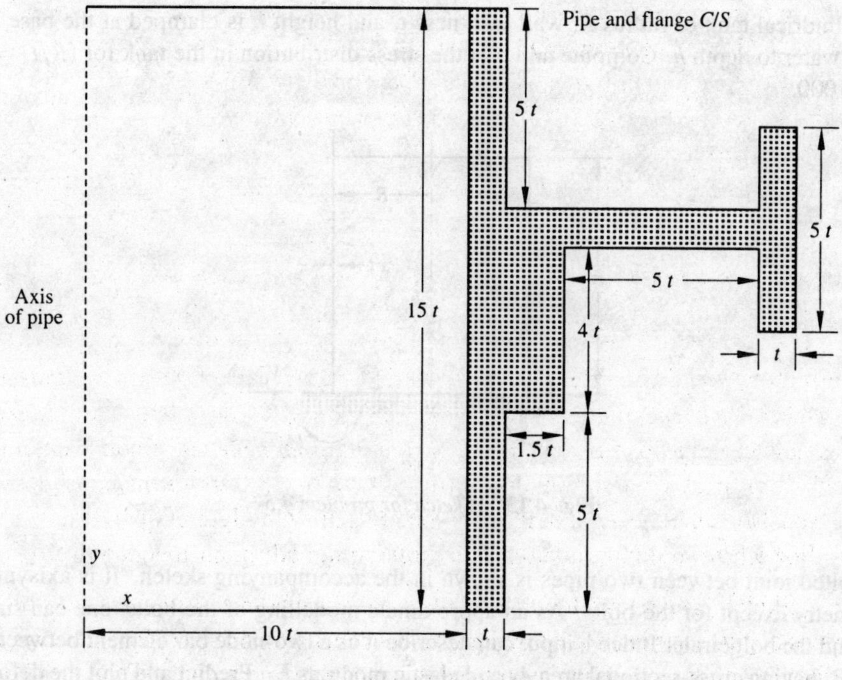

Fig. 4.11 *Sketch for problem 4.1*

4.4 The accompanying sketch shows the cross-section of a pipe with a supporting flange welded to the pipe. Predict and plot the deformed shape, and stress distribution in the assembly for loading by internal pressure.

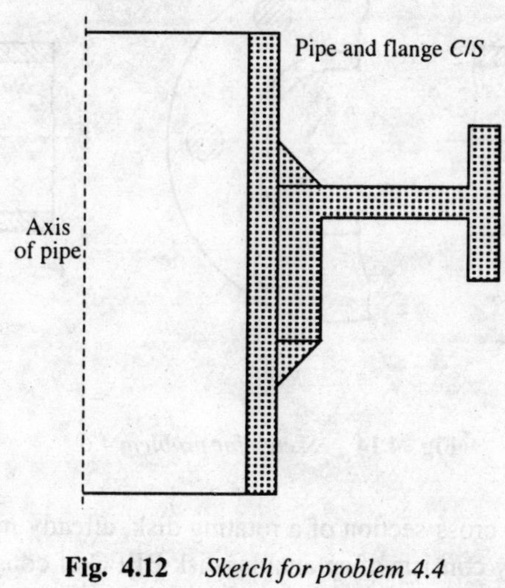

Fig. 4.12 *Sketch for problem 4.4*

4.5 A cylindrical tank of radius R, wall thickness t, and height h is clamped at the base and filled with water to depth h. Compute and plot the stress distribution in the tank for $(R/t) = 10, 100$, and 1000.

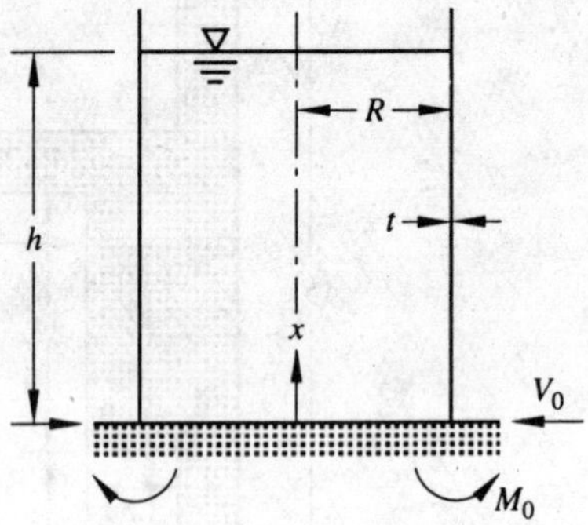

Fig. 4.13 *Sketch for problem 4.5*

4.6 A bolted joint between two pipes is shown in the accompanying sketch. It is axisymmetric in geometry except for the bolts. As an approximate modelling of the bolts, one can smear them around the bolt circle. Indeed, input can describe it as a two-node bar element between nodes A and B, having cross-sectional area A_s and elastic modulus E_s. Predict and plot the deformations, and stresses in the assembly for loading by internal pressure. The reader can consult [4.8] for help.

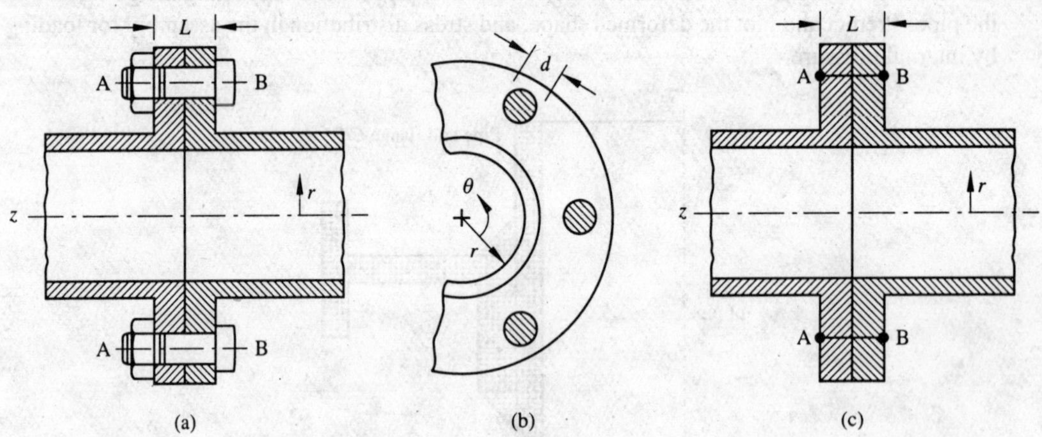

Fig. 4.14 *Sketch for problem 4.6*

4.7 The sketch below shows cross-section of a rotating disk, already meshed with the elements we propose to use. The body consists of an annular disk BEFC of constant thickness, attached to a tapered solid disk DABE of the same material by means of a shrink fit. Physically, the shrink fit is accomplished by heating the outer disk, slipping it over the inner disk, and allowing both disks to return to a uniform temperature. Dimensions are such that when the outer disk is 100°C hotter than the inner disk, the inner radius of the outer disk and the outer radius of the inner disk are both precisely 400 mm. We ask: (i) what contact pressures along BE are produced by the shrink fit? (ii) The resulting stresses in the two disks; (iii) if the assembly is set spinning about the z-axis, at what speed will the shrink fit loosen? The reader can consult Ref: [4.8] for help.

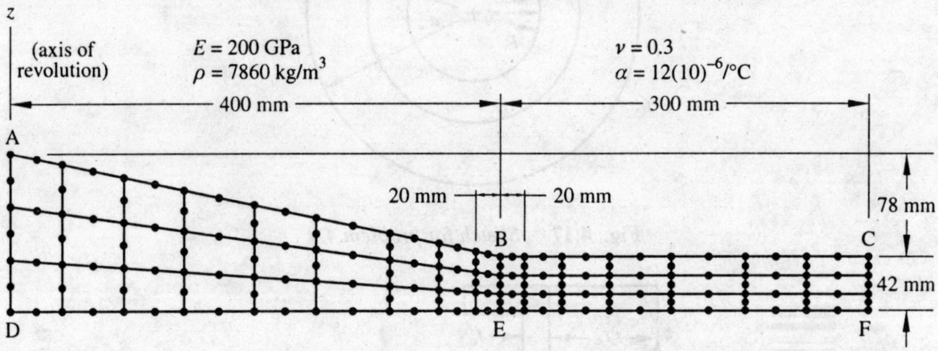

Fig. 4.15 *Sketch for problem 4.7*

4.8 The sketch shows the cross-section of a long cylinder capped by a hemisphere. The juncture at A contains quarter circle fillets of radius r. Loading is by internal pressure. Predict and plot the stress distribution in the vessel. Locate points at which the onset of yielding takes place.

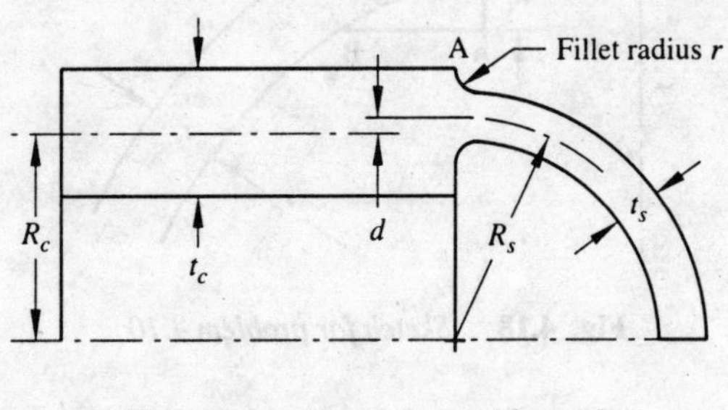

Fig. 4.16 *Sketch for problem 4.8*

4.9 Consider a compound cylinder, built by shrink fitting a jacket of inner radius c on an inner cylinder of outer radius $c + \Delta$, where $\Delta \ll c$. Internal pressure p is subsequently applied. For given values of p, a, b, c, what Δ is needed to induce the same maximum Von-Mises stress in the inner cylinder and in the jacket?

4.10 A hemispherical vessel has a radially directed nozzle, as shown in the cross-section. The local reinforcements are also shown. Let $t_c = 2t_s(r_c/r_s)$. Apply internal pressure p and compute the largest Von-Mises equivalent stress as a percentage of p for $r_s = 400$ mm, $t_s = 40$ mm, $r_c = 100$ mm, $h_O = 40$ mm, $h_i = 40$ mm.

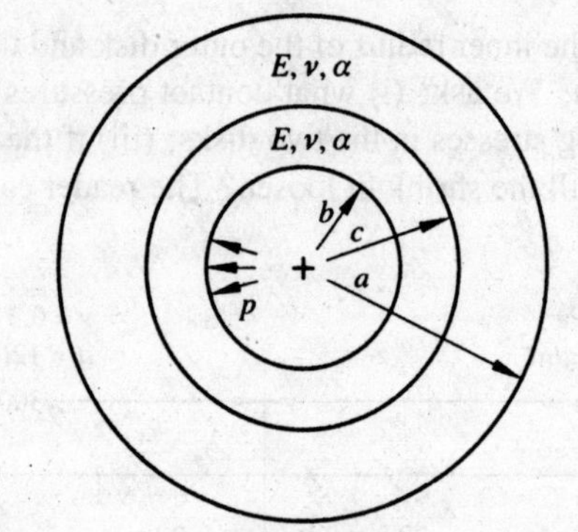

Fig. 4.17 *Sketch for problem 4.9*

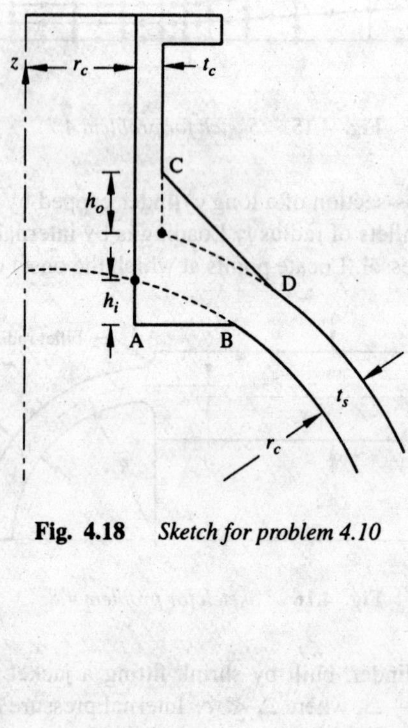

Fig. 4.18 *Sketch for problem 4.10*

5 Three-Dimensional Elements

Applications to Solid Mechanics Problems

5.1 Overview

Even trivial problems featuring three-dimensional geometry (Fig. 5.1) and/or three-dimensional stress state (Fig. 5.2) defy exact solution by continuum mechanics (analytical) methods. Experimental investigations are equally complex and prohibitively expensive in terms of facilities, model construction, instrumentation, and the actual test. However, there is only one further step to apply the finite element method to fully three-dimensional problems in continuum mechanics.

Theory of elasticity [5.1] provides the basic equations for the analysis of three-dimensional solid bodies. These equations are summarized and used to construct the potential energy functional.

A library of three-dimensional continuum elements of different shapes (tetrahedra, hexahedra, pentahedra) and increasing orders (linear, quadratic, cubic, etc.,) are available in commercial FEM systems. Among these the four-noded tetrahedral solid element (Fig. 5.3) and the eight-noded hexahedral solid element (Fig. 5.4) are formulated in detail in this chapter.

Higher order isoparametric elements with curved edges and curved faces that are implemented in commercial FEM systems are described.

Standard sets of test problems to assess element accuracy are given followed by practical applications.

The presentation also includes identification of topics for advanced study.

5.2 Basic Equations and Potential Energy Functional

Let us consider a three-dimensional solid body occupying a volume V and having surface(s) S as shown in Fig. 5.5. A point P in the body is located by its x, y, and z coordinates. The support conditions include point, roller, elastic, etc. The applied loads include concentrated forces, body forces, surface tractions and those due to thermal expansion. The thermal, physical and mechanical properties of the material(s) out of which the body is made are known. The deformation of a point P (x, y, z) is given by

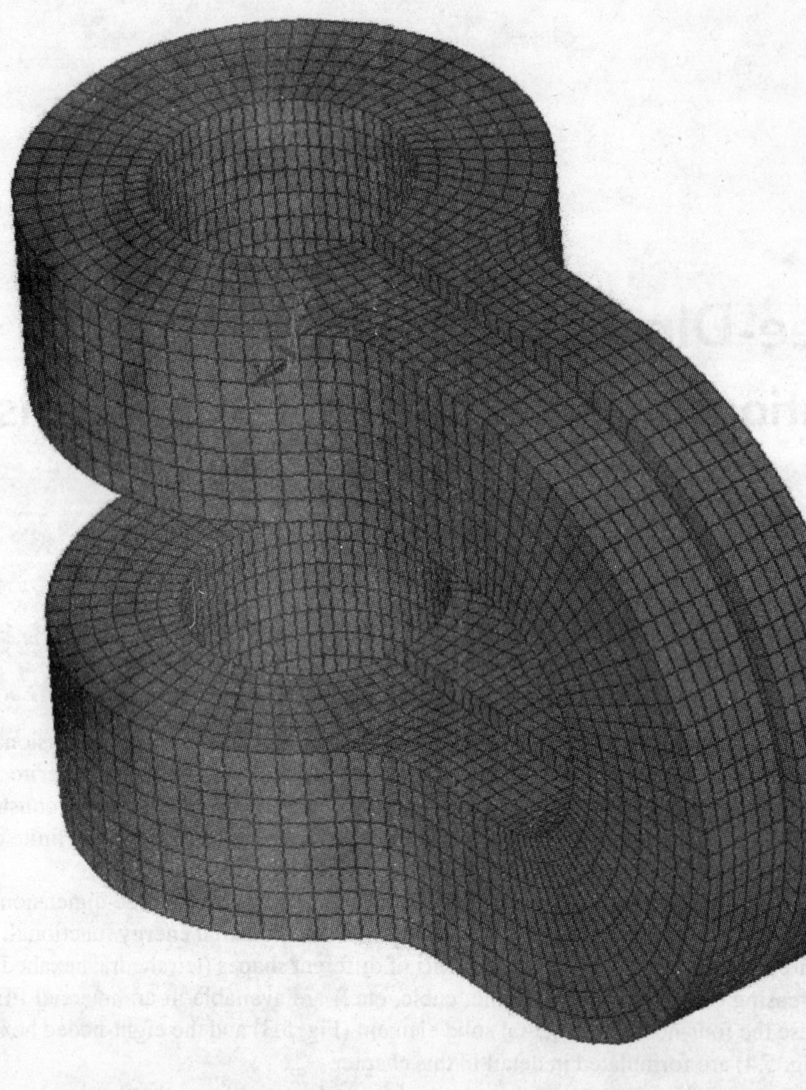

Fig. 5.1 *A machine component discretized using hexahedral elements*

the components of the displacement vector **u**.

$$\{\mathbf{u}\} = \begin{Bmatrix} u(x, y, z) \\ v(x, y, z) \\ w(x, y, z) \end{Bmatrix}. \tag{5.1}$$

The body force per unit volume is defined by the vector **b**, when

$$\{\mathbf{b}\} = \begin{Bmatrix} b_x(x, y, z) \\ b_y(x, y, z) \\ b_z(x, y, z) \end{Bmatrix}. \tag{5.2}$$

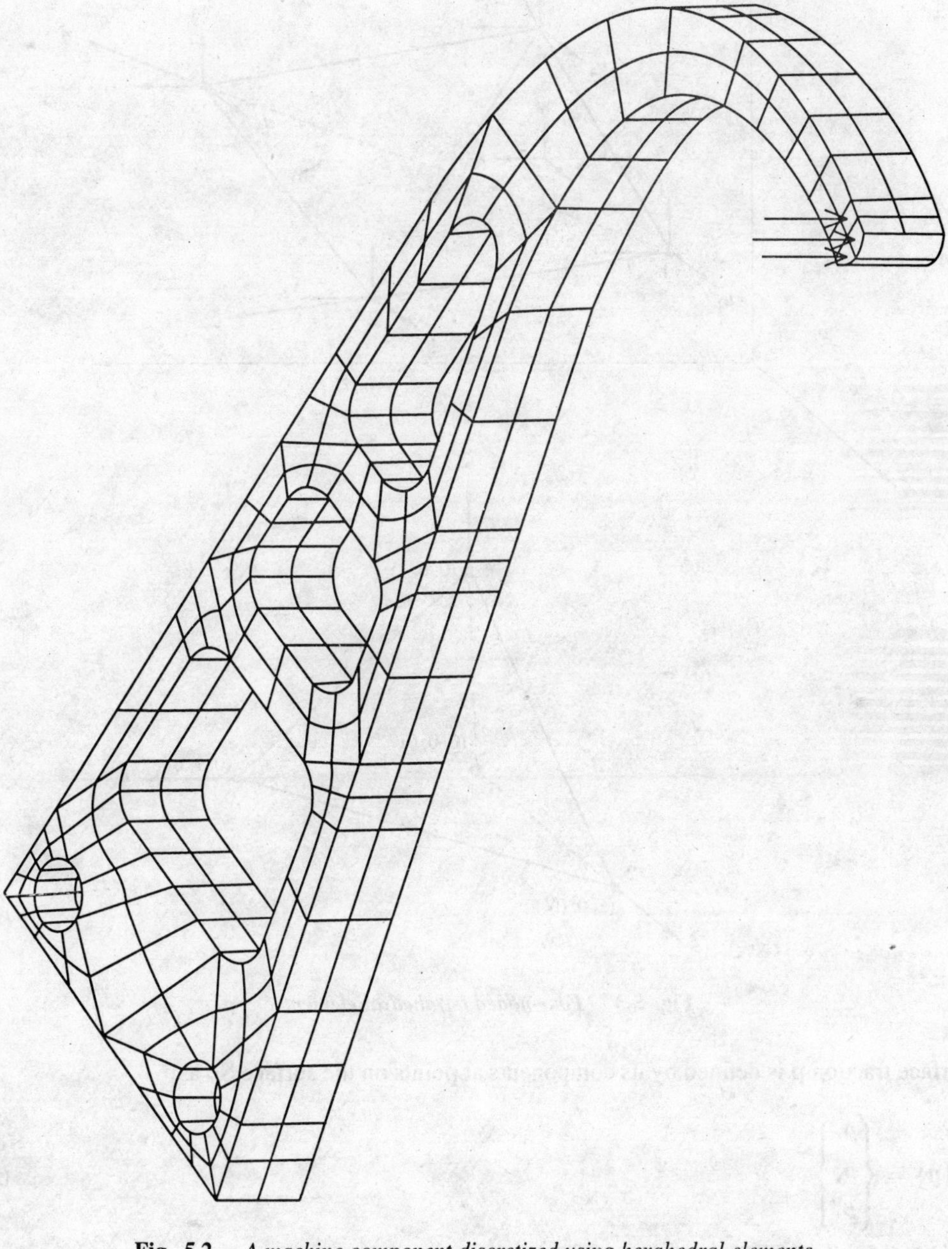

Fig. 5.2 *A machine component discretized using hexahedral elements*

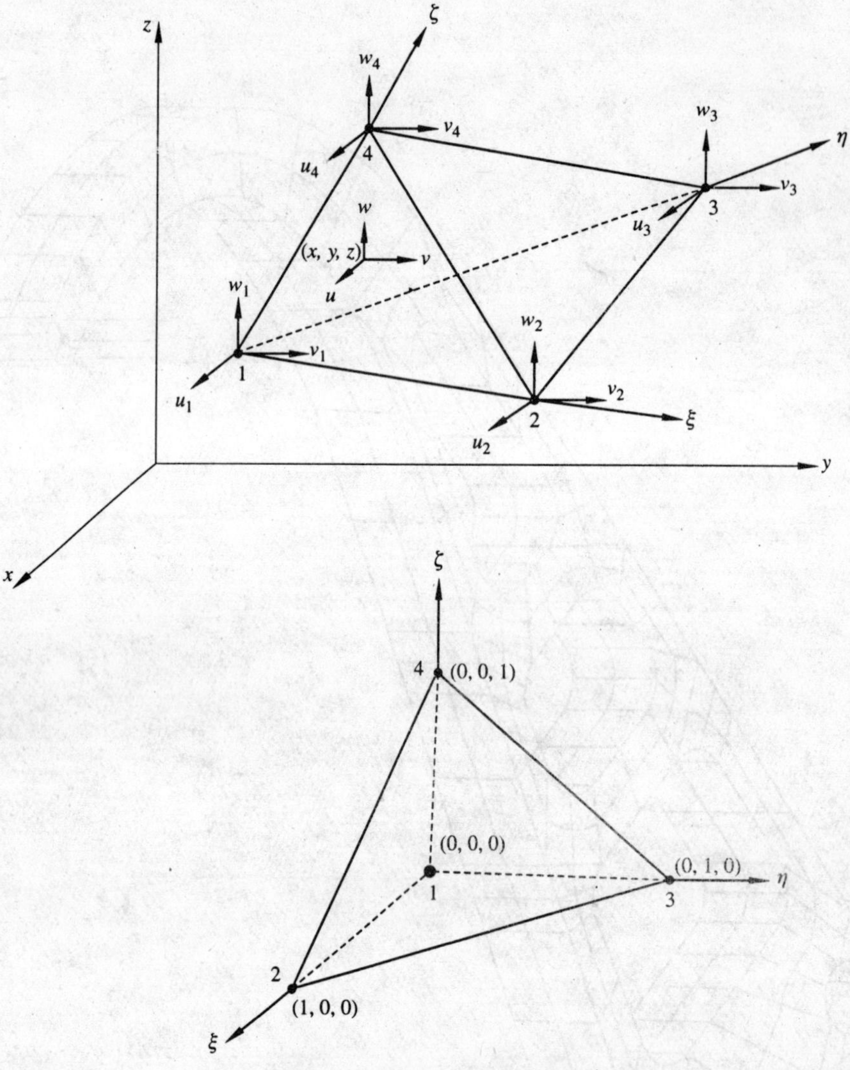

Fig. 5.3 *Four-noded tetrahedral element*

The surface traction **p** is defined by its components at points on the surface $\mathbf{S}_\sigma$ as

$$\{\mathbf{p}\} = \begin{Bmatrix} p_x \\ p_y \\ p_z \end{Bmatrix}. \tag{5.3}$$

The resultant pressure **p** acts in the direction of outward normal **n** to the surface $\mathbf{S}_\sigma$.

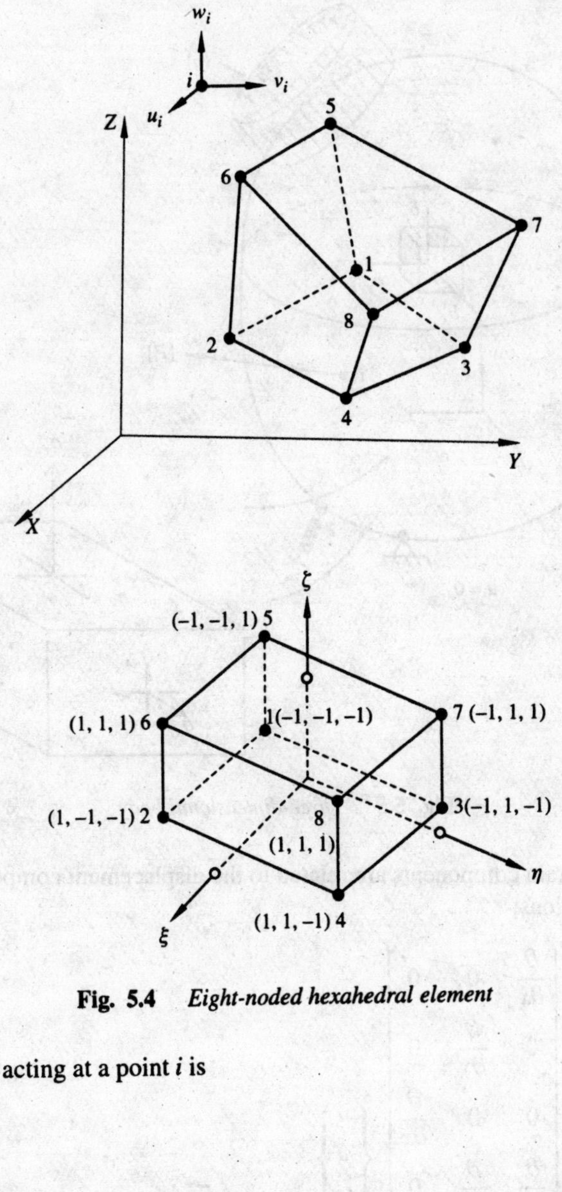

Fig. 5.4 *Eight-noded hexahedral element*

A concentrated load acting at a point i is

$$\{\mathbf{P}\}_i = \begin{Bmatrix} P_{xi} \\ P_{yi} \\ P_{zi} \end{Bmatrix}. \tag{5.4}$$

The stresses acting on the faces of an infinitesimal cube $dxdydz$ can be represented by the stress vector $\boldsymbol{\sigma}$:

$$\{\sigma\}^T = (\sigma_{xx}, \sigma_{yy}, \sigma_{zz}, \tau_{xy}, \tau_{yz}, \tau_{zx}). \tag{5.5}$$

In Eqn. (5.5) σ_{xx}, σ_{yy}, σ_{zz} denote normal stresses while τ_{xy}, τ_{yz}, τ_{zx} are shear stresses.

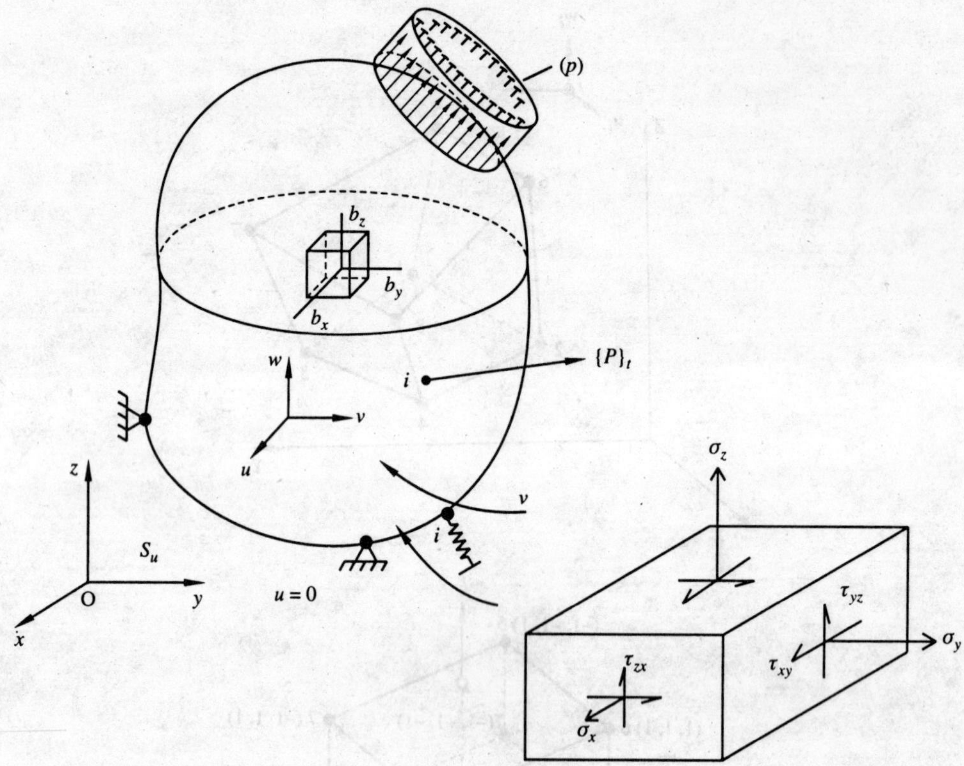

Fig. 5.5 *Three-dimensional body*

The corresponding strain components are related to the displacement components by the following strain displacement relations:

$$\{\boldsymbol{\varepsilon}\} = \begin{Bmatrix} \varepsilon_{xx} \\ \varepsilon_{yy} \\ \varepsilon_{zz} \\ \gamma_{xy} \\ \gamma_{yz} \\ \gamma_{zx} \end{Bmatrix} = \begin{bmatrix} \dfrac{\partial}{\partial x} & 0 & 0 \\ 0 & \dfrac{\partial}{\partial y} & 0 \\ 0 & 0 & \dfrac{\partial}{\partial x} \\ \dfrac{\partial}{\partial y} & \dfrac{\partial}{\partial x} & 0 \\ 0 & \dfrac{\partial}{\partial z} & \dfrac{\partial}{\partial y} \\ \dfrac{\partial}{\partial z} & 0 & \dfrac{\partial}{\partial x} \end{bmatrix} \begin{Bmatrix} u \\ v \\ w \end{Bmatrix}$$

or

$$\{\boldsymbol{\varepsilon}\} = [\mathbf{L}]\{\mathbf{u}\}. \tag{5.6}$$

The relations being linear hold good for small deformations. For large deformation problems, nonlinear terms are included in Eqn. (5.6). Such problems are called geometrically nonlinear problems.

The stresses are related to the strains using constitutive relations. Assuming the material to be modelled as homogeneous and isotropic, and its behaviour linear and elastic, the constitutive law is nothing but the generalized Hooke's law.

$$
\begin{aligned}
\varepsilon_{xx} &= \frac{\sigma_{xx}}{E} - \frac{\nu}{E}(\sigma_{yy} + \sigma_{zz}) + \alpha T, \\
\varepsilon_{yy} &= \frac{\sigma_{yy}}{E} - \frac{\nu}{E}(\sigma_{xx} + \sigma_{zz}) + \alpha T, \\
\varepsilon_{zz} &= \frac{\sigma_{zz}}{E} - \frac{\nu}{E}(\sigma_{xx} + \sigma_{yy}) + \alpha T, \\
\gamma_{xy} &= \frac{\tau_{xy}}{G}, \\
\gamma_{yz} &= \frac{\tau_{yz}}{G}, \\
\gamma_{zx} &= \frac{\tau_{zx}}{G}.
\end{aligned}
\tag{5.7}
$$

In Eqn. (5.7), E is the Young's modulus of elasticity, ν is the Poisson's ratio, G is the shear modulus, α is the coefficient of thermal expansion, and T the temperature change (measured from a stress free temperature).

Using Eqn. (5.7), the stresses $\boldsymbol{\sigma}$ are related to the strains $\boldsymbol{\varepsilon}$ as

$$\{\boldsymbol{\sigma}\} = [\mathbf{D}](\{\boldsymbol{\varepsilon}\} - \{\boldsymbol{\varepsilon}^{\text{init}}\}) \tag{5.8}$$

where the elasticity matrix **D** is given by

$$[\mathbf{D}] = \frac{E}{(1+\nu)(1-2\nu)} \begin{bmatrix} (1-\nu) & \nu & \nu & 0 & 0 & 0 \\ \nu & (1-\nu) & \nu & 0 & 0 & 0 \\ \nu & \nu & (1-\nu) & 0 & 0 & 0 \\ 0 & 0 & 0 & \frac{(1-2\nu)}{2} & 0 & 0 \\ 0 & 0 & 0 & 0 & \frac{(1-2\nu)}{2} & 0 \\ 0 & 0 & 0 & 0 & 0 & \frac{(1-2\nu)}{2} \end{bmatrix} \tag{5.9}$$

and the vector of initial strains $\boldsymbol{\varepsilon}^{\text{init}}$ is

$$\{\boldsymbol{\varepsilon}^{\text{init}}\}^T = (\alpha T, \alpha T, \alpha T, 0, 0, 0).$$

The stress strain relations given by Eqn. (5.8) being linear hold good for linear elastic behaviour. For nonlinear and/or inelastic behaviour, one has to use a nonlinear form of Eqn. (5.8) and what we have is called material nonlinearity.

Finally, the stresses σ should satisfy the differential equations of equilibrium.

$$\frac{\partial \sigma_{xx}}{\partial x} + \frac{\partial \tau_{xy}}{\partial y} + \frac{\partial \tau_{zx}}{\partial z} + b_x = 0.$$

$$\frac{\partial \tau_{xy}}{\partial x} + \frac{\partial \sigma_{yy}}{\partial y} + \frac{\partial \tau_{yz}}{\partial z} + b_y = 0.$$

$$\frac{\partial \tau_{zx}}{\partial x} + \frac{\partial \tau_{yz}}{\partial y} + \frac{\partial \sigma_{zz}}{\partial z} + b_z = 0. \tag{5.10}$$

In solid mechanics, the goal is to determine **u**, σ, and ε at any point in a given body. However, the displacements **u** should satisfy the kinematic boundary conditions and the stresses σ should satisfy the surface traction boundary conditions. The simultaneous solution of 15 equations (3 equilibrium equations, 6 stress-strain relations, and 6 strain-displacement relations) for 15 unknowns (3 displacement components, 6 strain components, and 6 stress components) is generally referred to as the exact solution. Such exact solutions are possible provided the geometry, loading and support conditions are very simple, even for linear elastic behaviour. Due to this limitation of the differential equations formulation, a variational formulation has been suggested.

In the variational formulation, the differential equations of equilibrium Eqn. (5.10) are omitted and the equilibrium conditions are satisfied by invoking a variational principle. Here, an introduction to the widely used minimum potential energy principle is in order. It can be stated as "Of all displacement functions satisfying the admissibility requirements, those which satisfy the equilibrium equations make the potential energy functional π_p assume a stationary value, and for stable equilibrium, π_p is a minimum".

By employing the variational formulation, we may find solutions of boundary value problems in solid mechanics (and structural mechanics) from another point of view. Instead of solving the governing differential equations together with the boundary conditions, a task often very difficult mathematically, we may seek displacement functions which satisfy the boundary conditions a priori and minimize the potential energy functional. One of the widely used variational methods is the Rayleigh–Ritz method. The FEM is, in fact a piece-wise Rayleigh-Ritz method!

The potential energy functional π_p is defined as the sum of strain energy (SE) and the work potential (WP).

$$\pi_p = SE + WP, \tag{5.11}$$

where

$$SE = \frac{1}{2}\int_v \{\sigma\}^T\{\boldsymbol{\varepsilon}\}dv \tag{5.12}$$

and

$$WP = -\int_v \{\mathbf{u}\}^T\{\mathbf{b}\}dv - \int_{S_\sigma} \{\bar{\mathbf{u}}\}^T\{\mathbf{p}\}dS - \sum_i \{\mathbf{u}\}_i^T\{\mathbf{P}\}_i. \tag{5.13}$$

In view of Eqns. (5.6) and (5.8), π_p is a function of **u** only, i.e.,

$$\pi_p = \pi_p(\mathbf{u}).$$

In order to evaluate the potential energy functional for complex domains, the concept of finite element discretization is very useful.

5.3 Finite Element Discretization

In order to apply the variational methods, the displacement functions **u** should be kinematically admissible. Kinematically admissible displacement functions are those that are finite, single-valued, and are continuous functions satisfying a priori the displacement boundary conditions. Creating such mathematical functions for practical problems featuring complex geometries and realistic support conditions is in fact impossible. Admitting even the availability of such functions, evaluation of the volume and surface integrals in Eqns. (5.12) and (5.13) is a formidable task. Both these hurdles are crossed by introducing the concept of discretization. The solid body in Fig. 5.5 is now considered as an assemblage of finite elements. A library of three-dimensional solid type elements of different shapes and orders has been developed for this purpose. Among these a four-noded tetrahedral element shown in Fig. 5.3 and an eight-noded hexahedral element shown in Fig. 5.4 are identified as workhorse elements and given explicit attention. However, an introduction to higher order elements is included.

5.4 Four-Noded Tetrahedral Element (TET 4)

The element shown in Fig. 5.3 is bounded by four triangular faces and has four nodes. The nodes are numbered 1, 2, 3 and 4 as shown. The geometry of the element is described by the Cartesian coordinates $(X_i\ Y_i\ Z_i)$ $(i = 1,4)$ of the nodes. For solid mechanics applications, each node i has three degrees of freedom u_i, v_i, w_i and the vector of nodal degrees of freedom **q** is

$$\{\mathbf{q}\}^T = (u_1, v_1, w_1, u_2, v_2, w_2, u_3, v_3, w_3, u_4, v_4, w_4). \tag{5.14}$$

The corresponding vector of nodal forces is

$$\{\mathbf{f}\}^T = (F_{x1}, F_{y1}, F_{z1}, F_{x2}, F_{y2}, F_{z2}, F_{x3}, F_{y3}, F_{z3}, F_{x4}, F_{y4}, F_{z4}). \tag{5.15}$$

The derivation of element equations, i.e., algebraic equations that relate **q** and **f** by the variational approach is described below.

5.4.1 Admissible displacement functions

The displacement components (u, v, w) at any point (x, y, z) within the element have to be interpolated in terms of the nodal degrees of freedom. The simplest representation is

$$\{\mathbf{u}\} = [\mathbf{N}]\{\mathbf{q}\}, \tag{5.16}$$

where

$$[\mathbf{N}] = \begin{bmatrix} N_1 & 0 & 0 & N_2 & 0 & 0 & N_3 & 0 & 0 & N_4 & 0 & 0 \\ 0 & N_1 & 0 & 0 & N_2 & 0 & 0 & N_3 & 0 & 0 & N_4 & 0 \\ 0 & 0 & N_1 & 0 & 0 & N_2 & 0 & 0 & N_3 & 0 & 0 & N_4 \end{bmatrix}. \tag{5.17}$$

In order to derive explicit expressions for the interpolation functions $N_i(i = 1,4)$, it is now a standard practice to use the concept of a parent element and the covariant coordinate system (ξ, η, ζ) illustrated in Fig. 5.3. Any variable Φ is approximated over the parent element domain using the polynomial (see the Pascal's pyramid in Fig. 5.8):

$$\Phi = \alpha_1 + \alpha_2 \xi + \alpha_3 \eta + \alpha_4 \zeta. \tag{5.18}$$

Let us denote

$\Phi = \Phi_1$ at node 1, i.e., at $\xi = 0, \eta = 0, \zeta = 0$,

$\Phi = \Phi_2$ at node 2, i.e., at $\xi = 1, \eta = 0, \zeta = 0$,

$\Phi = \Phi_3$ at node 3, i.e., at $\xi = 0, \eta = 1, \zeta = 0$,

$\Phi = \Phi_4$ at node 4, i.e., at $\xi = 0, \eta = 0, \zeta = 1$.

Using Eqn. (5.18), it is then possible to evaluate the polynomial coefficients $\alpha_i(i = 1,4)$ in terms of $\Phi_i(i = 1,4)$ and write

$$\Phi = \sum_{i=1}^{4} N_i(\xi, \eta, \zeta)\Phi_i, \tag{5.19}$$

with the interpolation functions given by

$$\begin{aligned} N_1(\xi, \eta, \zeta) &= 1 - \xi - \eta - \zeta, \\ N_2(\xi, \eta, \zeta) &= \xi, \\ N_3(\xi, \eta, \zeta) &= \eta, \\ N_4(\xi, \eta, \zeta) &= \zeta. \end{aligned} \tag{5.20}$$

These functions are such that

$N_i = 1$ at node i,

$N_i = 0$ at all other nodes,

$$\sum_{i=1}^{4} N_i = 1.0.$$

Let us now distort this parent element to the shape of the tetrahedral element. The required relation between the Cartesian and the natural coordinates to accomplish this is

$$\begin{Bmatrix} x \\ y \\ z \end{Bmatrix} = \sum_{i=1}^{4} N_i(\xi, \eta, \zeta) \begin{Bmatrix} X_i \\ Y_i \\ Z_i \end{Bmatrix}, \tag{5.21}$$

where (X_i, Y_i, Z_i) denote Cartesian coordinates of node i.

The geometry of the element is therefore represented using the four shape functions $N_i (i = 1, 4)$ and the Cartesian coordinates of the four nodes $(x, y, z)(i = 1, 4)$. The centroid of the element is located at $\xi = \eta = \zeta = 1/3$. The four faces of the element are identified as $\xi = 0$, $\eta = 0$, $\zeta = 0$, and $\xi + \eta + \zeta = 1$.

The same shape functions are also used as displacement functions. We can therefore write

$$\begin{Bmatrix} u \\ v \\ w \end{Bmatrix} = \sum_{i=1}^{4} N_i(\xi, \eta, \zeta) \begin{Bmatrix} u_i \\ v_i \\ w_i \end{Bmatrix}. \tag{5.22}$$

Since the shape functions are so chosen in the covariant natural coordinates space as to observe the usual rules for convergence (continuity within the element and across inter-element boundaries, and the presence of complete first-order polynomials), convergence with mesh refinement is assured. The constant derivative criterion is also satisfied for this isoparametric element since $N_1 + N_2 + N_3 + N_4 = 1.0$.

5.4.2 Element strain matrix

Combining Eqns. (5.6) and (5.16) we get

$$\{\boldsymbol{\varepsilon}\} = [\boldsymbol{B}]\{\boldsymbol{q}\} \tag{5.23}$$

where

$$[\mathbf{B}] = \begin{bmatrix} \frac{\partial N_1}{\partial x} & 0 & 0 & \frac{\partial N_2}{\partial x} & 0 & 0 & \frac{\partial N_3}{\partial x} & 0 & 0 & \frac{\partial N_4}{\partial x} & 0 & 0 \\ 0 & \frac{\partial N_1}{\partial y} & 0 & 0 & \frac{\partial N_2}{\partial y} & 0 & 0 & \frac{\partial N_3}{\partial y} & 0 & 0 & \frac{\partial N_4}{\partial y} & 0 \\ 0 & 0 & \frac{\partial N_1}{\partial z} & 0 & 0 & \frac{\partial N_2}{\partial z} & 0 & 0 & \frac{\partial N_3}{\partial z} & 0 & 0 & \frac{\partial N_4}{\partial z} \\ \frac{\partial N_1}{\partial y} & \frac{\partial N_1}{\partial x} & 0 & \frac{\partial N_2}{\partial y} & \frac{\partial N_2}{\partial x} & 0 & \frac{\partial N_3}{\partial y} & \frac{\partial N_3}{\partial x} & 0 & \frac{\partial N_4}{\partial y} & \frac{\partial N_4}{\partial x} & 0 \\ 0 & \frac{\partial N_1}{\partial z} & \frac{\partial N_1}{\partial y} & 0 & \frac{\partial N_2}{\partial z} & \frac{\partial N_2}{\partial y} & 0 & \frac{\partial N_3}{\partial z} & \frac{\partial N_3}{\partial y} & 0 & \frac{\partial N_4}{\partial z} & \frac{\partial N_4}{\partial y} \\ \frac{\partial N_1}{\partial z} & 0 & \frac{\partial N_1}{\partial x} & \frac{\partial N_2}{\partial z} & 0 & \frac{\partial N_2}{\partial x} & \frac{\partial N_3}{\partial z} & 0 & \frac{\partial N_3}{\partial x} & \frac{\partial N_4}{\partial z} & 0 & \frac{\partial N_4}{\partial x} \end{bmatrix} \tag{5.24}$$

The Cartesian derivatives of the shape functions N_i (ξ, η, ζ) $(i = 1, 4)$ which are required to compute the strain matrix **B** are determined as

$$\begin{bmatrix} \frac{\partial N_i}{\partial x} \\ \frac{\partial N_i}{\partial y} \\ \frac{\partial N_i}{\partial z} \end{bmatrix} = [\mathbf{J}]^{-1} \begin{bmatrix} \frac{\partial N_i}{\partial \xi} \\ \frac{\partial N_i}{\partial \eta} \\ \frac{\partial N_i}{\partial \zeta} \end{bmatrix} \quad (i = 1, 2, 3, 4). \tag{5.25}$$

The Jacobian matrix $\mathbf{J}$ is in turn evaluated using

$$[\mathbf{J}] = \begin{bmatrix} \frac{\partial x}{\partial \xi} & \frac{\partial y}{\partial \xi} & \frac{\partial z}{\partial \xi} \\ \frac{\partial x}{\partial \eta} & \frac{\partial y}{\partial \eta} & \frac{\partial z}{\partial \eta} \\ \frac{\partial x}{\partial \zeta} & \frac{\partial y}{\partial \zeta} & \frac{\partial z}{\partial \zeta} \end{bmatrix} = \sum_{j=1}^{4} \begin{bmatrix} \frac{\partial N_j}{\partial \xi} x_j & \frac{\partial N_j}{\partial \xi} y_j & \frac{\partial N_j}{\partial \xi} z_j \\ \frac{\partial N_j}{\partial \eta} x_j & \frac{\partial N_j}{\partial \eta} y_j & \frac{\partial N_j}{\partial \eta} z_j \\ \frac{\partial N_j}{\partial \zeta} x_j & \frac{\partial N_j}{\partial \zeta} y_j & \frac{\partial N_j}{\partial \zeta} z_j \end{bmatrix} \tag{5.26}$$

$$= \begin{bmatrix} J_{11} & J_{12} & J_{13} \\ J_{21} & J_{22} & J_{23} \\ J_{31} & J_{32} & J_{33} \end{bmatrix}.$$

The derivates of the shape functions $N_i\ (\xi, \eta, \zeta)$ with respect to ξ, η, and ζ are easily obtained from Eqn. (5.20). They are either ± 1 or zeroes. Therefore the Jacobian matrix $\mathbf{J}$ does not change from point-to-point within the element. The inverse of the Jacobian matrix is evaluated as

$$[\mathbf{J}]^{-1} = \frac{1}{|\mathbf{J}|} \begin{bmatrix} (J_{22}J_{33} - J_{32}J_{23}) & (J_{32}J_{13} - J_{12}J_{33}) & (J_{12}J_{23} - J_{22}J_{13}) \\ (J_{23}J_{31} - J_{33}J_{21}) & (J_{33}J_{11} - J_{13}J_{31}) & (J_{12}J_{21} - J_{23}J_{11}) \\ (J_{21}J_{32} - J_{31}J_{22}) & (J_{21}J_{32} - J_{31}J_{32}) & (J_{11}J_{22} - J_{21}J_{12}) \end{bmatrix} \tag{5.27}$$

The net result is that the nonzero elements of the strain matrix $\mathbf{B}$ are real constants and hence the popular name constant strain tetrahedron. It can be verified that the volume $v^{(e)}$ of the tetrahedral element is

$$v^{(e)} = \frac{1}{6} |\mathbf{J}|$$

and therefore $|\mathbf{J}| \neq 0$.

5.4.3 Element stress recovery

Combining Eqns. (5.8) and (5.23), the element stresses $\boldsymbol{\sigma}$ are evaluated as

$$\{\sigma\} = [\mathbf{D}][\mathbf{B}]\{\mathbf{q}\} - [\mathbf{D}]\{\boldsymbol{\varepsilon}^{\text{init}}\}. \tag{5.28}$$

The initial strains $\boldsymbol{\varepsilon}^{\text{init}}$ now due to thermal loading can be determined using

$$\left\{\boldsymbol{\varepsilon}^{\text{init}}\right\} = \begin{Bmatrix} \alpha \sum_{i=1}^{4} N_i\,(\xi, \eta, \zeta)\, T_i \\ \alpha \sum_{i=1}^{4} N_i\,(\xi, \eta, \zeta)\, T_i \\ \alpha \sum_{i=1}^{4} N_i\,(\xi, \eta, \zeta)\, T_i \\ 0 \\ 0 \\ 0 \end{Bmatrix}, \tag{5.29}$$

where T_i denotes the temperature change at node i.

By default, the stresses are recovered at the centroid ($\xi = \eta = \zeta = 1/3$). Using the stress components, the principal stresses $\sigma_1 > \sigma_2 > \sigma_3$ can then be determined. The principal stresses are used to evaluate an equivalent stress based on the appropriate failure criteria.

5.4.4 Element stiffness matrix

The strain energy stored in the element is

$$SE = \frac{1}{2}\int_{v^{(e)}} \boldsymbol{\varepsilon}^T \mathbf{D}\boldsymbol{\varepsilon} dv,$$

$$= \frac{1}{2}\{\mathbf{q}\}^T[\mathbf{k}]\{\mathbf{q}\},$$

$$[\mathbf{k}] = \frac{1}{2}\int_{v^{(e)}} [\mathbf{B}]^T\mathbf{D}[\mathbf{B}]dv. \tag{5.30}$$

The integral in Eqn. (5.31) can be explicitly evaluated leading to

$$[\mathbf{k}] = [\mathbf{B}]^T[\mathbf{D}][\mathbf{B}]\frac{1}{6}|\mathbf{J}|. \tag{5.31}$$

Eqn. (5.32) is the formula to compute the element stiffness matrix. The indicated matrix multiplication is performed on the computer.

5.4.5 Consistent nodal force vectors

The formulae to compute consistent nodal force vectors due to body force $\mathbf{f}^b$, surface traction, $\mathbf{f}^p$, and initial strains $\mathbf{f}^{\text{init}}$ are derived from the potential energy functional as

$$\left\{\mathbf{f}^b\right\} = \int_{v^{(e)}} [\mathbf{N}]^T \{\mathbf{b}\}\, dv. \tag{5.32}$$

$$\left\{\mathbf{f}^{\text{init}}\right\} = \int_{v^{(e)}} [\mathbf{B}]^T \{\mathbf{D}\} \left\{\boldsymbol{\varepsilon}^{\text{init}}\right\} dv. \tag{5.33}$$

$$\{\mathbf{f}^p\} = \int_{S_\sigma} [\bar{\mathbf{N}}]^T \{\mathbf{p}\}\, dS. \tag{5.34}$$

The volume and surface integrals in the above equations have to be evaluated using numerical integration methods. It will be convenient to evaluate these over the parent element domain using Gauss quadrature formulae. The required sampling point locations, and weighting factors are readily available [5.6].

The division of an arbitrary volume into individual tetrahedral elements sometimes presents difficulties of visualization. A more convenient subdivision of space is into brick shaped elements shown in Fig. 5.4. The design of a linear hexahedral element is described in the next section.

5.5 Eight-Noded Hexahedral Element (HEXA 8)

The element shown in Fig. 5.4 is bounded by six quadrilateral faces and has eight nodes. The nodes are numbered 1 to 8 as shown. The geometry of the element is described by the Cartesian coordinates (X_i, Y_i, Z_i) of the eight nodes. For solid mechanics applications, each node i has three degrees of freedom u_i, v_i, w_i and the nodal degrees of freedom vector **q** is

$$\{\mathbf{q}\}^T = (u_1, v_1, w_1, u_2, v_2, w_2, \ldots, \ldots, \ldots, u_8, v_8, w_8)\,. \tag{5.35}$$

The corresponding vector of nodal forces is

$$\{\mathbf{f}\}^T = \left(F_{x1}, F_{y1}, F_{z1}, F_{x2}, F_{y2}, F_{z2}, \ldots, \ldots, \ldots, F_{x8}, F_{y8}, F_{z8}\right). \tag{5.36}$$

The derivation of element equations, i.e., the necessary and sufficient number of algebraic equations to determine **q** using the assumed displacement functions and minimum potential energy functional approach is described below.

5.5.1 Admissible displacement functions

The displacement components (u, v, w) at any point (X, Y, Z) within the HEXA 8 element have to be interpolated in terms of the nodal degrees of freedom. The simplest representation is

$$\{\mathbf{u}\} = [\mathbf{N}]\{\mathbf{q}\}, \tag{5.37}$$

where

$$[\mathbf{N}] = \begin{bmatrix} N_1 & 0 & 0 & N_2 & 0 & 0 & \ldots & \ldots & \ldots & N_8 & 0 & 0 \\ 0 & N_1 & 0 & 0 & N_2 & 0 & \ldots & \ldots & \ldots & 0 & N_8 & 0 \\ 0 & 0 & N_1 & 0 & 0 & N_2 & \ldots & \ldots & \ldots & 0 & 0 & N_8 \end{bmatrix}. \tag{5.38}$$

In order to derive explicit expressions for the interpolation functions $N_i (i = 1, 8)$, it is now a standard practice to use the concept of a parent element in the covariant natural coordinate system ξ, η, ζ as shown in Fig. 5.4. The parent element is a bi-unit cube with nodes located at its eight corners. The local coordinates of these nodes are listed in Table 5.1. Any variable Φ is approximated over the parent element domain using the tri-linear polynomial:

$$\Phi = \alpha_1 + \alpha_2\xi + \alpha_3\eta + \alpha_4\zeta + \alpha_5\xi\eta + \alpha_6\eta\zeta + \alpha_7\xi\zeta + \alpha_8\xi\eta\zeta. \tag{5.39}$$

The reader is advised to justify this choice with reference to Pascal's pyramid given in Fig. 5.8.

Using Table 5.1, it is possible to evaluate the polynomial coefficients $\alpha_i (i = 1, 8)$ in terms of the nodal values of Φ namely, $\Phi_i (i = 1, 8)$. The result is

$$\Phi = \sum_{i=1}^{8} N_i(\xi, \eta, \zeta)\Phi_i, \tag{5.40}$$

where

$$N_i(\xi, \eta, \zeta) = \frac{1}{8}(1 + \xi\xi_i)(1 + \eta\eta_i)(1 + \zeta\zeta_i)\,. \tag{5.41}$$

Table 5.1 *Natural coordinates of nodes of the parent element (see Fig. 5.4(b))*

Node i	ξ_i	η_i	ζ_i
1	-1	-1	-1
2	$+1$	-1	-1
3	-1	$+1$	-1
4	$+1$	$+1$	-1
5	-1	-1	$+1$
6	$+1$	-1	$+1$
7	-1	$+1$	$+1$
8	$+1$	$+1$	$+1$

These are the shape functions for the HEXA 8 element and are listed in Table 5.2.

Table 5.2 *Shape functions for the HEXA 8 element*

$$N_1(\xi,\eta,\zeta) = \frac{1}{8}(1-\xi)(1-\eta)(1-\zeta)$$
$$N_2(\xi,\eta,\zeta) = \frac{1}{8}(1+\xi)(1-\eta)(1-\zeta)$$
$$N_3(\xi,\eta,\zeta) = \frac{1}{8}(1-\xi)(1+\eta)(1-\zeta)$$
$$N_4(\xi,\eta,\zeta) = \frac{1}{8}(1+\xi)(1+\eta)(1-\zeta)$$
$$N_5(\xi,\eta,\zeta) = \frac{1}{8}(1-\xi)(1-\eta)(1+\zeta)$$
$$N_6(\xi,\eta,\zeta) = \frac{1}{8}(1+\xi)(1-\eta)(1+\zeta)$$
$$N_7(\xi,\eta,\zeta) = \frac{1}{8}(1-\xi)(1+\eta)(1+\zeta)$$
$$N_8(\xi,\eta,\zeta) = \frac{1}{8}(1+\xi)(1+\eta)(1+\zeta)$$

These functions have the property

$N_i = 1$ at node i. $\quad N_i = 0$ at all other nodes.

$$\sum_{i=1}^{8} N_i = 1.0.$$

Let us now distort the parent element to the shape of the HEXA 8 element. The required relation between the Cartesian coordinates and the natural coordinates to accomplish this is

$$\begin{Bmatrix} x \\ y \\ z \end{Bmatrix} = \sum_{i=1}^{8} N_i(\xi,\eta,\zeta) \begin{Bmatrix} X_i \\ Y_i \\ Z_i \end{Bmatrix}, \tag{5.42}$$

where (X_i, Y_i, Z_i) denote the Cartesian coordinates of the node i.

The geometry of the HEXA 8 element is thereby represented by the shape functions $N_i (i = 1, 8)$ and the nodal coordinates $(X_i, Y_i, Z_i)(i = 1, 8)$. The centroid of the element is located at $\xi = \eta = \zeta = 0$; the six faces of the element are identified by $\xi = \pm 1$, $\eta = \pm 1$, and $\zeta = \pm 1$.

The same shape functions are also used as displacement functions. We can therefore write

$$\begin{Bmatrix} u \\ v \\ w \end{Bmatrix} = \sum_{i=1}^{8} N_i(\xi, \eta, \zeta) \begin{Bmatrix} u_i \\ v_i \\ w_i \end{Bmatrix}. \tag{5.43}$$

The HEXA 8 element can be of arbitrary shape since it is formulated as an isoparametric element. The assumed displacement functions will satisfy the continuity requirements at interfaces between elements. The constant derivative criterion will be satisfied for this isoparametric element since $\sum_{i=1}^{8} N_i = 1.0$. Therefore convergence with mesh refinement is assured.

5.5.2 Element strain matrix

Combining Eqns. (5.6) and (5.44), we can write

$$\{\boldsymbol{\varepsilon}\} = [\mathbf{B}]\{\mathbf{q}\}, \tag{5.44}$$

where

$$[\mathbf{B}] = \begin{bmatrix} \dfrac{\partial N_1}{\partial x} & 0 & 0 & -- & -- & -- & \dfrac{\partial N_8}{\partial x} & 0 & 0 \\ 0 & \dfrac{\partial N_1}{\partial y} & 0 & -- & -- & -- & 0 & \dfrac{\partial N_8}{\partial y} & 0 \\ 0 & 0 & \dfrac{\partial N_1}{\partial z} & -- & -- & -- & 0 & 0 & \dfrac{\partial N_8}{\partial z} \\ \dfrac{\partial N_1}{\partial y} & \dfrac{\partial N_1}{\partial x} & 0 & -- & -- & -- & \dfrac{\partial N_8}{\partial y} & \dfrac{\partial N_8}{\partial x} & 0 \\ 0 & \dfrac{\partial N_1}{\partial z} & \dfrac{\partial N_1}{\partial y} & -- & -- & -- & 0 & \dfrac{\partial N_8}{\partial z} & \dfrac{\partial N_8}{\partial y} \\ \dfrac{\partial N_1}{\partial z} & 0 & \dfrac{\partial N_1}{\partial x} & -- & -- & -- & \dfrac{\partial N_8}{\partial z} & 0 & \dfrac{\partial N_8}{\partial x} \end{bmatrix}. \tag{5.45}$$

The Cartesian derivatives of the shape functions, which are required to compute the element strain matrix **B**, are evaluated using

$$\begin{bmatrix} \dfrac{\partial N_i}{\partial x} \\ \dfrac{\partial N_i}{\partial y} \\ \dfrac{\partial N_i}{\partial z} \end{bmatrix} = [\mathbf{J}]^{-1} \begin{bmatrix} \dfrac{\partial N_i}{\partial \xi} \\ \dfrac{\partial N_i}{\partial \eta} \\ \dfrac{\partial N_i}{\partial \zeta} \end{bmatrix} \quad (i = 1, 8). \tag{5.46}$$

The Jacobian matrix **J** in turn is evaluated using

$$[\mathbf{J}] = \begin{bmatrix} \frac{\partial x}{\partial \xi} & \frac{\partial y}{\partial \xi} & \frac{\partial z}{\partial \xi} \\ \frac{\partial x}{\partial \eta} & \frac{\partial y}{\partial \eta} & \frac{\partial z}{\partial \eta} \\ \frac{\partial x}{\partial \zeta} & \frac{\partial y}{\partial \zeta} & \frac{\partial z}{\partial \zeta} \end{bmatrix} = \sum_{j=1}^{8} \begin{bmatrix} \frac{\partial N_j}{\partial \xi} x_j & \frac{\partial N_j}{\partial \xi} y_j & \frac{\partial N_j}{\partial \xi} z_j \\ \frac{\partial N_j}{\partial \eta} x_j & \frac{\partial N_j}{\partial \eta} y_j & \frac{\partial N_j}{\partial \eta} z_j \\ \frac{\partial N_j}{\partial \zeta} x_j & \frac{\partial N_j}{\partial \zeta} y_j & \frac{\partial N_j}{\partial \zeta} z_j \end{bmatrix} \tag{5.47}$$

$$= \begin{bmatrix} J_{11} & J_{12} & J_{13} \\ J_{21} & J_{22} & J_{23} \\ J_{31} & J_{32} & J_{33} \end{bmatrix}.$$

The derivatives of the shape functions with respect to ξ, η, ζ are easily obtained. However, these are known functions of ξ, η, ζ. Therefore the matrix **J** and its inverse $\mathbf{J}^{-1}$ are also functions of ξ, η, ζ. The net result is that the nonzero elements of the matrix **B** are highly complex functions of ξ, η, ζ. It can be shown that

$$dxdydz = |\mathbf{J}| d\xi d\eta d\zeta.$$

For the distortions of the HEXA 8 element to be reasonable, $|\mathbf{J}| \neq 0$ and the sign of $|\mathbf{J}|$ should not change from point-to-point within the element. Excessive distortions are unacceptable as they affect the accuracy of the finite element solutions.

5.5.3 Element stress recovery

The stresses $\boldsymbol{\sigma}$ at any point within the HEXA 8 element are evaluated using

$$\{\sigma\} = [\mathbf{D}][\mathbf{B}]\{\mathbf{q}\} - [\mathbf{D}]\{\boldsymbol{\varepsilon}^{\text{init}}\}, \tag{5.48}$$

where the initial strains due to thermal expansion are

$$\{\boldsymbol{\varepsilon}^{\text{init}}\} = \begin{Bmatrix} \alpha \sum_{i=1}^{8} N_i(\xi,\eta,\zeta)\, T_i \\ \alpha \sum_{i=1}^{8} N_i(\xi,\eta,\zeta)\, T_i \\ \alpha \sum_{i=1}^{8} N_i(\xi,\eta,\zeta)\, T_i \\ 0 \\ 0 \\ 0 \end{Bmatrix} \tag{5.49}$$

and T_i denotes the temperature change at node i.

By default, the stresses are recovered at the centroid ($\xi = \eta = \zeta = 0$) of the HEXA 8 element. However, the Gauss point locations ($\xi = 1/\pm\sqrt{3}$, $\eta = 1/\pm\sqrt{3}$, $\zeta = 1/\pm\sqrt{3}$) are optimal for accurate stress recovery. If nodal point stresses are requested, extrapolation of the Gauss point stresses to the nodal point locations and averaging of contributions from all elements connected to a node are essential.

Three-dimensional stress analysis is therefore feasible and geometric complexity is not an issue at all.

5.5.4 Element stiffness matrix

The stiffness matrix **k** for the HEXA 8 element is evaluated using

$$[\mathbf{k}] = \int_{v^{(e)}} [\mathbf{B}]^T [\mathbf{D}] [\mathbf{B}] \, dv.$$

The integration can now be performed over the parent element domain using

$$[\mathbf{k}] = \int_{\xi=-1}^{+1} \int_{\eta=-1}^{+1} \int_{\zeta=-1}^{+1} [\mathbf{B}]^T [\mathbf{D}] [\mathbf{B}] \, |\mathbf{J}| \, d\xi d\eta d\zeta. \tag{5.50}$$

We note here that both **B** and **J** are involved functions of ξ, η, ζ and so the evaluation of the integral in Eqn. (5.51) has to be performed numerically. The Gauss quadrature formula is ideally suited and hence exclusively used.

To evaluate numerically the integral of a function f(ξ, η, ζ) over the parent element domain, the Gauss quadrature formula is

$$I = \int_{\xi=-1}^{+1} \int_{\eta=-1}^{+1} \int_{\zeta=-1}^{+1} f(\xi, \eta, \zeta) \, d\xi d\eta d\zeta = \sum_{i=1}^{n} \sum_{j=1}^{n} \sum_{k=1}^{n} W_i W_j W_k f(\xi_i, \eta_j, \zeta_k). \tag{5.51}$$

In Eqn. (5.52), n denotes number of Gauss points, W_i, W_j, W_k are the weighting coefficients and (ξ_i, η_j, ζ_k) denotes the sampling point locations. Table 3.2 therefore gives the required sampling point locations and weighting coefficients. The number of Gauss point n to be chosen depends upon the nature of the integrand.

Choosing $n = 2$ and using Eqn. (5.52), we arrive at the following summation expression for the computation of the HEXA 8 element stiffness matrix.

$$[\mathbf{k}] = \sum_{i=1}^{2} \sum_{j=1}^{2} \sum_{k=1}^{2} W_i W_j W_k \left[\Phi(\xi_i, \eta_j, \zeta_k)\right] \tag{5.52}$$

where

$$[\Phi(\xi, \eta, \zeta)] = [\mathbf{B}]^T [\mathbf{D}] [\mathbf{B}] |\mathbf{J}|. \tag{5.53}$$

It is to be noted that the eight Gauss points are located at the corners of a cube whose faces are at $\xi = 1/\pm\sqrt{3}, \eta = 1/\pm\sqrt{3}, \zeta = 1/\pm\sqrt{3}$.

In practice, care must be taken to avoid severe distortions (i.e., deviations from the cubic shape). Therefore checks on the sign of $|\mathbf{J}|$ are necessary during the computation of stiffness matrix.

5.5.5 Consistent nodal force vector: Body force

The nodal force vector $\mathbf{f}^b$ due to body force $\mathbf{b}$ is evaluated using

$$\left\{\mathbf{f}^b\right\} = \int_{v^{(e)}} [\mathbf{N}]^T \{\mathbf{b}\}\, dv. \tag{5.54}$$

The integration can be performed over the parent element domain using

$$\left\{\mathbf{f}^b\right\} = \int_{\xi=-1}^{+1} \int_{\eta=-1}^{+1} \int_{\zeta=-1}^{+1} [\mathbf{N}]^T \{\mathbf{b}\}\, |\mathbf{J}|\, d\xi d\eta d\zeta. \tag{5.55}$$

Using Gauss quadrature formula and choosing $n = 2$, we arrive at the following summation expression for the computation of $\mathbf{f}^b$.

$$\left\{\mathbf{f}^b\right\} = \sum_{i=1}^{2}\sum_{j=1}^{2}\sum_{k=1}^{2} W_i W_j W_k \left\{\psi\left(\xi_i, \eta_j, \zeta_k\right)\right\}, \tag{5.56}$$

where

$$\{\psi(\xi, \eta, \zeta)\} = [\mathbf{N}]^T \{\mathbf{b}\} |\mathbf{J}|. \tag{5.57}$$

5.5.6 Consistent nodal force vector: Initial strain

The nodal force vector $\mathbf{f}^{\text{init}}$ due to initial strain $\boldsymbol{\varepsilon}^{\text{init}}$ is evaluated using

$$\{\mathbf{f}^{\text{init}}\} = \int_{v^{(e)}} [\mathbf{B}]^T [\mathbf{D}] \{\boldsymbol{\varepsilon}^{\text{init}}\} dv. \tag{5.58}$$

The initial strains due to thermal expansion are in turn evaluated using Eqn. (5.50).

The integration can be performed over the parent element domain using

$$\{\mathbf{f}^{\text{init}}\} = \int_{\xi=-1}^{+1} \int_{\eta=-1}^{+1} \int_{\zeta=-1}^{+1} [\mathbf{B}]^T [\mathbf{D}] \{\boldsymbol{\varepsilon}^{\text{init}}\} |\mathbf{J}| d\xi d\eta d\zeta. \tag{5.59}$$

Using Gauss quadrature formula and choosing $n = 2$, we arrive at the following summation expression for the computation of $\mathbf{f}^{\text{init}}$.

$$\{\mathbf{f}^{\text{init}}\} = \sum_{i=1}^{2}\sum_{j=1}^{2}\sum_{k=1}^{2} W_i W_j W_k \{\chi(\xi_i, \eta_j, \zeta_k)\}, \tag{5.60}$$

where

$$\{\chi(\xi,\eta,\zeta)\} = [\mathbf{B}]^T[\mathbf{D}]\{\boldsymbol{\varepsilon}^{\text{init}}\}|\mathbf{J}|. \tag{5.61}$$

5.5.7 Consistent nodal force vector: Surface traction

The HEXA 8 element is bounded by six arbitrary quadrilateral faces. These are identified as face (1) to face (6). On one or more of the faces, pressure load can be applied. We can derive a formula to compute the consistent nodal force vector $\mathbf{f}^p$ due to surface traction $\{\mathbf{p}\}^T = (p_x, py, p_z)$ as:

$$\{\mathbf{f}^p\} = \int_{S_\sigma} [\bar{\mathbf{N}}]^T\{\mathbf{p}\}dS. \tag{5.62}$$

In Eqn. (5.63), S_σ denotes the surface area on which the pressure **p** is applied, $\bar{\mathbf{N}}$ denotes the shape functions now evaluated on the face under consideration. The pressure **p** acting in a direction **n** normal to the face can be resolved into components p_x, p_y, p_z. In practice, the integral in Eqn. (5.63) is evaluated numerically using a two-point Gauss quadrature formula. It is to be noted that a pressure applied to a face will contribute forces only to those four nodes associated with that face. The calculation of the direction cosines of the outward normal vector to a face is an important pre-requisite.

5.6 Six-Noded Pentahedral Element

The element shown in Fig. 5.6 is bounded by five faces and has six nodes. For solid mechanics applications, each node i has three degrees of freedom u_i, v_i, w_i.

This element is created by further distortions of the HEXA 8 element as indicated in Fig. 5.6. Specifically, it involves collapsing a face and forcing the nodes that are co-located to have identical degrees of freedom. Collapsing of a face is achieved by editing the coordinates of nodes 4 and 8 to be the same as those of nodes 3 and 7. Multi-point constraint (MPC) equations are used to tie the nodes 3 to 4 and 7 to 8.

5.7 Higher Order Elements

It is obvious that the number of linear tetrahedral and hexahedral elements to achieve adequate modelling of geometry and a desired degree of accuracy of the solution (displacements, stresses etc.) has to be very large. This will result in a very large number of simultaneous equations in practical applications. Further, the bandwidth of the resulting system of equations will also be large demanding huge computer storage requirements. Given that with usual solution techniques, the computational effort is roughly proportional to the product of the number of equations and the square of the bandwidth, the magnitude of the problem can be appreciated. It is not surprising therefore that efforts are on to improve geometric modelling and ensure faster convergence with mesh refinement. Higher accuracy for a given mesh, and computational efficiency have been the strongest in the area of three-dimensional elements. This in fact is the motivation for the development of higher order elements.

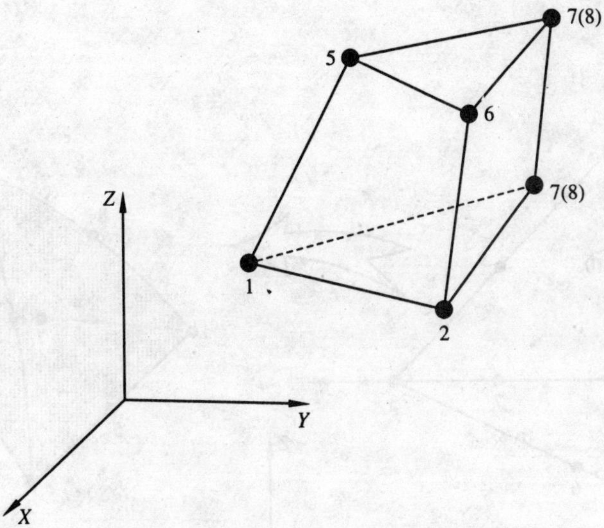

Fig. 5.6 *Six-noded pentahedral element*

5.7.1 Tetrahedral elements

A family of tetrahedral elements can be developed on a pattern indicated in Fig. 5.7. The number of nodes in each member coincides with the number of polynomial terms employed to approximate a variable over the element domain. This follows by comparison with Pascal's pyramid shown in Fig. 5.8. A complete polynomial of the order needed to ensure compatibility is used.

The *quadratic order element* will have ten nodes. Four nodes are located at the vertices and the other six are at mid-side locations. Any variable Φ is approximated over the parent element domain using

$$\Phi = \alpha_1 + \alpha_2\xi + \alpha_3\eta + \alpha_4\zeta + \alpha_5\xi^2 + \alpha_6\xi\eta + \alpha_7\eta^2 + \alpha_8\eta\zeta + \alpha_9\zeta^2 + \alpha_{10}\xi\zeta.$$

Explicit shape functions $N_i(\xi, \eta, \zeta)$ $(i = 1, 10)$ derived from this polynomial are given in Table 5.3.

Using isoparametric formulation, the parent element can be distorted to have curved edges and curved faces as illustrated in Fig. 5.7. The resulting geometric representation is

$$\begin{Bmatrix} X \\ Y \\ Z \end{Bmatrix} = \sum_{i=1}^{10} N_i(\xi, \eta, \zeta) \begin{Bmatrix} X_i \\ Y_i \\ Z_i \end{Bmatrix},$$

where X_i, Y_i, Z_i denote the Cartesian coordinates of node i. The admissible displacement functions are written as

$$\begin{Bmatrix} u \\ v \\ w \end{Bmatrix} = \sum_{i=1}^{10} N_i(\xi, \eta, \zeta) \begin{Bmatrix} u_i \\ v_i \\ w_i \end{Bmatrix},$$

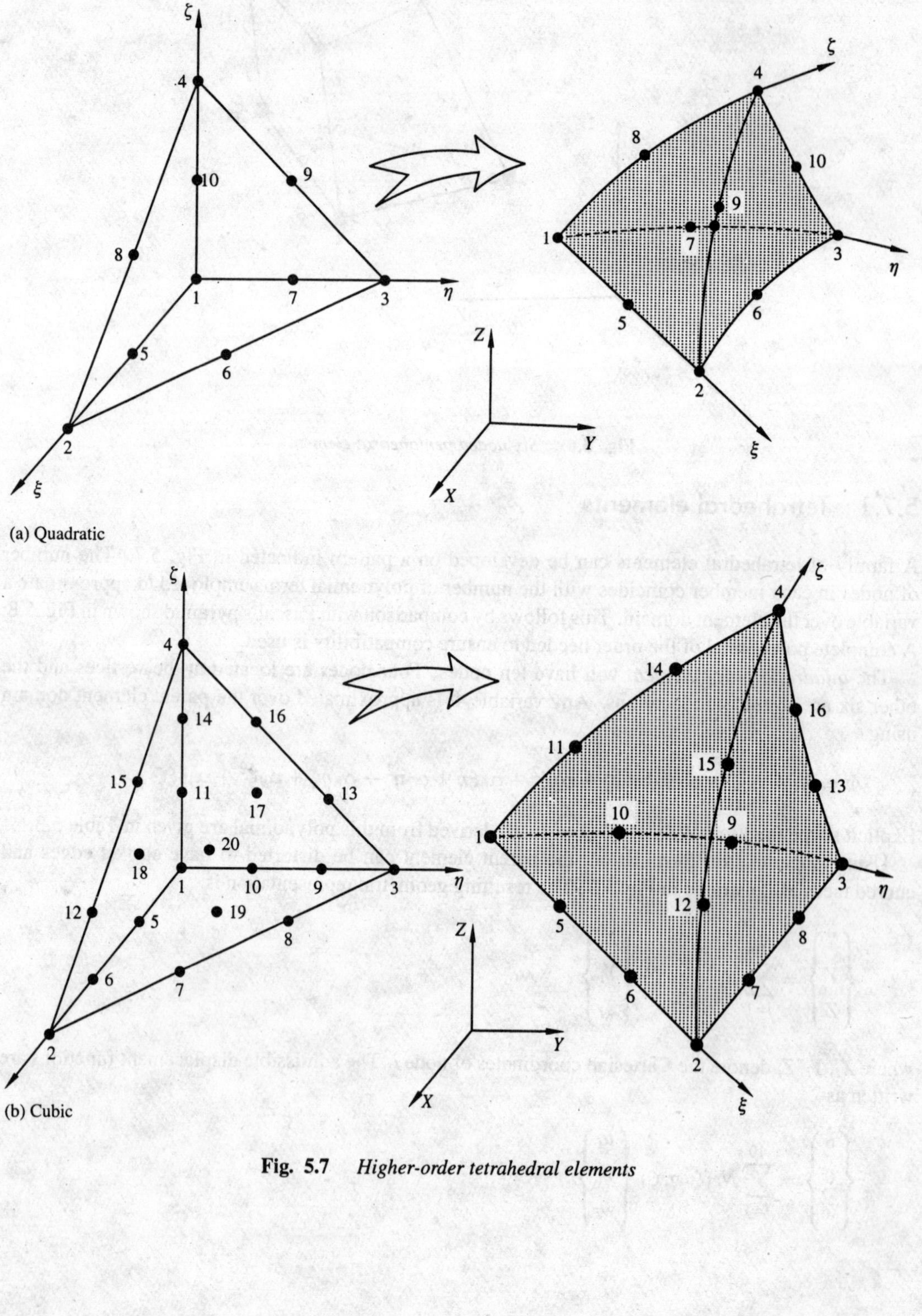

Fig. 5.7 *Higher-order tetrahedral elements*

Table 5.3 *Shape functions for a three dimensional element, tetrahedral in shape and of quadratic order*

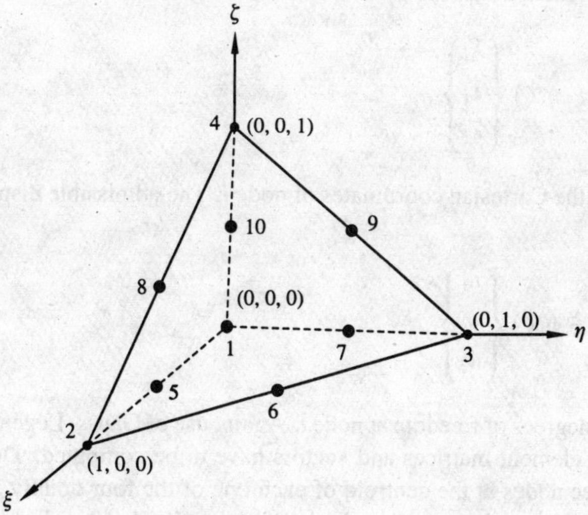

$$N_1(\xi,\eta,\zeta) = (1-\xi-\eta-\zeta) - \frac{1}{2}N_5 - \frac{1}{2}N_7 - \frac{1}{2}N_{10}$$
$$N_2(\xi,\eta,\zeta) = \xi - \frac{1}{2}N_5 - \frac{1}{2}N_6 - \frac{1}{2}N_8$$
$$N_3(\xi,\eta,\zeta) = \eta - \frac{1}{2}N_6 - \frac{1}{2}N_7 - \frac{1}{2}N_9$$
$$N_4(\xi,\eta,\zeta) = \zeta - \frac{1}{2}N_8 - \frac{1}{2}N_9 - \frac{1}{2}N_{10}$$
$$N_5(\xi,\eta,\zeta) = 4\xi(1-\xi-\eta-\zeta)$$
$$N_6(\xi,\eta,\zeta) = 4\xi\eta$$
$$N_7(\xi,\eta,\zeta) = 4\eta(1-\xi-\eta-\zeta)$$
$$N_8(\xi,\eta,\zeta) = 4\xi\zeta$$
$$N_9(\xi,\eta,\zeta) = 4\eta\zeta$$
$$N_{10}(\xi,\eta,\zeta) = 4\zeta(1-\xi-\eta-\zeta)$$

where u_i, v_i, w_i are the degrees of freedom at node i. Use of Gauss–Legendre quadrature formula will be required when element matrices and vectors have to be computed.

The *cubic order element* will have twenty nodes; four corner nodes, twelve side nodes and 4 mid-surface nodes as shown in Fig. 5.7. Any variable Φ is approximated over the parent element domain using a complete cubic polynomial:

$$\Phi = \alpha_1 + \alpha_2\xi + \alpha_3\eta + \alpha_4\zeta + \alpha_5\xi^2 + \alpha_6\xi\eta + \alpha_7\eta^2 + \alpha_8\eta\zeta + \alpha_9\zeta^2 + \alpha_{10}\xi\zeta + \alpha_{11}\xi^3$$
$$+ \alpha_{12}\xi^2\eta + \alpha_{13}\xi\eta^2 + \alpha_{14}\eta^3 + \alpha_{15}\eta^2\zeta + \alpha_{16}\eta\zeta^2 + \alpha_{17}\zeta^3 + \alpha_{18}\zeta^2\xi + \alpha_{19}\zeta\xi^2 + \alpha_{20}\xi\eta\zeta.$$

Explicit shape functions $N_i(\xi,\eta,\zeta)$ $(i = 1, 20)$ can be derived from this polynomial basis.

Using isoparametric formulation, the parent element can be distorted to have straight or curved edges and flat or curved faces as illustrated in Fig. 5.7. The geometric representation therefore is

$$\begin{Bmatrix} X \\ Y \\ Z \end{Bmatrix} = \sum_{i=1}^{20} N_i(\xi, \eta, \zeta) \begin{Bmatrix} X_i \\ Y_i \\ Z_i \end{Bmatrix},$$

where X_i, Y_i, Z_i denote the Cartesian coordinates of node i. The admissible displacement functions are written as

$$\begin{Bmatrix} u \\ v \\ w \end{Bmatrix} = \sum_{i=1}^{20} N_i(\xi, \eta, \zeta) \begin{Bmatrix} u_i \\ v_i \\ w_i \end{Bmatrix},$$

where u_i, v_i, w_i are the degrees of freedom at node i. Again, use of Gauss–Legendre quadrature formula will be essential when element matrices and vectors have to be computed. Due to the difficulties in locating the mid-surface nodes at the centroid of each one of the four doubly curved faces, the cubic order element is not widely used in practice. In fact, it is not implemented in many commercial FEM systems.

The derivation of element equations and computation of element matrices for the ten-noded tetrahedral element is left to the reader as a home assignment.

5.7.2 Hexahedral elements: Serendipity family

Figure 5.9 shows quadratic and cubic order three-dimensional hexahedral elements of the so-called Serendipity family.

The *quadratic order element* has twenty nodes. Eight nodes are located at the vertices and the others are at mid-side points of the parent element which is a bi-unit cube. Any variable Φ is approximated over the parent element domain using the following incomplete quadratic order polynomial:

$$\begin{aligned}\Phi = {} & \alpha_1 + \alpha_2\xi + \alpha_3\eta + \alpha_4\zeta + \alpha_5\xi^2 + \alpha_6\xi\eta + \alpha_7\eta^2 + \alpha_8\eta\zeta + \alpha_9\zeta^2 + \alpha_{10}\xi\zeta \\ & + \alpha_{11}\xi^2\eta + \alpha_{12}\xi\eta^2 + \alpha_{13}\eta^2\zeta + \alpha_{14}\eta\zeta^2 + \alpha_{15}\xi\zeta^2 + \alpha_{16}\xi^2\zeta + \alpha_{17}\xi\eta\zeta \\ & + \alpha_{18}\xi^2\eta\zeta + \alpha_{19}\xi\eta^2\zeta + \alpha_{20}\xi\eta\zeta^2\end{aligned}$$

The reader is advised to justify this choice with reference to Pascal's pyramid given in Fig. 5.8. Explicit interpolation functions $N_i(\xi, \eta, \zeta)$ $(i = 1, 20)$ have been derived from this polynomial basis and are readily available in [5.3]. The reader is advised to list them using the current notation and node numbering scheme.

Using isoparametric formulation, the parent element can be distorted to have straight or curved edges and flat or curved faces as illustrated in Fig. 5.9. The geometric representation for the element is

$$\begin{Bmatrix} X \\ Y \\ Z \end{Bmatrix} = \sum_{i=1}^{20} N_i(\xi, \eta, \zeta) \begin{Bmatrix} X_i \\ Y_i \\ Z_i \end{Bmatrix},$$

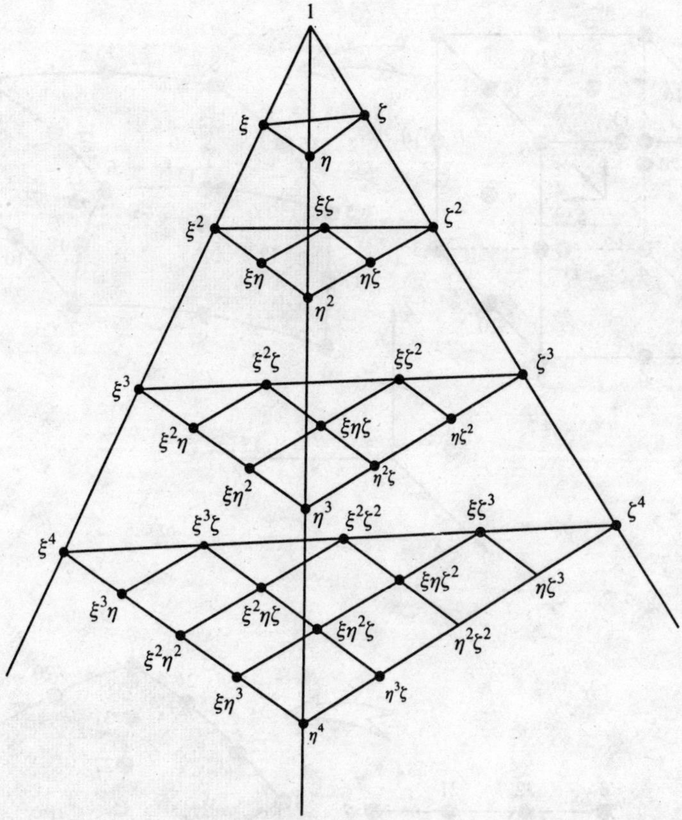

Fig. 5.8 *Pascal's pyramid*

where X_i, Y_i, Z_i denote the Cartesian coordinates of node i. The admissible displacement functions are written as

$$\begin{Bmatrix} u \\ v \\ w \end{Bmatrix} = \sum_{i=1}^{20} N_i\,(\xi, \eta, \zeta) \begin{Bmatrix} u_i \\ v_i \\ w_i \end{Bmatrix},$$

where u_i, v_i, w_i are the degrees of freedom at node i.

Use of a three-point Gauss quadrature formula in each of the ξ, η, ζ coordinate directions is recommended to compute the element matrices and vectors. The required Gauss point locations and weighting factors are readily available (see Chapter 3).

The design of the twenty-noded hexahedral element is left to the reader as a home assignment.

The *cubic order hexahedral element* (see Fig. 5.9) will have thirty two nodes. Eight nodes are located at the vertices of the parent element, namely a bi-unit cube. There are two equally spaced nodes on each side of the element. Any variable Φ is approximated over the parent element domain

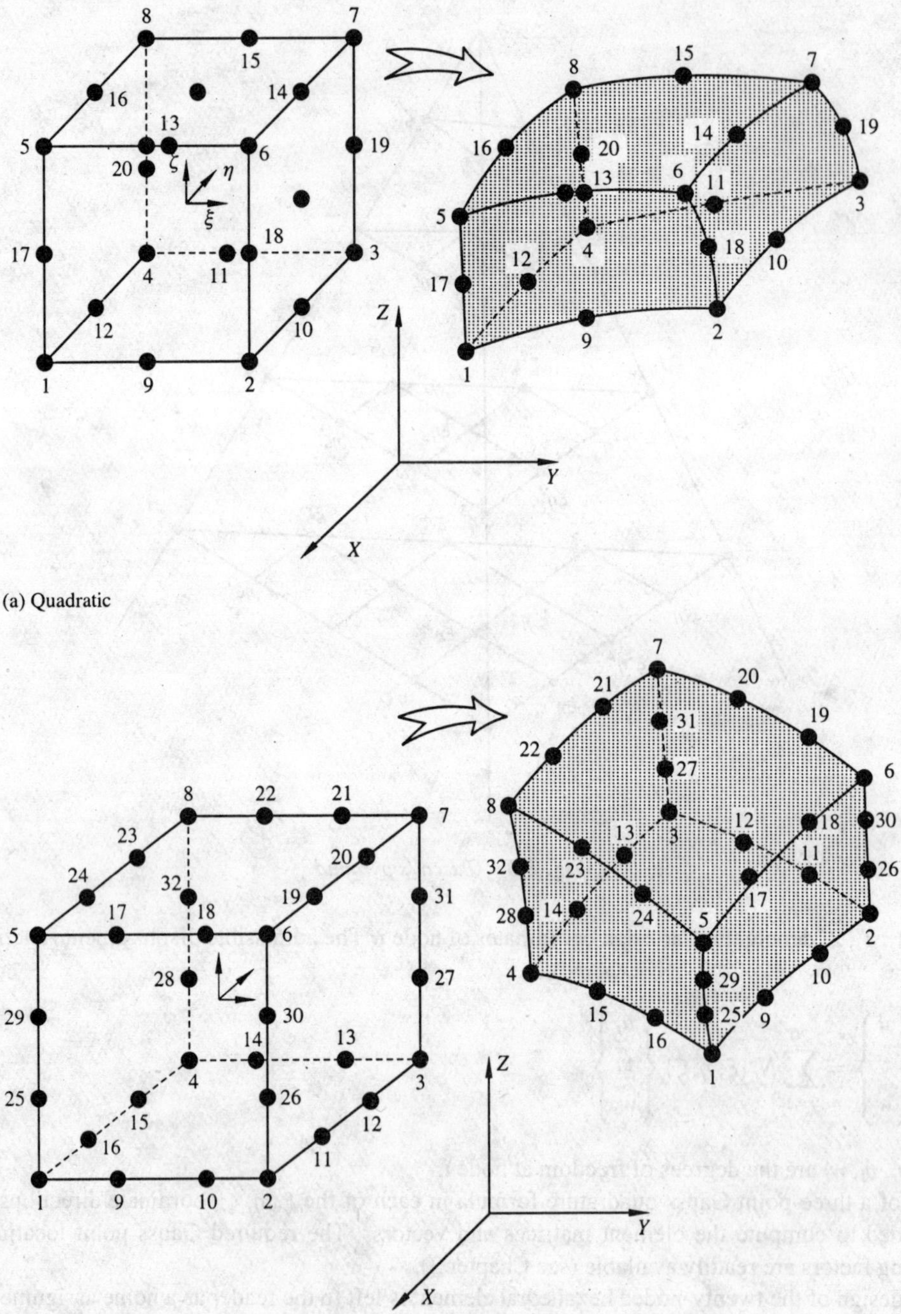

Fig. 5.9 *Higher-order hexahedral elements – Serendipity family*

using the following incomplete cubic order polynomial.

$$\begin{aligned}\Phi = {} & \alpha_1 + \alpha_2\xi + \alpha_3\eta + \alpha_4\zeta + \alpha_5\xi^2 + \alpha_6\xi\eta + \alpha_7\eta^2 + \alpha_8\eta\zeta + \alpha_9\zeta^2 + \alpha_{10}\xi\zeta \\ & + \alpha_{11}\xi^3 + \alpha_{12}\xi^2\eta + \alpha_{13}\xi\eta^2 + \alpha_{14}\eta^3 + \alpha_{15}\eta^2\zeta + \alpha_{16}\eta\zeta^2 + \alpha_{17}\zeta^3 + \\ & + \alpha_{18}\xi\zeta^2 + \alpha_{19}\xi^2\zeta + \alpha_{20}\xi\eta\zeta + \alpha_{21}\xi^3\eta + \alpha_{22}\xi^2\eta^2 + \alpha_{23}\xi\eta^3 + \alpha_{24}\eta^3\zeta + \\ & + \alpha_{25}\eta^2\zeta^2 + \alpha_{26}\eta\zeta^3 + \alpha_{27}\xi\zeta^2 + \alpha_{28}\xi^2\zeta^2 + \alpha_{29}\xi^3\zeta + \alpha_{30}\xi^2\eta\zeta + \alpha_{31}\xi\eta^2\zeta + \alpha_{32}\xi\eta\zeta^2.\end{aligned}$$

The reader is advised to substantiate this choice with reference to Pascal's pyramid given in Fig. 5.8. Explicit interpolation functions $N_i(\xi, \eta, \zeta)$ $(i = 1, 32)$ have been derived from this polynomial basis and are readily available in [5.6]. The reader is advised to list them using the current notation and node-numbering scheme.

Using isoparametric formulation, the parent element can be distorted to have straight or curved edges and flat or curved faces as illustrated in Fig. 5.9. The geometric representation for the element is

$$\begin{Bmatrix} X \\ Y \\ Z \end{Bmatrix} = \sum_{i=1}^{32} N_i(\xi, \eta, \zeta) \begin{Bmatrix} X_i \\ Y_i \\ Z_i \end{Bmatrix},$$

where X_i, Y_i, Z_i denote the Cartesian coordinates of node i.

The admissible displacement functions are

$$\begin{Bmatrix} u \\ v \\ w \end{Bmatrix} = \sum_{i=1}^{32} N_i(\xi, \eta, \zeta) \begin{Bmatrix} u_i \\ v_i \\ w_i \end{Bmatrix},$$

where u_i, v_i, w_i are the nodal degrees of freedom at node i.

Use of a four-point Gauss quadrature formula in each of the ξ, η, ζ coordinate directions is recommended to compute the element matrices and vectors. The required Gauss point locations and weighting factors are readily available (see Chapter 3).

The design of the thirty-two noded hexahedral element is left to the reader as a home assignment.

The HEXA 20 and HEXA 32 elements are widely used in practice and are implemented in every commercial FEM system. However, for complex geometries, auto-mesh capability using the HEXA 8, HEXA 20, and HEXA 32 family is desirable.

Explicit shape functions for HEXA 20 and HEXA 32 elements have been originally derived by inspection, and progression to yet higher orders is quite difficult and requires some ingenuity. It was therefore appropriate to name this family as Serendipity.

5.7.3 Hexahedral elements: Lagrange family

Hexahedral elements of the Lagrange family of quadratic (27 nodes) and cubic (64 nodes) orders have also been developed. Explicit shape functions for these elements are easy to derive as products of Lagrange polynomials in the three natural coordinates ξ, η and ζ. Progression to yet higher orders is therefore feasible. However, these elements are difficult to use in practice due to the presence of surface nodes and internal nodes, and are not implemented in commercial FEM systems.

5.7.4 Pentahedral elements

Pentahedral elements of the serendipity family of quadratic (15 nodes) and cubic (24 nodes) orders are shown in Fig. 5.10. They are created by further distorting the HEXA 20 and HEXA 32 elements as indicated in Fig. 5.10. Specially, it involves collapsing a face and forcing the nodes that are collocated to have identical degrees of freedom.

One more distortion of these elements creates "singular" elements for Computational Fracture Mechanics applications. The reader is advised to refer Newmann and Raju [5.7] for details.

5.8 Test Problems

Mac-Neal and Harder [5.5] have proposed a standard set of problems to test finite element accuracy. A description of each of these problems in terms of geometry, boundary conditions, applied loads, material properties, and finite element meshes is available in [5.5]. Theoretical results for each problem are also tabulated in [5.5] and these have to be used to present the results in a normalized form. The test results are also reduced to a letter grade. These test problems help to find and correct element errors and identify their weaknesses. Almost every test problem is capable of evoking results ranging from excellent (error $< 2\%$) to poor (error $> 50\%$). Therefore, a low grade in one or more problems should not disqualify the element under test for practical applications.

The reader is advised to test the three-dimensional elements in a commercial FEA software using these as test problems.

An effort to establish a rational set of test cases has also been undertaken in the United Kingdom by NAFEMS (National Agency for Finite Element Methods and Standards). The aims of NAFEMS include: setting FE standards, creating test procedures and coordinating evaluation of FEA software. The numerous NAFEMS test cases are called benchmarks and have the following characteristics: each uses a single element type, data preparation is simple, geometry, loading, and boundary conditions are unambiguous and each has a single well-defined target known from theory or perhaps from soundly justified computation [5.9]. Commercial FEM systems can be compared by applying them to a chosen benchmark. Such comparisons occasionally appear in NAFEMS publications.

Commercial FEA software developers also maintain an extensive set of test cases. They are called verification problems. Users can often access the verification problems manual and use it to learn the software.

5.9 Practical Applications

5.9.1 A curved cantilever beam

A curved cantilever beam loaded by a bending moment is shown in Fig. 5.11. Deformations, stresses and strains are sought. A typical mesh of linear hexahedral solid elements used is also shown in Fig. 5.11. The bending moment M is applied indirectly by appropriate nodal forces as shown in Fig. 5.11. The clamped boundary conditions is enforced by suppressing all three degrees of freedom at the nodes on this boundary surface. The solution of governing matrix equation provides nodal displacements and hence strains and stresses at points within each element.

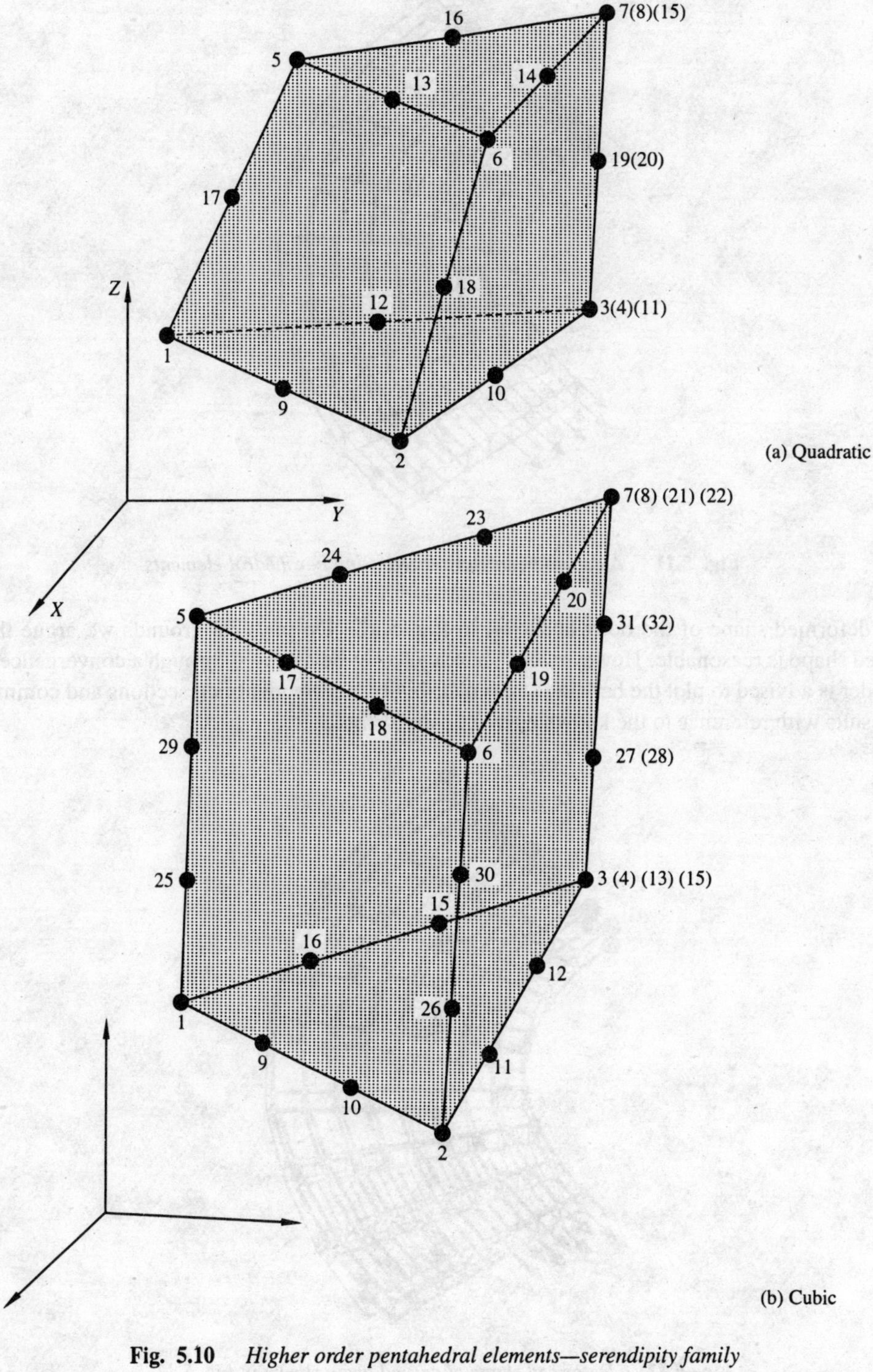

Fig. 5.10 *Higher order pentahedral elements—serendipity family*

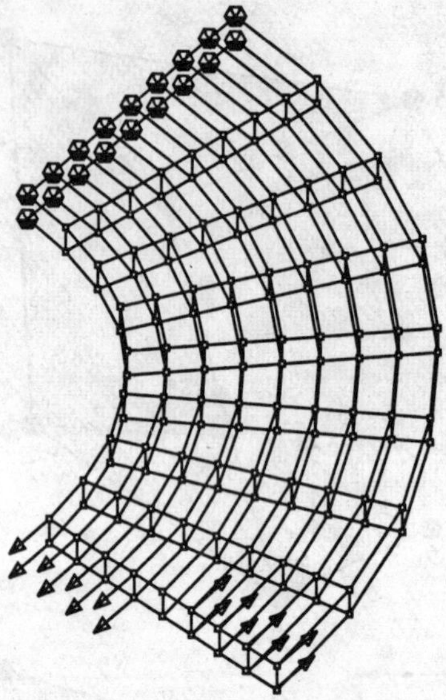

Fig. 5.11 *A curved beam discretized using hexahedral elements*

The deformed shape of the body is shown in Fig. 5.12. On physical grounds we argue that the deformed shape is reasonable. However, the accuracy has to be assessed through a convergence study. The reader is advised to plot the bending stress distribution across identified sections and comment on these results with reference to the known curved beam formulae.

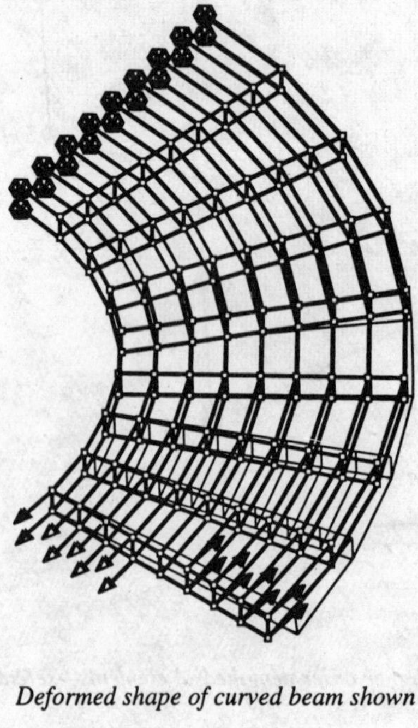

Fig. 5.12 *Deformed shape of curved beam shown in Fig. 5.11*

The reader can experiment with different cross-sections and other load cases, and report in each case the converged solutions for displacements, strains and stresses.

5.9.2 Aircraft wing attachment

Wing attachment to the fuselage is one of the critical areas in an aircraft prone to static and fatigue failures. Figure 5.13 shows a typical wing root joint. The outer wing and center wing structures are connected to forgings, which in turn are bolted together. The forgings have pockets to accommodate either the bolt head or the nut. The geometric complexities of a typical center wing and outer wing pockets are illustrated in Figs. 5.13(b) and (c). The outer wing pocket houses the bolt head whereas the nut is located in the center wing pocket. Three-dimensional circular fillets of approximately 1.5 mm radius are used to provide a smooth transition between the web and flange of each pocket. The goal of finite element analysis here is to predict the stress concentration around the fillet in each pocket.

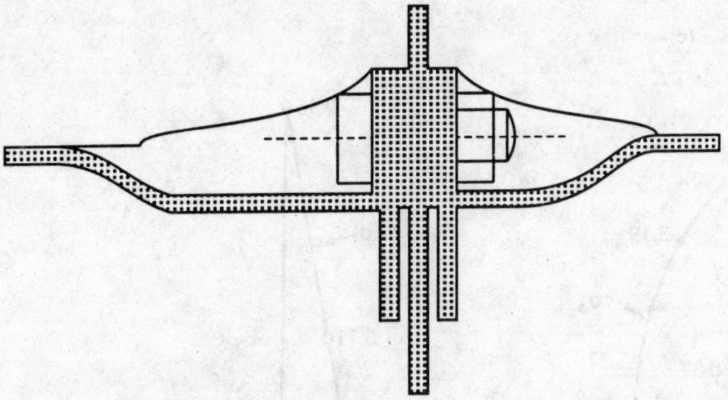

(a)

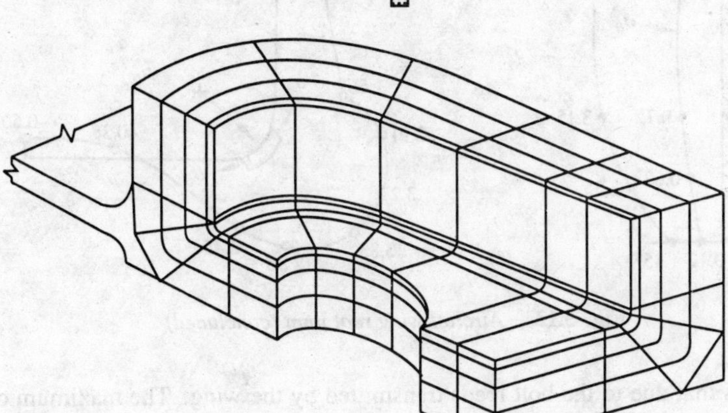

(b)

Fig. 5.13 *Aircraft wing root joint (continues)*

The center wing pocket was chosen for the analysis and was discretized using a coarse mesh of twenty-noded hexahedral elements shown in Fig. 5.13(d).

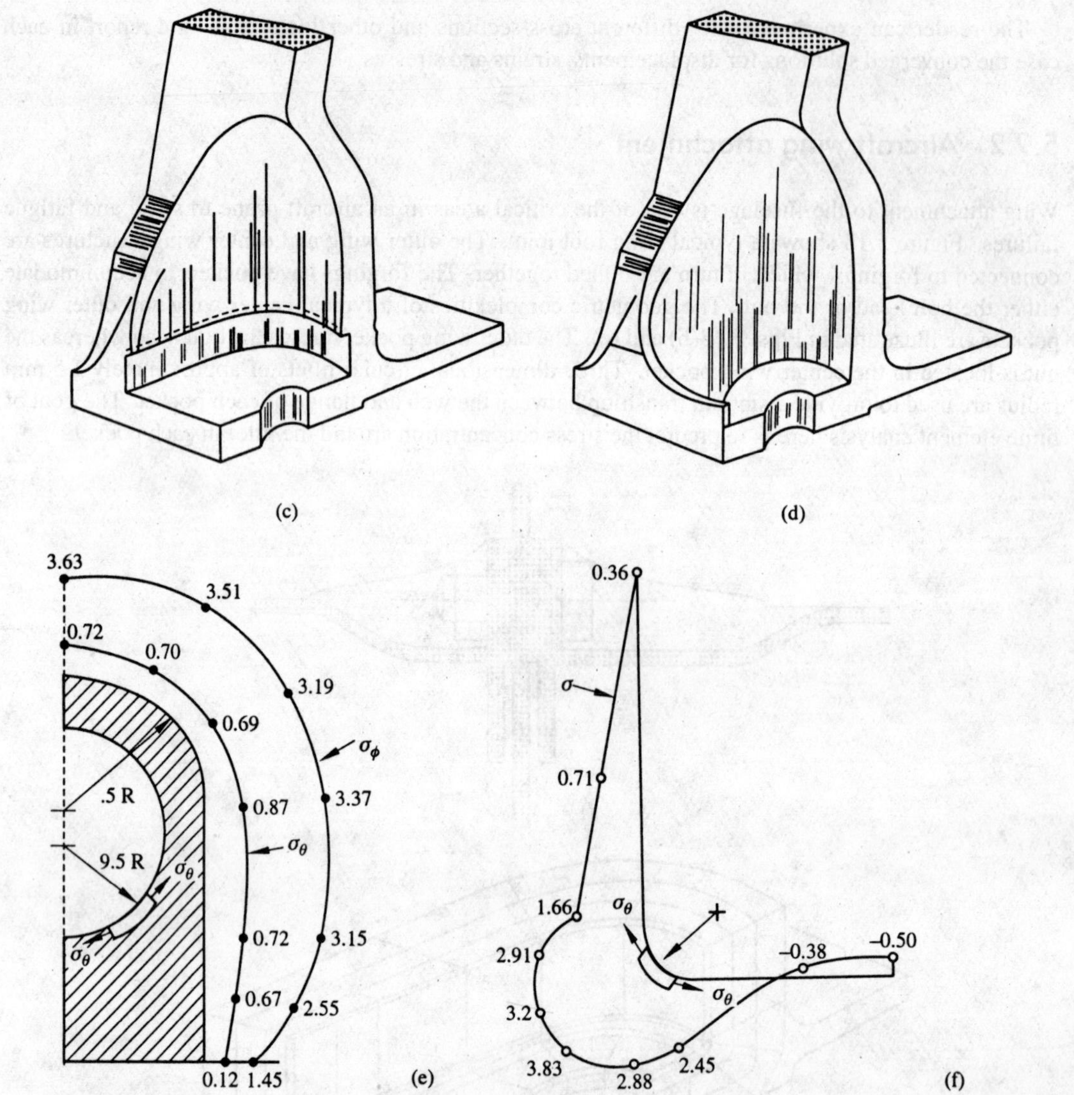

Fig. 5.13 *Aircraft wing root joint (concluded)*

The loading was that due to the bolt loads transmitted by the wing. The maximum of these loads was applied as a uniform pressure on the bearing area of the nut.

Figure 5.13(e) shows the predicted variation of the normalized tangential stress components σ_θ along the boundary of the fillet. The stresses were normalized with reference to the bolt stress. Figure 5.14(f) shows the variation of the normalized tangential component σ_ϕ. Therefore, the stress σ_ϕ is predominant when compared to σ_θ. The maximum value of σ_ϕ is 3.63 times the bolt stress.

This application highlights the use of higher order hexahedral elements to model such intricate geometric details and the capability of the FEM to predict with accuracy the stress concentrations.

5.10 Closure

A library of three-dimensional finite elements of different shapes (tetrahedra, hexahedra and pentahedra) and increasing orders (linear, quadratic, and cubic) are presented in this chapter. Majority of these are implemented in commercial FEM systems and extensively used in practice.

The ability of these elements to model complex three-dimensional parts created using geometric modelling systems and the availability of finite element analysis modules in integrated CAD/CAM/CAE systems makes the dream of design automation a reality.

Solid mechanics applications and linear static analysis procedure was stressed during the presentation. However, buckling, vibration and transient response analysis of solids as well as structures (beams, plates and shells) is possible. In addition to linear, elastic behaviour, it is also possible to predict nonlinear and/or inelastic behaviour of complex components. These are identified as topics for advanced study.

The elements and procedures described are also useful in damage tolerance analysis of solids and structures using fracture mechanics approaches. The reader should refer [5.7].

Assessment of criticality of weld defects is a burning problem in mechanical engineering design. The modelling of such defects as surface cracks and predicting their growth demands three-dimensional finite element analysis.

Three-dimensional elements are useful in the solution of heat transfer, fluid flow and other field problems also. However, derivation of element equations and design of acceptable elements for this purpose deserves an in-depth study. The reader is advised to refer Baker and Pepper [5.8] and Bathe [5.3] for details.

The practical issues of integration of three-dimensional elements into a general-purpose finite element analysis program that can be executed in the PC environment is a much larger challenge not only for element matrix computation, but also for the solution of the governing matrix equations. Frontal solver, super elements, parallel-processing systems are the contributions in this direction. These are topics for advanced study.

5.11 References

[5.1] Timoshenko, S.P. and Goodier, J.N. *Theory of Elasticity*. McGraw-Hill, New York, 3rd ed., 1970.

[5.2] Zienkiewicz, O.C. and Taylor, R.L. The Finite Element Method. 5th ed., Vol 2: *Solid Mechanics*. Arnold Publishers, London, 2000.

[5.3] Bathe, K.J. *Finite Element Procedures*. Prentice Hall of India, New Delhi, 1996.

[5.4] Cook, R.D. *Finite Element Modeling for Stress Analysis*. John Wiley, New York, 1995.

[5.5] MacNeal, R.H. and Harder, R.L. A proposed standard set of problems to test finite element accuracy. *Finite Elements in Analysis and Design*, 1(1):3–20, 1985.

[5.6] Zienkiewicz, O.C. and Taylor, R.L. The Finite Element Method. 5th ed., Vol 2: *The Basis*. Arnold Publishers, London, 2000.

[5.7] Newmann, J.C. and Raju, I.S. Methods for analysis of cracks in three dimensional solids. *Journal of Aeronautical Society of India*, 36(3), Aug 1984.

[5.8] Baker, A.J. and Pepper, D.W. *Finite Elements 1-2-3*. McGraw Hill, New York, 1991.

[5.9] National Agency for Finite Element Methods and Standards, Glasgow, UK. *The standard NAFEMS Benchmarks (Revision)*, 1989.

5.12 Computational Problems

Compute significant results, displacements, stresses, etc. as appropriate. Exploit symmetry if possible. Choose convenient numbers and consistent units for material properties, dimensions, and loads. Unless directed otherwise, assume that the material is homogeneous and isotropic and the behaviour is linear and elastic.

5.1 The sketch shows a massive coil spring, formed by bending a bar of rectangular cross section into a helix. Loading is by axial forces F at the ends (ends are not shown). Assume the spring to be closely coiled, which means that angle α is small. Predict and plot load deflection characteristics, stress distributions across critical sections.

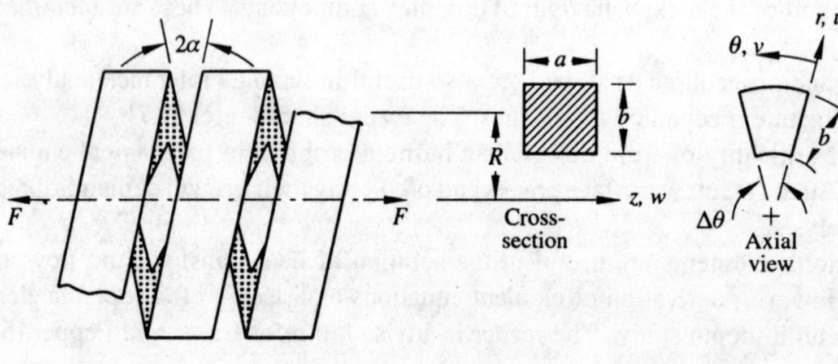

Fig. 5.14 *Sketch for Problem 5.1*

5.2 Repeat Problem 5.1 with a bar of solid circular section.

5.3 Consider a small circular hole in a plate, such that $R = 0.5t$ (see sketch). Well away from the hole, the plate is in a state of pure shear. Investigate the stress concentration around the hole.

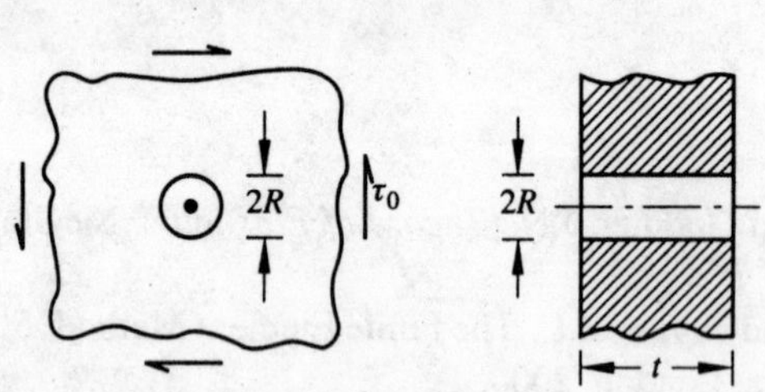

Fig. 5.15 *Sketch for Problem 5.3*

5.4 Repeat Problem 5.3 for out-of-plane bending load.

5.5 The accompanying sketch shows a prismatic beam whose cross-section is a right triangle. The beam is clamped at the end $z = 0$. Apply x-direction force to the plane $z = L$. place the force at $y = 0$, then at $y = b$. In addition to stress analysis, determine the y location of the force for which the cross-sections have no rotation about the z-axis.

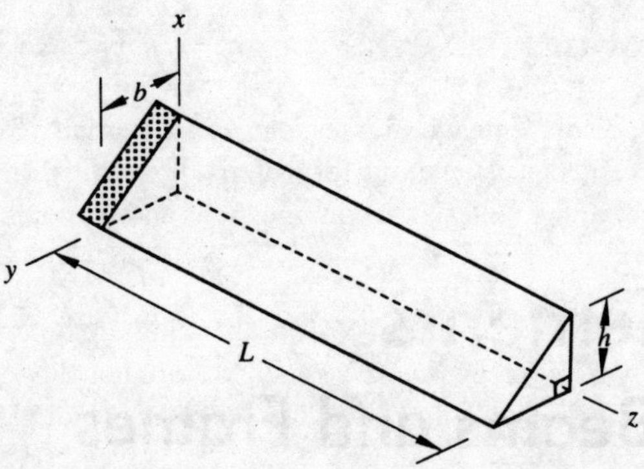

Fig. 5.16 *Sketch for Problem 5.5*

5.6 A curved cantilever beam is bent in its own plane as shown in the sketch. A rigorous stress analysis is sought resulting in contour plot of the Von-Mises equivalent stress on the critical section.

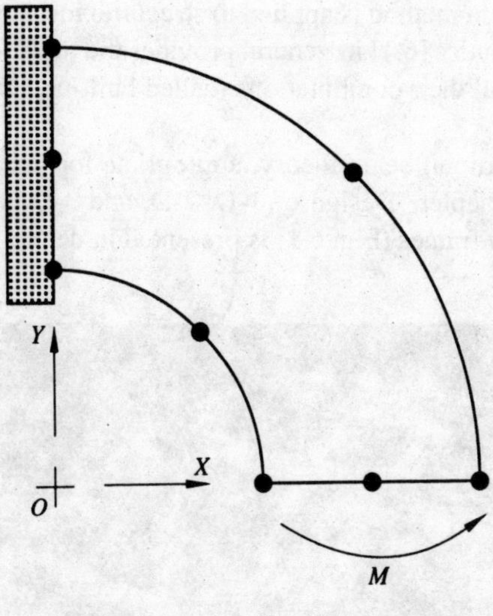

Fig. 5.17 *Sketch for Problem 5.6*

5.7 Repeat problem 5.6 with a load applied in the *x*-direction.

5.8 Create a finite element model for the machine tool component shown in Fig. 5.1. Load the same through a bolt and nut combination. Predict and plot the load deflection characteristics and stress distribution across critical section.

5.9 Create a finite element model of the automobile part shown in Fig. 5.2. Hold it at the two-hole locations and load it as shown. Perform linear static analysis and report the results. Critically comment on the component behaviour.

5.10 Repeat Problem 5.9. Now perform free vibration analysis, harmonic response analysis, and transient response analysis.

6 Beam Elements
Analysis of Beams and Frames

6.1 Overview

In this chapter, the finite element method is applied to structural mechanics problems to further expose its versatility. Structural mechanics [6.1] in general provides the methodology to predict the behaviour of beams, plates, and shells and their combinations (called built-up structures) subjected to static and dynamic loads.

The equations of Euler–Bernouli beam theory, appropriate for linear static analysis of beams and frames are presented in this chapter. Design of 1-D, 2-D, and 3-D beam elements for modelling of beams, plane frames and space frames (Fig. 6.1) is presented in detail.

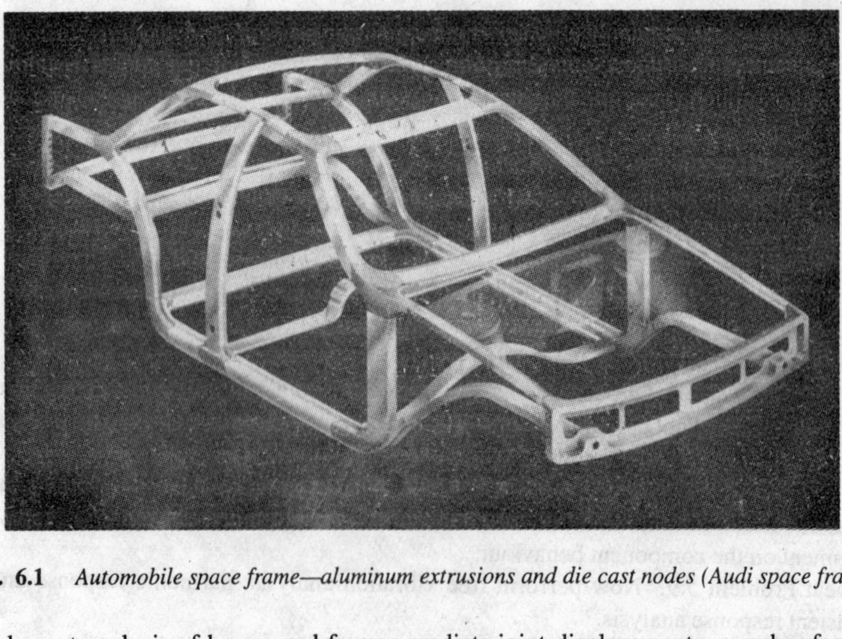

Fig. 6.1 *Automobile space frame—aluminum extrusions and die cast nodes (Audi space frame)*

Finite element analysis of beams and frames predicts joint displacements, member forces and/or stresses, member strains, and support reactions accurately. This is demonstrated using standard test problems. Finally, example applications are presented to demonstrate how beam elements are used in practice.

The limitations of Euler–Bernouli beam theory are apparent. Theories proposed by Reissner, and Mindlin, which include shear deformation effects are therefore preferred. Shear flexible beam elements, implemented in commercial FEM systems and widely used in practice, are also described.

The presentation includes identification of topics for advanced study.

6.2 1-D Beam Element

Beam is a structural member that can resist only in-plane bending moment and transverse shear force. A general straight beam shown in Fig. 6.2 has cross-sections that are symmetric with respect to the plane of bending. Restricting to linear elastic behaviour, we recall from the Euler–Bernouli beam theory (engineering beam theory) that at any section, $X =$ constant; $V(x)$ denotes the displacement of the centroid in the y-direction;

$$\theta(x) = \frac{dV}{dx} = \theta_z$$

is the slope of the deflection curve or the rotation of the cross-section about the z-axis;

$$\chi(x) = \frac{d\theta}{dx} = \frac{d^2V}{dx^2} \text{ is the curvature;}$$

$BM = EI\chi$ is the bending moment;

$$SF = \frac{d(BM)}{dx} \text{ is the shear force;}$$

$$BM = \int \sigma_{xx} y dA, \quad \text{where } \sigma_{xx} \text{ is the bending stress; and}$$

$$SF = \int \tau_{xy} dA, \quad \text{where } \tau_{xy} \text{ is the transverse shear stress.}$$

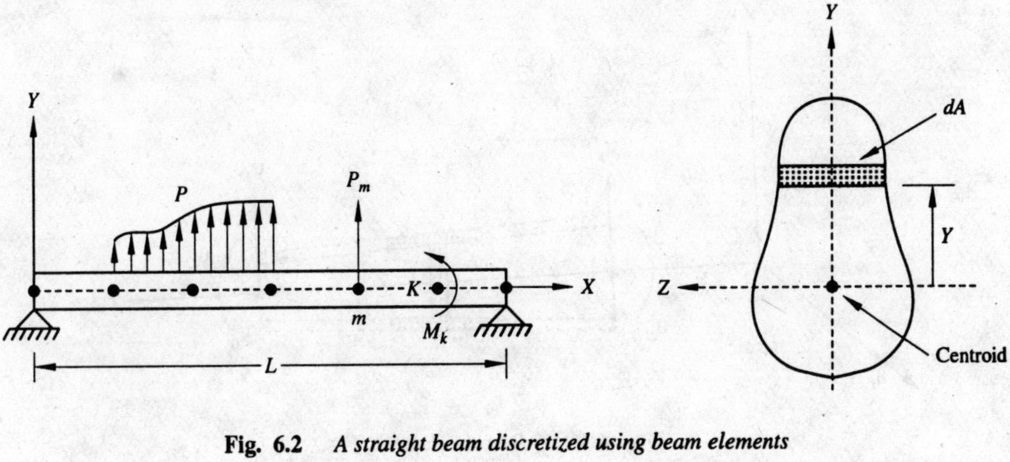

Fig. 6.2 *A straight beam discretized using beam elements*

We note here that **E** is the Young's modulus of elasticity of the beam material, and **I** is the second moment of area of the cross-section about the z-axis passing through the centroid. If the cross-section is varying along the length of the beam, then $I = I(x)$.

Identifying $V(x)$ as the primary variable, instead of deriving the governing differential equation, we construct the potential energy functional π_p.

$$\pi_p = \frac{1}{2}\int_0^L \chi^T (EI)\, \chi dx - \int_0^L p(x)\, V(x)\, dx - \sum_i P_i V_i - \sum_j M_j \theta_j. \tag{6.1}$$

In Eqn. (6.1), the first term is the strain energy due to bending of the beam; $p(x)$ denotes distributed load per unit length; P_i is the concentrated load applied at i; M_j is the moment of the couple applied at j; V_i denotes the deflection of point i; and θ_j is the slope of the deflection curve at point j.

Since the potential energy functional contains $\frac{d^2V}{dx^2}$, kinematic admissibility demands continuity of V as well as $\frac{dV}{dx}$. Therefore C^1 continuous mathematical functions have to be chosen as trial functions in order to apply the Rayleigh–Ritz method. To rationalize the construction of C^1continuous trial functions, the concept of discretization is introduced.

The beam is discretized using finite elements, as shown in Fig. 6.2. A typical element isolated from the mesh is shown in Fig. 6.3 and is called 1-D beam element hereafter. The element mathematically is a line between two nodes numbered 1 and 2 in Fig. 6.3. The user inputs the flexural stiffness $(EI)_i$ at each node i. To predict the bending behaviour, each node i now has two degrees of freedom V_i and θ_{zi} as shown in Fig. 6.3. The vector of nodal degrees of freedom **q** is

$$\{\mathbf{q}\}^T = (V_1, \theta_{z1}, V_2, \theta_{z2})\,. \tag{6.2}$$

The corresponding nodal force vector **f** is

$$\{\mathbf{f}\}^T = (F_{y1}, F_{\theta z1}, F_{y2}, F_{\theta z2})\,. \tag{6.3}$$

The derivation of algebraic equations relating **q** and **f** is accomplished by the assumed displacement function, minimum potential energy functional approach.

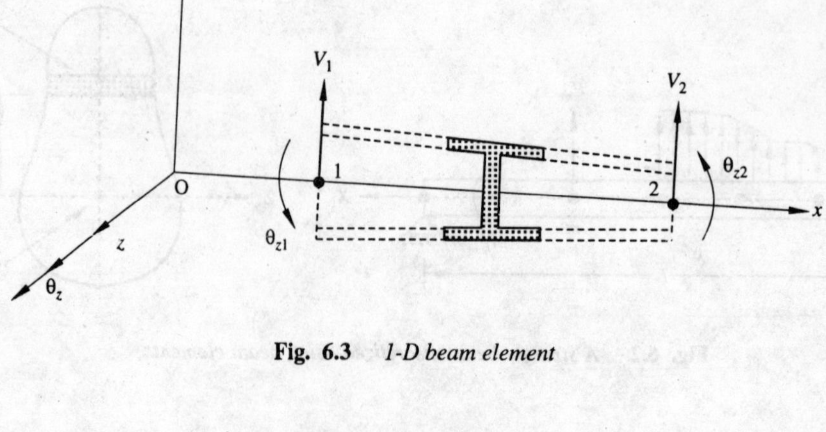

Fig. 6.3 *1-D beam element*

6.2.1 Admissible displacement function

The displacement V at any point within the 1-D beam element has to be interpolated in terms of the nodal degrees of freedom. The simplest representation is

$$V = [\mathbf{N}]\{\mathbf{q}\}, \tag{6.4}$$

where

$$[\mathbf{N}] = [N_1, N_2, N_3, N_4]. \tag{6.5}$$

In order to derive explicit expressions for the interpolation functions $N_i (i = 1,4)$, it is now a standard practice to use the concept of a parent element and the natural coordinate ξ as shown in Fig. 6.4. The parent element is a bi-unit line and the coordinate ξ is normalized to be -1 at node 1 and $+1$ at node 2. Any variable Φ within the parent element domain is approximated using the cubic polynomial

$$\Phi(\xi) = \alpha_1 + \alpha_2\xi + \alpha_3\xi^2 + \alpha_4\xi^3. \tag{6.6}$$

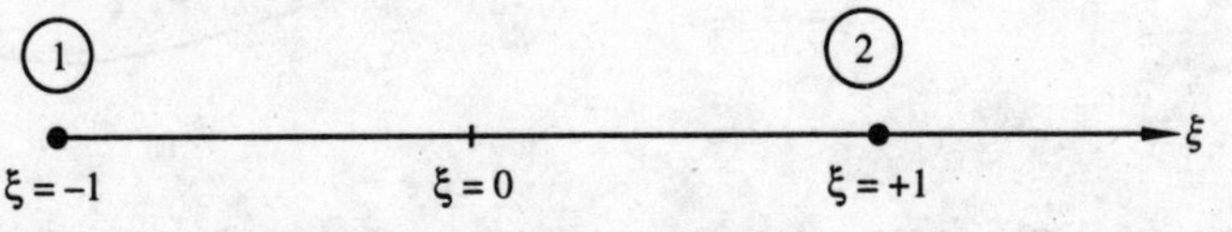

Fig. 6.4 *Parent element and natural coordinate for the 1-D beam element*

Denoting the values of Φ and $\dfrac{d\Phi}{d\xi}$ at node i as Φ_i and $\left(\dfrac{d\Phi}{d\xi}\right)_i$, we can determine the polynomial coefficients $\alpha_i (i = 1,4)$. We can then write Eqn. (6.6) as

$$\Phi(\xi) = H_1(\xi)\,\Phi_1 + H_2(\xi)\left(\frac{d\Phi}{d\xi}\right)_1 + H_3(\xi)\,\Phi_2 + H_4(\xi)\left(\frac{d\Phi}{d\xi}\right)_2, \tag{6.7}$$

where

$$H_1(\xi) = \frac{1}{4}(1-\xi)^2(2+\xi);$$

$$H_2(\xi) = \frac{1}{4}(1-\xi)^2(1+\xi);$$

$$H_3(\xi) = \frac{1}{4}(1+\xi)^2(2-\xi);$$

$$H_4(\xi) = \frac{1}{4}(1+\xi)^2(\xi-1). \tag{6.8}$$

A plot of these C^1 continuous functions is shown in Fig. 6.5. These functions are also known as Hermite polynomials in the mathematics literature.

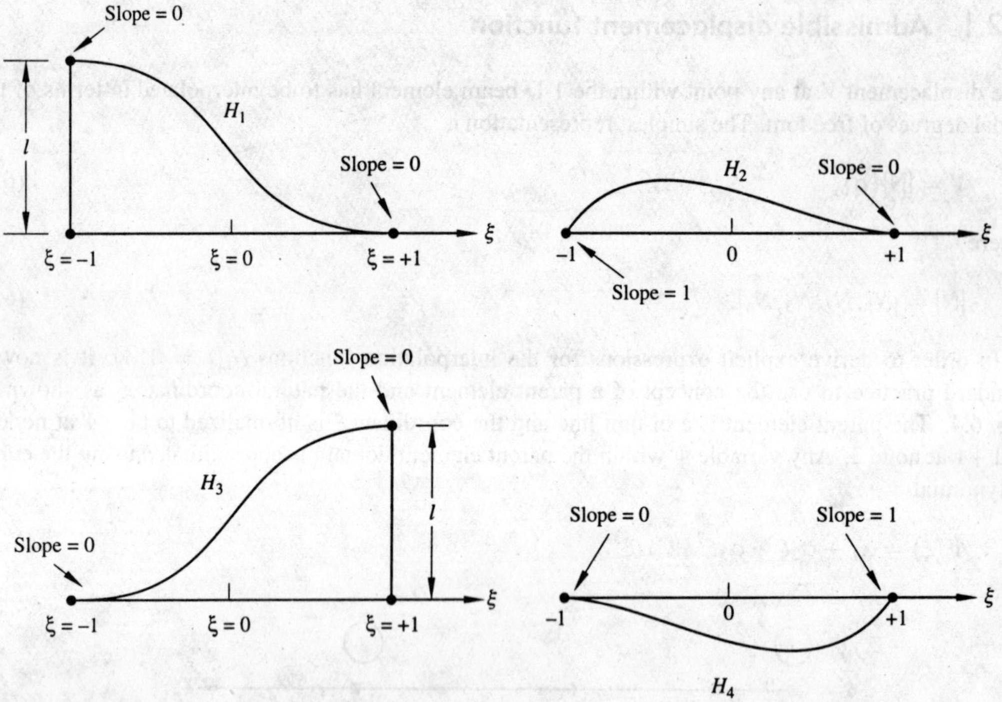

Fig. 6.5 *Plots of Hermite polynomials (C^1 continuous shape functions)*

Let us now map the parent element into the 1-D beam element. The relationship between the Cartesian coordinate X and the natural coordinate ξ to accomplish this is

$$X = \sum_{i=1}^{2} \bar{N}_i(\xi)X_i, \tag{6.9}$$

where

$$\bar{N}_1(\xi) = \frac{1}{2}(1-\xi),$$

$$\bar{N}_2(\xi) = \frac{1}{2}(1+\xi). \tag{6.10}$$

In Eqn. (6.9) X_i denotes the Cartesian coordinates of node i of the 1-D beam element.

Denoting $\ell = (X_2 - X_1)$ as the length of the element in the Cartesian space, we have from Eqn. (6.9)

$$\frac{dx}{d\xi} = \frac{\ell}{2} \tag{6.11}$$

and by chain rule of differentiation

$$\frac{d\Phi}{d\xi} = \frac{d\Phi}{dx}\frac{dx}{d\xi} = \frac{\ell}{2}\frac{d\Phi}{dx}. \tag{6.12}$$

In view of the above relations, the admissible displacement function for the 1-D beam element is

$$V = N_1(\xi)\, V_1 + N_1(\xi)\, \theta_{z1} + N_3(\xi)\, V_2 + N_4(\xi)\, \theta_{z2}, \tag{6.13}$$

where

$$\begin{aligned} N_1(\xi) &= H_1(\xi). \\ N_2(\xi) &= \frac{\ell}{2} H_2(\xi). \\ N_3(\xi) &= H_3(\xi). \\ N_4(\xi) &= \frac{\ell}{2} H_4(\xi). \end{aligned} \tag{6.14}$$

It is instructive to check that the beam element formulated using these interpolation functions satisfy the convergence criteria, namely, continuity, compatibility and completeness.

6.2.2 Element curvature matrix

The curvature χ at any point within the 1-D beam element is evaluated using

$$\chi = \frac{d^2V}{dX} = \left(\frac{2}{\ell}\right)^2 \frac{d^2V}{d\xi^2}. \tag{6.15}$$

Using Eqns. (6.13) and (6.14), we get

$$\chi = [\mathbf{B}]\{\mathbf{q}\}, \tag{6.16}$$

where the element curvature matrix $\mathbf{B}$ is

$$[\mathbf{B}] = \frac{4}{\ell^2}\left[\frac{3}{2}\xi, \frac{(-1+3\xi)}{2}\frac{\ell}{2}, \frac{-3}{2}\xi, \frac{(1+3\xi)}{2}\frac{\ell}{2}\right]. \tag{6.17}$$

6.2.3 Bending moment and shear force recovery

The bending moment (BM) at any point within the 1-D beam element is evaluated using

$$\mathrm{BM} = \{\bar{N}_1(\xi)(EI)_1 + \bar{N}_2(\xi)(EI)_2\}[\mathbf{B}]\{\mathbf{q}\}. \tag{6.18}$$

The shear force (SF) at any point within the 1-D beam element is evaluated as

$$\mathrm{SF} = \frac{2}{\ell}\frac{d\,(\mathrm{BM.})}{d\xi} \tag{6.19}$$

By default, the BM and SF are recovered at the centroid ($\xi = 0$). The Gauss point locations ($\xi = \pm 1/\sqrt{3}$) are optimal for accurate BM and SF recovery. If nodal point values are required, extrapolation to the nodes and averaging of contributions from adjacent elements have to be used.

For a prismatic beam, $(EI)_1 = (EI)_2 = EI$, Eqns. (6.18) and (6.19) simplify to

$$\text{BM} = \frac{EI}{\ell^2}\left[6\xi, (3\xi - 1)\,\ell, -6\xi, -(3\xi + 1)\,\ell\right]\{\mathbf{q}\}. \tag{6.20}$$

$$\text{SM} = \frac{6EI}{\ell^3}\left[2, \ell, -2, \ell\right]\{\mathbf{q}\}. \tag{6.21}$$

6.2.4 Element stiffness matrix

The formula for the stiffness matrix **k** of the 1-D beam element is

$$[\mathbf{k}] = \int_{X_1}^{X_2} [\mathbf{B}]^T EI(x) [\mathbf{B}]\, dx. \tag{6.22}$$

In order to evaluate the integral in Eqn. (6.22), the variation of the flexural stiffness EI over the element length must be known. Assuming linear tapering of the cross-section we can write

$$EI(\xi) = \bar{N}_1(\xi)(EI)_1 + \bar{N}_2(\xi)(EI)_2, \tag{6.23}$$

where $(EI)_i$ denotes the flexural stiffness at node i.

Eqn. (6.22) then becomes

$$[\mathbf{k}] = \int_{-1}^{+1} [\mathbf{B}]^T EI(\xi) [\mathbf{B}] \frac{\ell}{2}\, d\xi. \tag{6.24}$$

It is appropriate to evaluate the integral in Eqn. (6.24) numerically. Using the two-point Gauss quadrature formula, we get the following summation expression to compute the 1-D beam element stiffness matrix:

$$[\mathbf{k}] = \sum_{i=1}^{2} W_i [\Phi(\xi_i)], \tag{6.25}$$

where

$$\Phi(\xi) = [\mathbf{B}]^T EI(\xi) [\mathbf{B}] \frac{\ell}{2}. \tag{6.26}$$

The weighting factors W_i and Gauss point locations ξ_i are listed in Table 3.2.

For the particular case of a prismatic beam, $(EI)_1 = (EI)_2 = EI$, the element stiffness matrix is

$$[\mathbf{k}] = \frac{EI}{\ell^3}\begin{bmatrix} 12 & 6\ell & -12 & 6\ell \\ & 4\ell^2 & -6\ell & 2\ell^2 \\ & & 12 & -6\ell \\ \text{sym} & & & 4\ell^2 \end{bmatrix}. \tag{6.27}$$

6.2.5 Consistent nodal force vector: Distributed load

The formula for the consistent nodal force vector $\mathbf{f}^p$ due to distributed load $p(x)$ is

$$\{\mathbf{f}^p\} = \int_{X_1}^{X_2} [\mathbf{N}]^T p(x) dx \tag{6.28}$$

Assuming linearly varying load, we can write

$$p(\xi) = \bar{N}_1(\xi)p_1 + \bar{N}_2(\xi)p_2, \tag{6.29}$$

where p_i denotes the value of p at node i.

With $dx = \frac{\ell}{2}d\xi$, Eqn. (6.28) becomes

$$\{\mathbf{f}^p\} = \int_{-1}^{+1} [\mathbf{N}]^T p(\xi) \frac{\ell}{2} d\xi. \tag{6.30}$$

It is appropriate to evaluate the integral in Eqn. (6.30) using the two-point Gauss quadrature formula. The result is the following summation expression to compute $\mathbf{f}^p$:

$$\{\mathbf{f}^p\} = \sum_{i=1}^{2} W_i \{\psi(\xi_i)\}, \tag{6.31}$$

where

$$\{\psi(\xi)\} = [\mathbf{N}]^T p(\xi) \frac{\ell}{2}. \tag{6.32}$$

For the particular case of uniformly distributed load, $p_1 = p_2 = p$, (see Fig. 6.6).

$$\{\mathbf{f}^p\}^T = \left(\frac{p\ell}{2}, \frac{p\ell^2}{2}, \frac{p\ell}{2}, \frac{-p\ell^2}{2} \right) \tag{6.33}$$

The nodal forces are shown in Fig. 6.6.

6.2.6 Element equations

The potential energy functional, Eqn. (6.1), for the 1-D beam element can be written as

$$\pi_p(\mathbf{q}) = \frac{1}{2}\{\mathbf{q}\}^T[\mathbf{K}]\{\mathbf{q}\} - \{\mathbf{q}\}^T\{\mathbf{f}^p\} - \{\mathbf{q}\}^T\{\mathbf{f}^c\}, \tag{6.34}$$

where $\mathbf{f}^c$ includes concentrated nodal forces and moments.

We now invoke the variational principle of minimum potential energy to satisfy the equilibrium condition.

$$\frac{\partial \pi_p}{\partial \{\mathbf{q}\}} = 0 \Rightarrow [\mathbf{K}]\{\mathbf{q}\} = \{\mathbf{f}^p\} + \{\mathbf{f}^c\}. \tag{6.35}$$

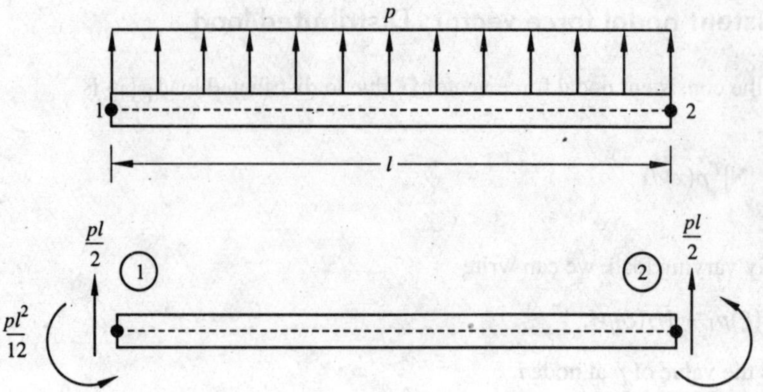

Fig. 6.6 *Consistent nodal forces—distributed load*

Eqns. (6.35) are the element equations.

For linear static analysis of the beam shown in Fig. 6.2, now modelled as an assemblage of 1-D beam elements, leads us to the global matrix statement

$$[\mathbf{K}]\{\mathbf{Q}\} = \mathbf{F}. \tag{6.36}$$

Solution of Eqn. (6.36) provides the total solution to the problem including support reactions, deflection curve, bending moment and shear force diagrams, bending and shear stress distributions,etc.

6.3 2-D Beam Element

Members in a plane frame resist axial force, transverse shear force and in-plane bending moment. For modelling and analysis of plane frames, the 2-D beam element shown in Fig. 6.7 is proposed. This element is mathematically a line between two nodes 1 and 2. The user inputs the Cartesian coordinates $(x_i, y_i, 0)$, axial stiffness $(EA)_i$, and flexural stiffness $(EI)_i$ at each node i. In order to recover the axial force (AF), the shear force (SF), and the bending moment (BM) (see Fig. 6.7), each node i will have u_i, v_i, θ_{zi} as degrees of freedom. The nodal degrees of freedom vector $\mathbf{q}$ in the global Cartesian coordinate system is

$$\{\mathbf{q}\}^T = (u_1, v_1, \theta_{z1}, u_2, v_2, \theta_{z2}). \tag{6.37}$$

In order to derive formulae to compute the 2-D beam element matrices and vectors, we define an element local coordinate system $(\bar{X}, \bar{Y}, \bar{Z})$ as shown in Fig. 6.7. The $\bar{X}$-axis is oriented along the line joining nodes 1 and 2 and $\bar{Z}$ is parallel to Z. The nodal degrees of freedom vector $\bar{\mathbf{q}}$ in the $(\bar{X}, \bar{Y}, \bar{Z})$ system is

$$\{\bar{\mathbf{q}}\}^T = (\bar{u}_1, \bar{v}_1, \theta_{z1}, \bar{u}_2, \bar{v}_2, \theta_{z2}). \tag{6.38}$$

Obviously, $\bar{\mathbf{q}}$ is uniquely related to $\mathbf{q}$. The required local–global transformation can be written as

$$\{\bar{\mathbf{q}}\} = [\mathbf{L}]\{\mathbf{q}\}, \tag{6.39}$$

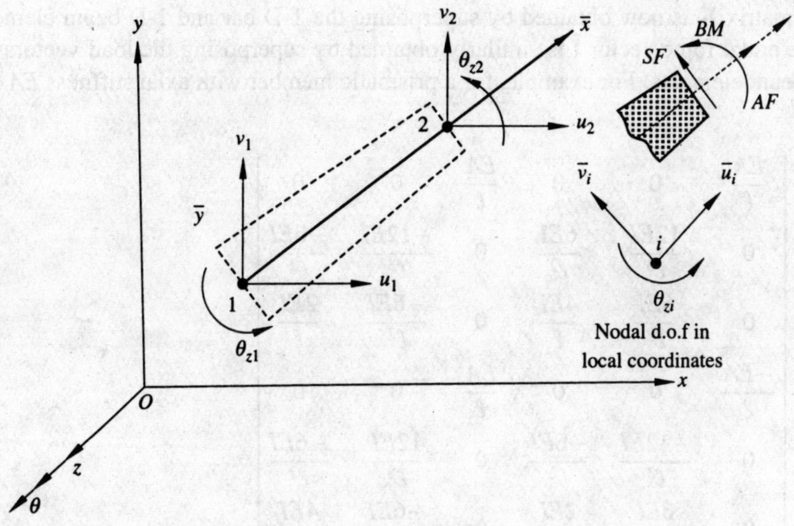

Fig. 6.7 *2-D beam element*

where the transformation matrix **L** is

$$[\mathbf{L}] = \begin{bmatrix} \cos(\bar{x},x) & \cos(\bar{x},y) & 0 & 0 & 0 & 0 \\ \cos(\bar{y},x) & \cos(\bar{y},y) & 0 & 0 & 0 & 0 \\ 0 & 0 & 1 & 0 & 0 & 0 \\ 0 & 0 & 0 & \cos(\bar{x},x) & \cos(\bar{x},y) & 0 \\ 0 & 0 & 0 & \cos(\bar{y},x) & \cos(\bar{y},y) & 0 \\ 0 & 0 & 0 & 0 & 0 & 1 \end{bmatrix}. \tag{6.40}$$

In Eqn. (6.40), $\cos(\bar{x},x)$, $\cos(\bar{x},y)$ and $\cos(\bar{y},x)$, $\cos(\bar{y},y)$ are the direction cosines of the $\bar{x}$- and $\bar{y}$-axes respectively. They are evaluated as

$$\cos(\bar{x},x) = \frac{(X_2 - X_1)}{\ell},$$

$$\cos(\bar{x},y) = \frac{(Y_2 - Y_1)}{\ell},$$

$$\cos(\bar{y},x) = -\cos(\bar{x},y),$$

$$\cos(\bar{y},y) = -\cos(\bar{x},x),$$

$$\ell = \sqrt{(X_2 - X_1)^2 + (Y_2 - Y_1)^2}. \tag{6.41}$$

In Eqns. (6.41), $(X_i,\ Y_i)$ are the Cartesian coordinates of node i, and ℓ denote the length of the line between nodes 1 and 2.

The element equations in the local coordinate system $(\bar{X}, \bar{Y}, \bar{Z})$ are

$$[\bar{\mathbf{K}}]\{\bar{\mathbf{q}}\} = \{\bar{\mathbf{f}}\}. \tag{6.42}$$

The stiffness matrix $\bar{\mathbf{K}}$ is now obtained by superposing the 1-D bar and 1-D beam element stiffness matrices. The nodal force vector $\bar{\mathbf{f}}$ is similarly obtained by superposing the load vectors for the 1-D bar and 1-D beam elements. For example, for a prismatic member with axial stiffness EA and flexural stiffness EI,

$$[\bar{\mathbf{K}}] = \begin{bmatrix} \frac{EA}{\ell} & 0 & 0 & \frac{EA}{\ell} & 0 & 0 \\ 0 & \frac{12EI}{\ell^3} & \frac{6EI}{\ell^2} & 0 & \frac{-12EI}{\ell^3} & \frac{-6EI}{\ell^2} \\ 0 & \frac{6EI}{\ell^2} & \frac{4EI}{\ell} & 0 & \frac{-6EI}{\ell^2} & \frac{2EI}{\ell} \\ \frac{-EA}{\ell} & 0 & 0 & \frac{EA}{\ell} & 0 & 0 \\ 0 & \frac{-12EI}{\ell^3} & \frac{-6EI}{\ell^2} & 0 & \frac{12EI}{\ell^3} & \frac{-6EI}{\ell^2} \\ 0 & \frac{6EI}{\ell^2} & \frac{2EI}{\ell} & 0 & \frac{-6EI}{\ell^2} & \frac{4EI}{\ell} \end{bmatrix}. \tag{6.43}$$

For linear taper, one can assume

$$EA = \frac{[(EA)_1 + (EA)_2]}{2},$$

$$EI = \frac{[(EI)_1 + (EI)_2]}{2}.$$

For the 2-D beam element, the strain energy (SE) is evaluated as

$$\text{SE} = \frac{1}{2}\{\bar{\mathbf{q}}\}^T[\bar{\mathbf{K}}]\{\bar{\mathbf{q}}\}. \tag{6.44}$$

Using Eqn. (6.39), we get

$$SE = \frac{1}{2}\{\mathbf{q}\}^T[\mathbf{K}]\{\mathbf{q}\}, \tag{6.45}$$

where

$$[\mathbf{K}] = [\mathbf{L}]^T[\bar{\mathbf{K}}][\mathbf{L}]. \tag{6.46}$$

Eqn. (6.46) is used to compute the stiffness matrix $\mathbf{K}$ for the 2-D beam element.

The work potential of the nodal forces is the same in the element local and global Cartesian coordinate systems. Therefore,

$$\{\bar{\mathbf{q}}\}^T\{\bar{\mathbf{f}}\} = \{\mathbf{q}\}^T\{\mathbf{f}\}. \tag{6.47}$$

Using Eqn. (6.39), we get

$$\{\mathbf{f}\} = [\mathbf{L}]^T\{\bar{\mathbf{f}}\}. \tag{6.48}$$

Eqn. (6.48) is used to compute the consistent nodal force vectors for the 2-D beam element.

For any plane frame, now discretized using 2-D beam elements, the usual assembly procedure leads to the global matrix statement

$$[\mathbf{K}]\{\mathbf{Q}\} = \{\mathbf{F}\},$$

and the solution of which gives the joint deflections as well as support reactions.

The axial force, bending moment and shear force can then be recovered at any point within each element. These stress resultants are all defined in the element local coordinate system $(\bar{X}, \bar{Y}, \bar{Z})$ only. Therefore, for each element, Eqn. (6.39) is used to calculate $\bar{\mathbf{q}}$ from the now known $\mathbf{q}$. Using $\bar{\mathbf{q}}$, it is now possible to recover AF, BM, and SF within the element as indicated in Fig. 6.7. Finally, one can calculate and plot the distribution of the normal stress $\sigma_{\bar{x}\bar{x}}$ and transverse shear stress $\tau_{\bar{x}\bar{y}}$ across the cross-section.

For a given frame structure to be modelled and analyzed using 2-D beam elements, the cross-sections of all its members should have a plane of symmetry parallel to the x–y plane. If this condition is violated, then, even though the structure is a plane frame, it can undergo deformations in the z-direction also. In such cases, the 2-D beam element is inadequate.

6.4 3-D Beam Element

Members in a space frame resist axial force, torsional moment, shear forces and bending moments in the two principal planes. For modelling and analysis of space frames, the 3-D beam element shown in Fig. 6.8 is proposed. This element is mathematically a line between two nodes 1 and 2. The user inputs the global Cartesian coordinates (X_i, Y_i, Z_i) of each node $i(i = 1, 2)$. The nodes are located at the centroid of the cross-section at each end. Between the nodes, a linear tapering of the cross-section is permitted. The required section properties to be furnished by the user are area of cross-section A, orientations of the principal axes ($\bar{Y}$and $\bar{Z}$ in Fig. 6.8), second moment of area about the principal axes ($I_{\bar{y}\bar{y}}$ and $I_{\bar{z}\bar{z}}$), and torsional constant j. The nodal degrees of freedom include global translations (u_i, v_i, w_i) and global rotations (θ_{xi}, θ_{yi}, θ_{zi}) at each node $i(i = 1, 2)$. Therefore, the nodal degrees of freedom vector $\mathbf{q}$ is

$$\{\mathbf{q}\}^T = (u_1, v_1, w_1, \theta_{x1}, \theta_{y1}, \theta_{z1}, u_2, v_2, w_2, \theta_{x2}, \theta_{y2}, \theta_{z2}) \tag{6.49}$$

and the corresponding vector of nodal forces $\mathbf{f}$ is

$$\{\mathbf{f}\}^T = (F_{x1}, F_{y1}, F_{z1}, F_{\theta x1}, F_{\theta y1}, F_{\theta z1}, F_{x2}, F_{y2}, F_{z2}, F_{\theta x2}, F_{\theta y2}, F_{\theta z2}). \tag{6.50}$$

In order to derive the formulae to compute the stiffness matrix $\mathbf{k}$ and consistent nodal force vector $\mathbf{f}$ for the 3-D beam element, it is convenient to define an element local coordinate system $(\bar{X}, \bar{Y}, \bar{Z})$ as shown in Fig. 6.8. The $\bar{X}$-axis is oriented along the line joining the nodes 1 and 2. $\bar{Y}$ and $\bar{Z}$-axes are oriented along the principal axes for the cross-section of the beam. The vector of nodal degrees of freedom in the $(\bar{X}, \bar{Y}, \bar{Z})$ system is $\bar{\mathbf{q}}$.

$$\{\bar{\mathbf{q}}\}^T = (\bar{u}_1, \bar{v}_1, \bar{w}_1, \theta_{\bar{x}1}, \theta_{\bar{y}1}, \theta_{\bar{z}1}, \bar{u}_2, \bar{v}_2, \bar{w}_2, \theta_{\bar{x}2}, \theta_{\bar{y}2}, \theta_{\bar{z}2}). \tag{6.51}$$

Obviously, $\bar{\mathbf{q}}$ is uniquely related to $\mathbf{q}$. The required local–global transformation can be written as

$$\{\bar{\mathbf{q}}\} = [L]\{\mathbf{q}\} \tag{6.52}$$

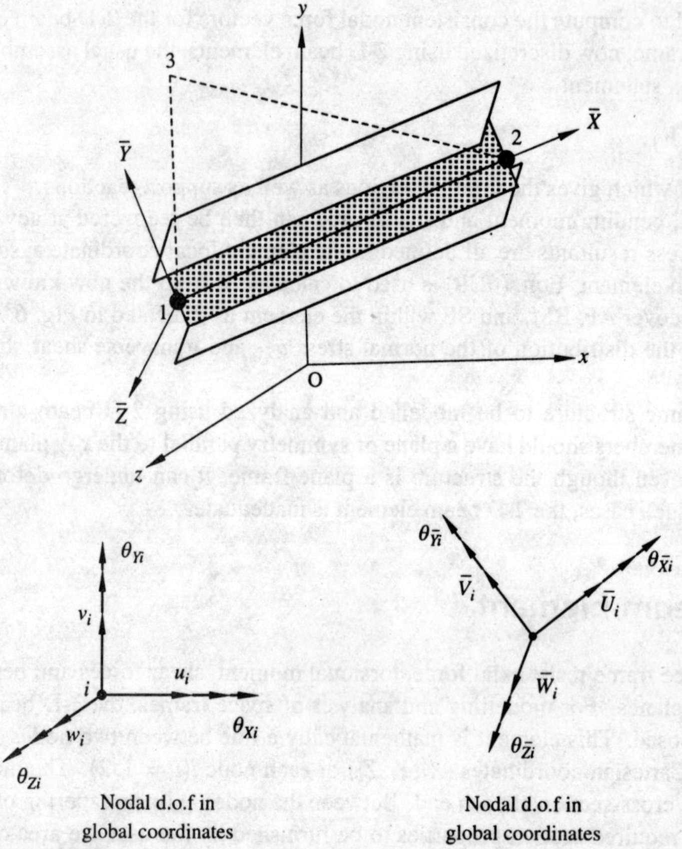

Fig. 6.8 *3-D beam element arbitrarily oriented in global coordinates X, Y, Z with local coordinates $\bar{x}$, $\bar{y}$, $\bar{z}$*

where the transformation matrix **L** is

$$[\mathbf{L}] = \begin{bmatrix} [\lambda] & [0] & [0] & [0] \\ [0] & [\lambda] & [0] & [0] \\ [0] & [0] & [\lambda] & [0] \\ [0] & [0] & [0] & [\lambda] \end{bmatrix} \tag{6.53}$$

and the direction cosine matrix $\boldsymbol{\lambda}$ is

$$[\lambda] = \begin{bmatrix} \cos(\bar{x},x) & \cos(\bar{x},y) & \cos(\bar{x},z) \\ \cos(\bar{y},x) & \cos(\bar{y},y) & \cos(\bar{y},z) \\ \cos(\bar{z},x) & \cos(\bar{z},y) & \cos(\bar{z},z) \end{bmatrix} \tag{6.54}$$

In Eqn. (6.54), $\cos(\bar{x},x)$, $\cos(\bar{x},y)$ and $\cos(\bar{x},z)$ are the direction cosines of the $\bar{X}$-axis direction. They are evaluated as

$$\cos(\bar{x},x) = (X_2 - X_1)/\ell,$$

$$\cos(\bar{x}, y) = (Y_2 - Y_1)/\ell,$$
$$\cos(\bar{x}, z) = (Z_2 - Z_1)/\ell,$$
$$\ell = \sqrt{(X_2 - X_1)^2 + (Y_2 - Y_1)^2 + (Z_2 - Z_1)^2}. \tag{6.55}$$

In general-purpose finite element analysis programs, one popular method to orient the $\bar{Y}$-axis is to specify a third node, node 3, so as to lie in the $\bar{X} - \bar{Y}$ plane. Node 3 is either a dummy node or another node of the structure whose global Cartesian coordinates (X_3, Y_3, Z_3) serve to compute the direction cosines of the $\bar{Y}$-axis. First, the plane formed by the three points 1, 2 and 3 is found by determining two vectors **R** and $\mathbf{S}^l$:

$$\mathbf{R} = (X, Y, Z)_2 - (X, Y, Z)_1.$$
$$\mathbf{S}^l = (X, Y, Z)_3 - (X, Y, Z)_1.$$

The vector **T** normal to this plane is then found and used to find the vector **S** that completes the orthogonal system.

$$\mathbf{T} = \mathbf{R} \times \mathbf{S}^l.$$
$$\mathbf{S} = \mathbf{T} \times \mathbf{R}.$$

Normalizing we get

$$\mathbf{t} = \frac{\mathbf{T}}{|\mathbf{T}|}, \ \mathbf{r} = \frac{\mathbf{R}}{|\mathbf{R}|}, \ \mathbf{s} = \frac{\mathbf{S}}{|\mathbf{S}|}.$$

The $\bar{X}$-axis is now fixed to coincide with **r**; the $\bar{Z}$-axis is fixed to coincide with **t** and this allows the $\bar{Y}$-axis to be aligned with **s**. Thus the $(\bar{X}, \bar{Y}, \bar{Z})$ system is formed in an orthogonal sense and their direction cosines to be used in Eqn. (6.54) are thus determined.

The finite element equations in the $(\bar{X}, \bar{Y}, \bar{Z})$ system are

$$[\bar{\mathbf{k}}]\{\bar{\mathbf{q}}\} = \{\bar{\mathbf{f}}\} \tag{6.56}$$

where the stiffness matrix $\bar{k}$ is obtained by superposition of appropriate bar and beam element stiffness matrices. For example, for a prismatic beam we have

$$[\bar{\mathbf{k}}] = \begin{bmatrix} a & 0 & 0 & 0 & 0 & 0 & -a & 0 & 0 & 0 & 0 & 0 \\ 0 & 12c & 0 & 0 & 0 & 6c\ell & 0 & -12c & 0 & 0 & 0 & 6c\ell \\ 0 & 0 & 12d & 0 & -6d\ell & 0 & 0 & 0 & -12d & 0 & -6c\ell & 0 \\ 0 & 0 & 0 & b & 0 & 0 & 0 & 0 & 0 & -b & 0 & 0 \\ 0 & 0 & -6d\ell & 0 & 4d\ell^2 & 0 & 0 & 0 & 6d\ell & 0 & 2d\ell^2 & 0 \\ 0 & 6c\ell & 0 & 0 & 0 & 4c\ell^2 & 0 & -6c\ell & 0 & 0 & 0 & 2c\ell^2 \\ -a & 0 & 0 & 0 & 0 & 0 & a & 0 & 0 & 0 & 0 & 0 \\ 0 & -12c & 0 & 0 & 0 & -6c\ell & 0 & 12c & 0 & 0 & 0 & -6c\ell \\ 0 & 0 & -12d & 0 & 6d\ell & 0 & 0 & 0 & 12d & 0 & 6d\ell & 0 \\ 0 & 0 & 0 & -b & 0 & 0 & 0 & 0 & 0 & b & 0 & 0 \\ 0 & 0 & -6d\ell & 0 & 2d\ell^2 & 0 & 0 & 0 & 6d\ell & 0 & 4d\ell^2 & 0 \\ 0 & 6c\ell & 0 & 0 & 0 & 2c\ell^2 & 0 & -6c\ell & 0 & 0 & 0 & 4c\ell^2 \end{bmatrix} \tag{6.57}$$

In Eqn. 6.57,

$$a = \frac{EA}{\ell},$$

$$b = \frac{GJ}{\ell},$$

$$c = \frac{EI_{\bar{y}\bar{y}}}{\ell^3},$$

$$d = \frac{EI_{\bar{z}\bar{z}}}{\ell^3}. \tag{6.58}$$

It is very important to note that in Eqns. (6.57) and (6.58),

- ℓ is the length of the 3-D Beam element;
- A is the area of cross-section, a section property;
- J is the torsional constant, a section property;
- $I_{\bar{y}\bar{y}}$ is the second moment of area of the cross-section about the centroidal $\bar{Y}$ axis, a section property;
- $I_{\bar{z}\bar{z}}$ is the second moment of area of the cross-section about the centroidal $\bar{Z}$ axis, a section property;
- E is the Young's modulus of elasticity, a material property; and
- G is the shear modulus, a material property.

$I_{\bar{y}\bar{y}}$ and $I_{\bar{z}\bar{z}}$ are also the principal moments of inertia of A. If the cross-section is non-circular, J is not the polar moment of inertia of the cross-section but is a section property such that GJ is the torsional stiffness.

The nodal force vector $\bar{\mathbf{f}}$ in Eqn. (6.56) is similarly obtained by superposing the appropriate bar and beam element vectors.

For the 3-D beam element, the strain energy (SE) is evaluated as

$$\text{SE} = \frac{1}{2}\{\bar{\mathbf{q}}\}^T[\bar{\mathbf{k}}]\{\bar{\mathbf{q}}\}. \tag{6.59}$$

Using Eqn. (6.52), we get

$$\text{SE} = \frac{1}{2}\{\mathbf{q}\}^T[\mathbf{k}]\{\mathbf{q}\}, \tag{6.60}$$

where

$$[\mathbf{k}] = [\mathbf{L}]^T[\bar{\mathbf{k}}][\mathbf{L}]. \tag{6.61}$$

Eqn. (6.61) is used to compute the stiffness matrix for the 3-D beam element.

The work potential of the nodal forces is the same in the element local and global Cartesian coordinate systems. Therefore

$$\{\mathbf{q}\}^T\{\mathbf{f}\} = \{\bar{\mathbf{q}}\}^T\{\bar{\mathbf{f}}\}. \tag{6.62}$$

Using Eqn. (6.52), we get

$$\{\mathbf{f}\} = [\mathbf{L}]^T\{\bar{\mathbf{f}}\}. \tag{6.63}$$

Eqn. (6.63) is used to compute the consistent nodal force vectors for the 3-D beam element.

Any space frame can be descretized using the 3-D beam elements. Usual assembly procedures lead us to the global matrix statement

$$[\mathbf{K}]\{\mathbf{Q}\} = \{\mathbf{F}\}. \tag{6.64}$$

Solution of Eqn. (6.64) provides the nodal degrees of freedom vector $\mathbf{q}$ for each element in the finite element model.

The stress resultants are defined in the element local coordinate system $(\bar{X}, \bar{Y}, \bar{Z})$ only. Therefore, for each element in the mesh, Eqn. (6.52) is used to calculate $\bar{\mathbf{q}}$ from the now known $\mathbf{q}$. Using $\bar{\mathbf{q}}$, it is now possible to recover the axial force (AF) (in the $\bar{X}$-direction), torsional moment (TM) about the $\bar{X}$-axis, transverse shear force (SF1) in the $\bar{Y}$-direction, transverse shear force (SF2) in the $\bar{Z}$-direction, bending moment (BM1) for bending about the $\bar{Y}$-axis, and bending moment (BM2) for bending about the $\bar{Z}$-axis. Finally one can calculate and plot the distribution of the normal stress $\sigma_{\bar{x}\bar{x}}$ and shear stresses $\tau_{\bar{x}\bar{z}}$ and $\tau_{\bar{x}\bar{y}}$ across the cross-section using the theory of beam bending and twisting [6.2].

If a 3-D beam element is used to model a Cantilever beam oriented along the global X-axis, what degrees of freedom must be suppressed at the support? The answer is all six of them, including θ_{xi} to prevent the beam from being free to spin about its centroidal axis (even if no torque is applied).

To model a continuous beam, either straight or curved in space, adjacent elements are rigidly connected together at shared nodes. Commercial FEA software allows the user to a release of one or more degree of freedom at a node, so that specified degrees of freedom are not connected. For example, two 3-D beam elements are connected end to end. Release of all three rotational degrees of freedom at the shared node makes this node a ball and socket joint. In effect, degrees of freedom are not released but simply left unconnected when adjacent elements are put together.

6.5 Shear Flexible Beam Elements

The Euler–Bernouli beam theory (EBT) assumes that plane sections before deformation remain plane even after deformation. However, when shear forces are present, plane sections do not remain plane after deformation. Warping of adjacent sections will be different in problems involving varying shear forces and this will invariably induce axial strains and stresses. For engineering analysis of beams and frames, the limitations of EBT are thereby apparent. Improved beam theories to include transverse shear deformation effects have been proposed. The Shear Deformation Theories (SDT) due to Reissner and Mindlin are widely used in practice to design the so-called shear flexible beam elements.

In the SDT, a priori assumptions are made regarding the displacement field. For example, for bending analysis of the beam in Fig. 6.2, the assumed displacement field in the SDT is

$$\begin{aligned} U(x,y,z) &= -y\beta(x), \\ V(x,y,z) &= v(x), \\ W(x,y,z) &= 0. \end{aligned} \tag{6.65}$$

In Eqn. 6.65, V and β are two independent variables. V is the lateral deflection and β denotes the rotation of the cross-section about the Z-axis. It is to be noted that $\beta = \dfrac{dv}{dx}$ for the EBT.

The bending curvature χ and transverse shear strain γ are then calculated as

$$\chi = \frac{d\beta}{dx}, \tag{6.66}$$

$$\gamma = \frac{dv}{dx} - \beta. \tag{6.67}$$

The potential energy functional π_p is then evaluated as

$$\pi_p(v, \beta) = \frac{1}{2}\int_0^L \chi^T EI(x)\chi dx + \frac{1}{2}S\int_0^L \gamma^T GA(x)\gamma dx - \int_0^L pvdx - \int_0^L m\beta dx. \tag{6.68}$$

In Equation (6.68), the first term is the strain energy due to bending and the second term is the strain energy due to shear deformation. S is a shear correction factor introduced to account for the difference in the constant shear stress in this theory and the parabolic variation of the actual shear stress through the thickness. The value of S is either 5/6 or $\pi^2/12$. p and m denote distributed transverse load and moment per unit length.

We can redesign the two-node beam element shown in Fig. 6.2 based on the shear deformation theory (SDT) described above. We now interpolate the two variables V and β independently as

$$V = \sum_{i=1}^{2} N_i(\xi)V_i, \tag{6.69}$$

$$\beta = \sum_{i=1}^{2} N_i(\xi)\theta_{zi}, \tag{6.70}$$

where

$$N_1(\xi) = \frac{1}{2}(1 - \xi),$$

$$N_2(\xi) = \frac{1}{2}(1 + \xi).$$

We can directly employ these assumed displacement functions and the minimum potential energy functional approach to derive element equations and use the Gauss quadrature formula to compute element matrices. The design of shear flexible beam elements of linear (two nodes), quadratic (three nodes) and cubic (four nodes) orders is assigned as term projects to the students. The reader can refer Bathe [6.3] for details.

We can identify a parameter

$$\alpha = \frac{G}{E}\frac{A}{I}S$$

and note that $\alpha \to \infty$ for slender beams and the constraint of zero shear deformation (i.e. $\frac{dv}{dx} = \beta$ and $\gamma = 0$) will be approached.

A phenomenon known as shear locking is often observed when the two-node shear flexible beam element is used for modelling and for analysis of slender beams. If interpolation functions used for V and β over the element domain should admit that for large values of α the transverse shear strain γ may not be small and indeed zero everywhere, then an erroneous shear strain energy, which can be large compared with the bending strain energy is included in the finite element analysis of beams. Hence, the resulting finite element model is much too stiff and predicts much smaller deflections than the exact solution based on EBT.

Various techniques have been proposed to alleviate the shear locking effects: reduced integration, selective reduced integration, assumed strain formulation, etc. The reader should refer Bathe [6.3] for further details.

6.6 Test Problems

***Problem* 1** Consider the beam shown in Fig. 6.9. The problem requires the use of 1-D beam elements. The stiffness matrix and consistent nodal force vectors are computed for each element in the mesh. The point loads F_0 and M_0 are included in the global nodal force vector. The global matrix equations are solved to get the nodal degrees of freedom after enforcing the clamped boundary conditions. The reaction forces and element stress resultants are then evaluated. Plots of the finite element solution for lateral deflection $V(x)$, section rotation $\theta_z(x)$, bending moment. BM(x), and shear force SF(x) are prepared. These are compared with the exact solution at the nodes. The predicted BM and SF at the Gauss points will be within 2% of the exact solution. The reader is advised to verify the above using commercial FEA software.

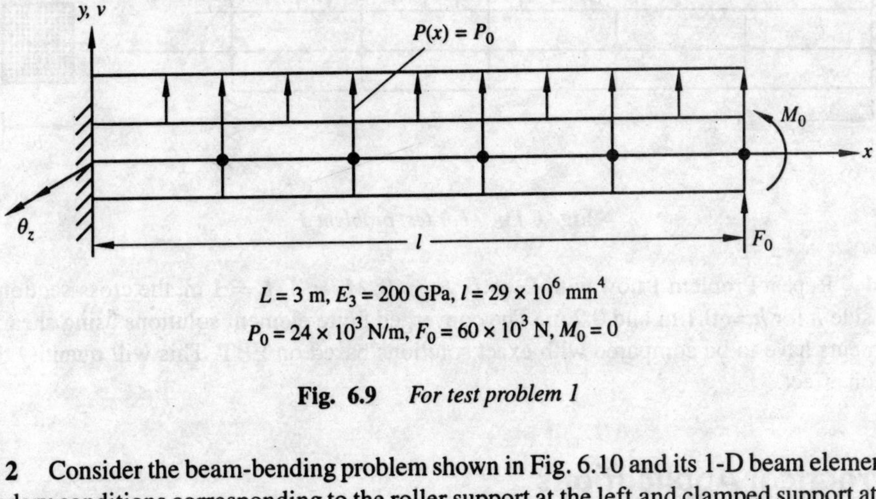

$L = 3$ m, $E_3 = 200$ GPa, $I = 29 \times 10^6$ mm^4

$P_0 = 24 \times 10^3$ N/m, $F_0 = 60 \times 10^3$ N, $M_0 = 0$

Fig. 6.9 *For test problem 1*

***Problem* 2** Consider the beam-bending problem shown in Fig. 6.10 and its 1-D beam element model. The boundary conditions corresponding to the roller support at the left and clamped support at the right are enforced. Plots of the finite element solution for the lateral deflection $V(x)$, section rotation $\theta_z(x)$, bending moment BM(x), have to be prepared. These can be compared with the exact solution. The reader can experiment with typical lengths, cross-sections (shapes and sizes), materials and loads.

***Problem* 3** Let us consider the beam that has a clamped support at one end and is spring-supported at the other (see Fig. 6.11), modelled using 1-D beam elements. The finite element solution for different values of EI and $K_S/(EI)$ can be obtained. The solutions for $K_S/(EI) = 0$ (no spring), $K_S/(EI) = 10^{-2}$

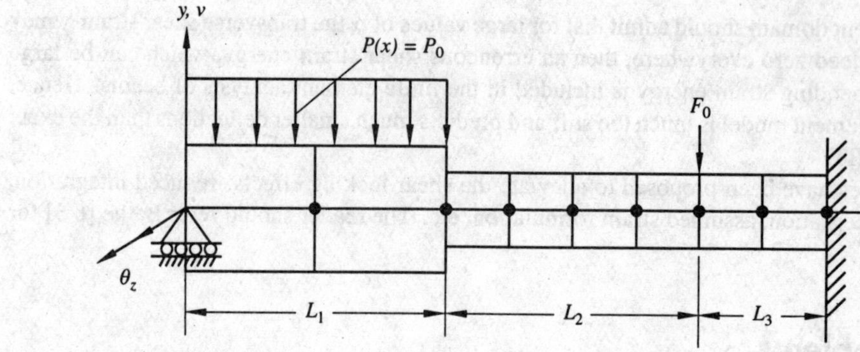

Fig. 6.10 *For test problem 2*

(a soft spring) and $K_S/(EI) = 10^2$ (a hard spring) have to be compared with the exact solution. Plots of the transverse deflection $V(x)$, section rotation $\theta_z(x)$, and bending moment BM(x) have to be prepared. The reader can experiment with appropriate lengths, cross-section dimensions, material properties, load magnitudes and spring stiffnesses.

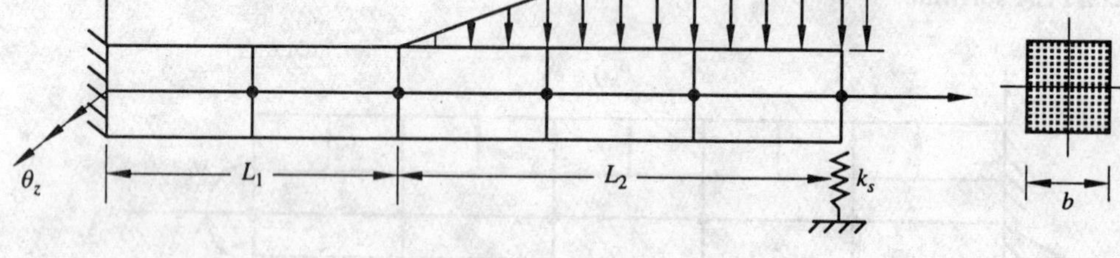

Fig. 6.11 *For test problem 3*

***Problem* 4** Repeat Problem 1 now with $P_0 = 0$, $F_0 = 0$, $M_0 = 1$, $L = 1$ m, the cross-section being a square of side h for $h = 0.1$ m and 0.2 m. The converged finite element solutions using shear flexible beam elements have to be compared with exact solutions based on EBT. This will quantity the shear deformation effect.

6.7 Practical Applications

6.7.1 Machine tool spindle

A simplified section of the spindle is shown in Fig. 6.12(a). The bearing at B has a radial stiffness of 60 N/μm and rotational stiffness (against bending) of 8×10^5 N-m/rad. The bearing at C has a radial stiffness of 20 N/μm and its rotational stiffness can be neglected. The cutting force $F = 1000$ N is applied at A. The spindle is made of steel with $E = 200$ GPa and $\nu = 0.3$.

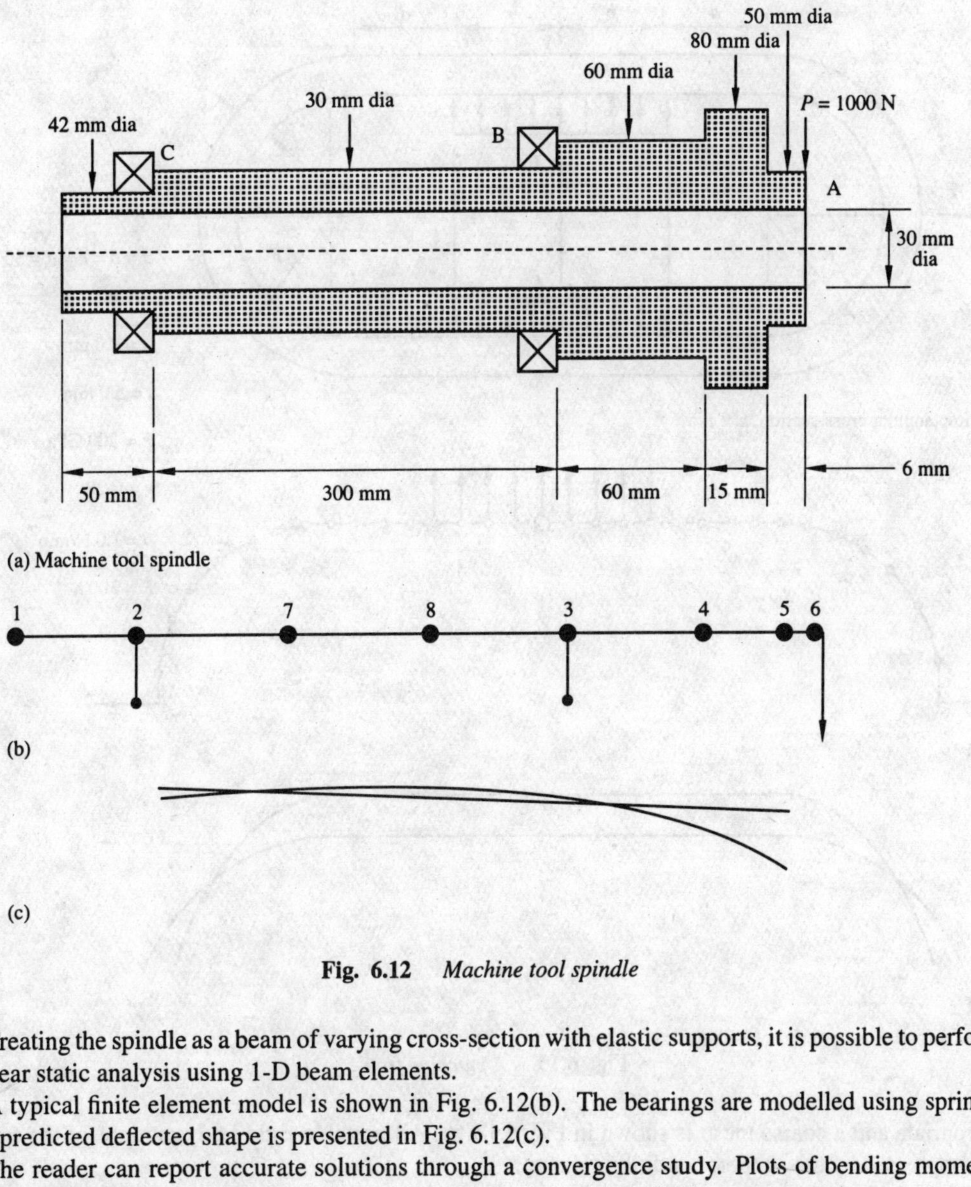

Fig. 6.12 *Machine tool spindle*

Treating the spindle as a beam of varying cross-section with elastic supports, it is possible to perform a linear static analysis using 1-D beam elements.

A typical finite element model is shown in Fig. 6.12(b). The bearings are modelled using springs. The predicted deflected shape is presented in Fig. 6.12(c).

The reader can report accurate solutions through a convergence study. Plots of bending moment, shear force, extreme fiber stress variations along the length of the spindle are expected.

It will be informative to predict the first ten natural frequencies and associated mode shapes for the spindle.

6.7.2 A machine frame

A machine frame made out of steel is shown in Fig. 6.13(a). The deformed shape and stress distribution are sought from the analyst.

There is a symmetry of geometry, loading and elastic properties about the horizontal centerline. Therefore only the upper half of this plane frame needs to be discretized. 2-D beam elements are

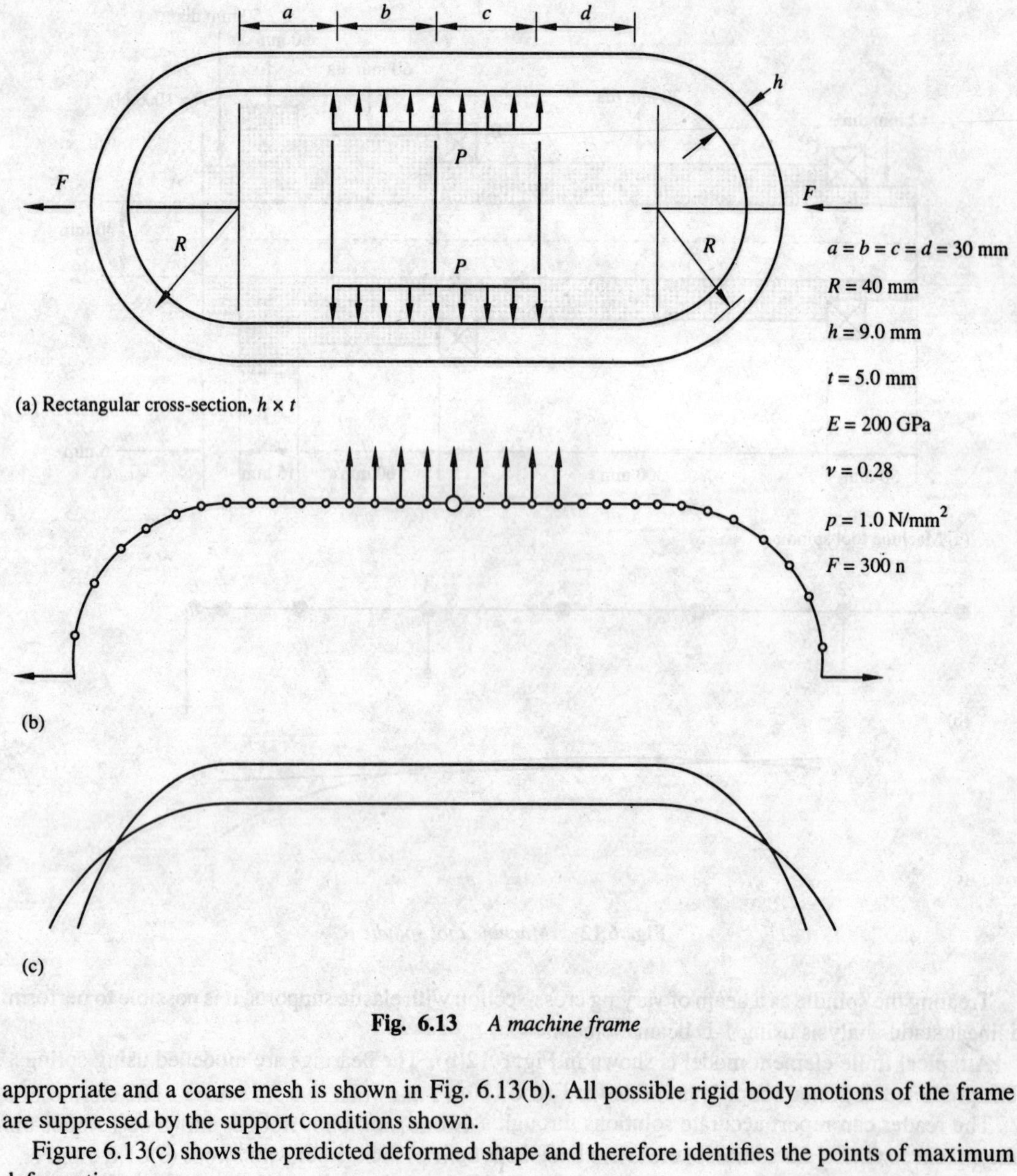

Fig. 6.13 *A machine frame*

appropriate and a coarse mesh is shown in Fig. 6.13(b). All possible rigid body motions of the frame are suppressed by the support conditions shown.

Figure 6.13(c) shows the predicted deformed shape and therefore identifies the points of maximum deformation.

How good are these results? The answer demands a convergence study. The reader is advised to report converged solutions for the distribution of normal stress ($\sigma_{\bar{x}\bar{x}}$) and shear stress ($\tau_{\bar{x}\bar{y}}$) across critical sections.

Some questions about the validity of the finite element modelling remain to be answered. The ratio (h/R) may be such that the Euler–Bernouli beam theory is not valid. It is then necessary to use shear flexible beam elements. If h is comparable to R, beam elements should be replaced by 2-D plane stress or 3-D solid elements.

We have assumed from the outset that the deformations are small, the material behaviour is linear and elastic and the applied load is static. However, depending upon the dimensions and magnitudes of applied loads, the deformations can become large and the stresses induced can be greater than the yield strength of the material. The problem then demands a nonlinear finite element analysis (combined geometric and material nonlinearities). For the direction of the applied load F considered, buckling is a possible mode of failure and therefore demands a stability analysis. If the applied load F is fluctuating, as for example $F(t) = f_0 \sin \Omega t$ where t denotes time, then dynamic analysis becomes essential to check the possibility of resonance. For impact loads, $F(t)$ is non-periodic and a transient response analysis is necessary to predict the behaviour.

The finite element method in general can take into account each one of the above complicating effects. Commercial FEM systems in particular provide a number of material models, and a procedure library in addition to a library of finite elements. Engineers can therefore perform static, dynamic, stability, and transient response analysis—either linear or nonlinear, of any plane frame!

6.7.3 Multi-storey building frame

Figure 6.14 shows the frame of a 26-storey building where the details of walls and slabs are omitted. Natural modes of vibration of this frame are sought from the analyst. 3-D beam elements are appropriate for modelling this space frame. A coarse mesh will model each member (column or beam) as an element and each joint as a node. The boundary conditions include suppressing all the degrees of freedom at each node at the foundation level.

Fig. 6.14 *Vibration mode shapes of a 26-storey building*

Figure 6.14 also shows the deformations corresponding to the first two modes of vibration predicted using NISA, a commercial FEM system. The corresponding natural frequencies (in Hz) are not reported here. They depend on the member lengths, cross-sections (shape and size) and material properties. The reader can report converged solutions obtained using any general-purpose finite element analysis software for typical member lengths, cross-sections and materials encountered in practice.

6.8 Closure

This chapter has thoroughly detailed the implementation of the finite element modelling for a class of structural mechanics problem, namely, beams and frames. The restriction to EBT is to give the reader a feeling of dexterity with the finite element method and its many details and attributes. Use of SDT is essential for practical applications and shear flexible beam elements are in fact available in commercial FEM systems. The design of such elements, therefore, is assigned as a term project to the students. The self-imposed restrictions to linear static analysis were solely for the sake of simplicity. Advanced applications to dynamic, stability and transient response analyses, either linear or nonlinear, of beams and frames are therefore included as computational problems. These problems can be solved using any commercial FEM system. The reader can refer Crisfield [6.4, 6.5] for details of element design as well as solution procedures.

Structural mechanics in general is concerned with not only beams and frames but also plates, and shells. The design of simple and robust finite elements for modelling plate and shell type structures has been and continues to be a problem challenging researchers for more than three decades. Finite elements for plates and shells are addressed in Chapters 7 and 8 respectively.

Finally, the FEM is essentially dependent for its success on its implementation in the form of a general-purpose program [6.6] on a computer. General-purpose programs (GPP) range from diskettes accompanying textbooks to commercial FEM systems. The range of computers include personal computers, workstations, mainframes, supercomputers and parallel processing machines. At present, commercial FEM systems are affordable, widely distributed, well documented, user-friendly, and ready to execute even on PCs! A brief description of the organization and desirable features of a general-purpose FEA software is presented in Chapter 9.

The application of the FEM in general and a GPP in particular is not limited to solid and structural mechanics problems. Advanced applications of FEM/GPP are identified in Chapter 10. Many other continuum mechanics problems arise in engineering and usually these problems are posed by appropriate differential equation (s) and boundary conditions to be imposed on the unknown function (s). It is possible to extend the FEM to deal with all such field problems. For example, the FEM is extensively used for the numerical simulation of problems in heat transfer [6.7], fluid flows [6.8], and electromagnetic fields [6.9]. These are obviously topics for advanced study.

6.9 References

[6.1] Meek, J.L. *Matrix Structural Analysis*. McGraw Hill, New York, 1971.
[6.2] Cook, R.D. and Young, W.C. *Advanced Mechanics of Materials*. Macmillan, New York, 1985.
[6.3] Bathe, K.J. *Finite Element Procedures*. Prentice Hall of India, New Delhi, 1996.

[6.4] Crisfield, M.A. Finite elements and solution procedures for structural analysis. *Essentials*, 1, 1986.
[6.5] Crisfield, M.A. *Nonlinear Finite Element Analysis of solids and structures*. John Wiley and Sons, Chichester, U.K., 1991.
[6.6] Smith, I.M. and Griffiths, D.V. *Programming the Finite Element Method*. John Wiley and Sons, Chichester, U.K., 1988.
[6.7] Baker, A.J. and Pepper, D.W. *Finite Elements 1-2-3*. McGraw Hill, New York, 1991.
[6.8] Zienkiewicz, O.C. and Taylor, R.L. The finite element method. *Fluid Mechanics*, 3, 2000.
[6.9] Silvester, P.P. and Ferrari, R.L. *Finite Elements for Electrical Engineers*. Cambridge University Press, second edition, 1990.

6.10 Computational Problems

In the following problems predict peak values of displacement, bending moment, shear force and/or stress. Exploit symmetry if possible. Use material properties of steel, when not stated.

6.1 The beam shown is uniformly tapered in width. Create FE model using beam elements. Choose convenient numbers for length L and force P. Repeat using a tip moment rather than a tip force. Use thickness $h = L/100$ and $L/10$. If the FEA software used permits, compare the solutions based on EBT and SDT for $L/10$.

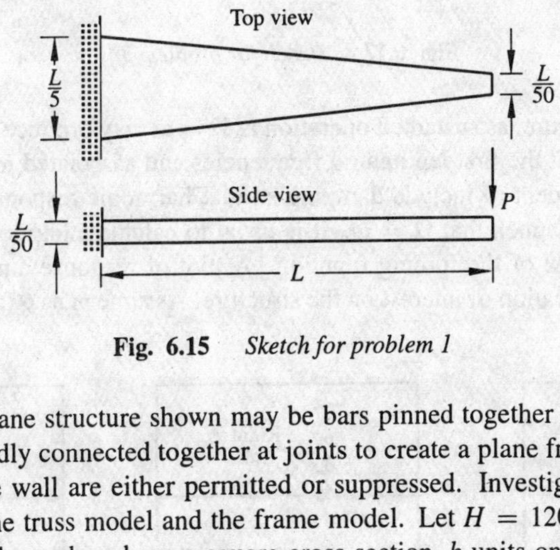

Fig. 6.15 *Sketch for problem 1*

6.2 Members of the plane structure shown may be bars pinned together at joints to create a plane truss or beams rigidly connected together at joints to create a plane frame. For the plane frame the rotations at the wall are either permitted or suppressed. Investigate how much difference there is between the truss model and the frame model. Let $H = 120$ mm, $L = 160$ mm, and $P = 1.0$ N. Let all members have a square cross-section, b units on side. Consider the cases $b = 5$ mm, $b = 15$ mm, and $b = 30$ mm.

6.3 Members of the space frame shown in the sketch are thin-walled tubes of outer diameter d and wall thickness t. All members are of equal length $h = 3$ m. Investigate the behaviour with a unit load in the X, Y, and Z directions applied at a joint on the top surface for $d = 60$ mm and $t = 10$ mm; $a = b = 6$ m.

6.4 A heavy machine of mass m is attached to a building frame as shown in the sketch. Let all members have square cross-section, b units on side. Consider the cases: $b = 0.1$ m, $b = 0.2$ m. The speed Ω of the rotating machinery should pass quickly through a resonant frequency ω of

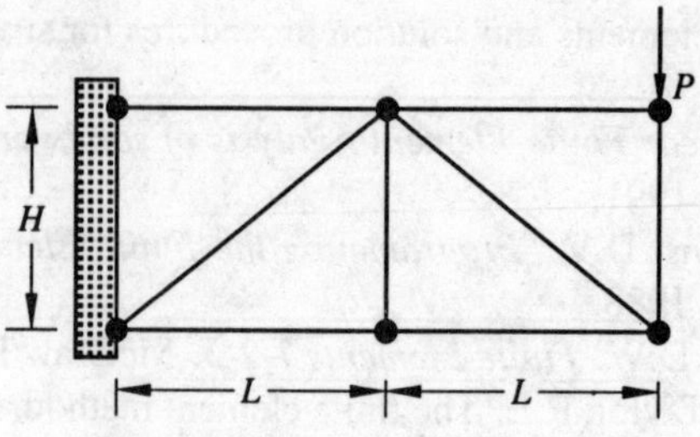

Fig. 6.16 *Sketch for problem 2*

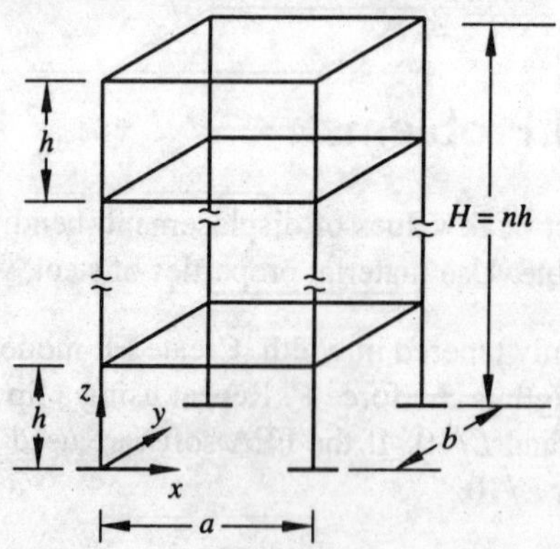

Fig. 6.17 *Sketch for problem 3*

the supporting structure, as sustained operation at $\Omega = \omega$ may produce destructive vibrations. It is for this reason that the first ten natural frequencies and associated mode shapes be predicted accurately. It is prudent to include damping so that harmonic response computations will not blow-up, should it happen that $\Omega = \omega$. It is usual to calculate harmonic response for each of several frequencies Ω of the forcing function. A plot of response amplitude versus Ω can be prepared for each location of interest on the structure. Assume *m* as a fraction of the mass of the frame.

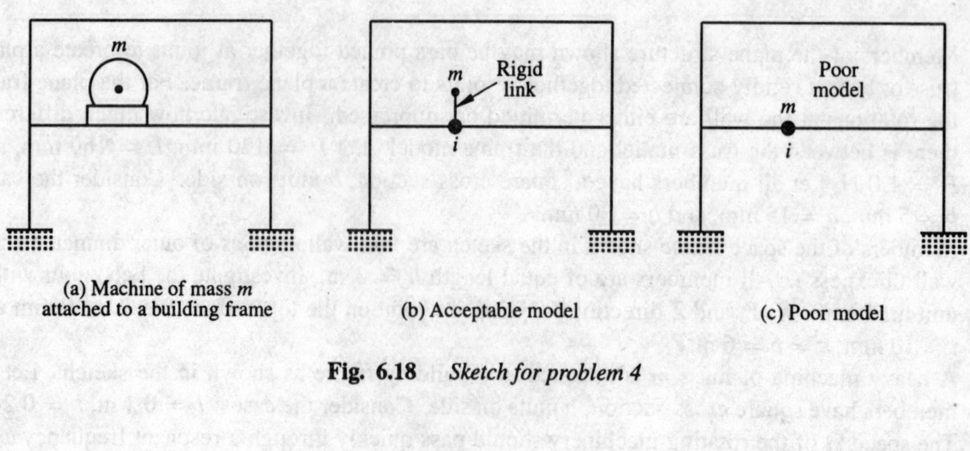

Fig. 6.18 *Sketch for problem 4*

6.5 The plane frame whose data appear in the sketch is subjected to an impulse loading. As shown, a horizontal force of 1×10^5 N is applied at node 51 for 0.01 seconds. The frame is initially undeformed and at rest. We will calculate selected displacements, velocities, and accelerations as functions of time, and also seek maximum values of these quantities. A comparison of dynamic response by mode superposition and direct integration methods is also requested.

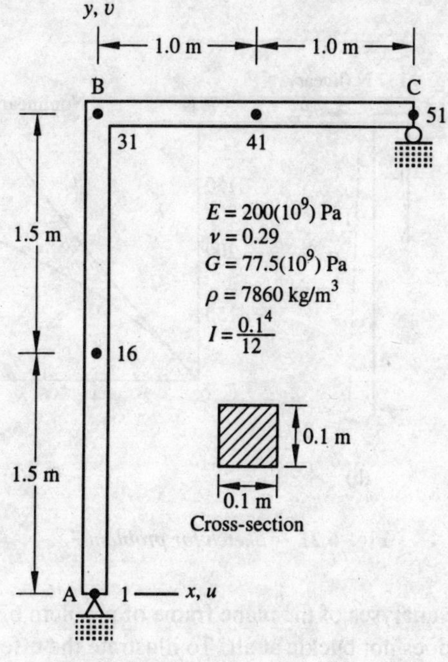

Fig. 6.19 *Sketch for problem 5*

6.6 Figure shows a simple buckling problem. But if there is any imperfection, for example, an off-center load or a slight initial curvature of the column, it does not buckle but instead exhibits a nonlinear load versus deflection response. You have to accurately predict P_{cr} and $P - D$ plot.

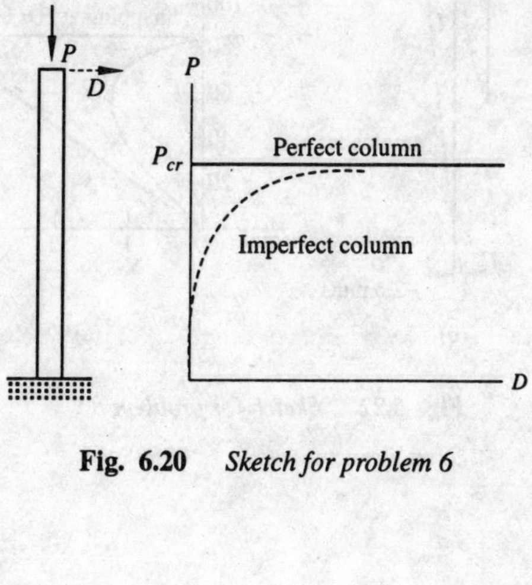

Fig. 6.20 *Sketch for problem 6*

6.7 Members of a frame ABCD shown in the sketch have a square cross-section 8 mm on a side. The material is assumed to be linearly elastic at all times, with $E = 200$ GPa. Beam elements are used for the finite element modelling: 5 along AB, 2 along BC, and 12 along CD. Node at A is hinged, node D may displace horizontally, but is confined to the *XY* plane. The behaviour produced by a horizontal load *P* at point C is to be investigated. Buckling appears possible, so predict the buckling load and mode.

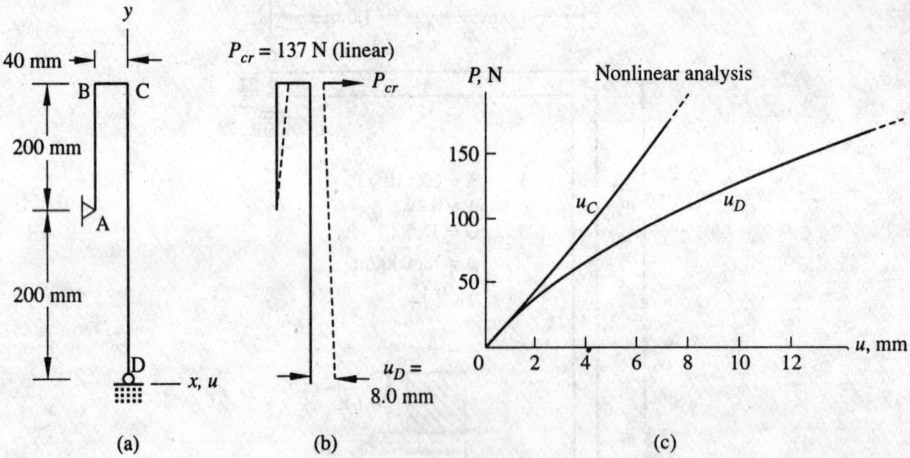

Fig. 6.21 *Sketch for problem 7*

6.8 Perform nonlinear static analyses of the plane frame of problem 6.7. Plot *P* vs. u_c and *P* vs. u_D. Confirm that the frame does not buckle at all. To illustrate the effect of geometric imperfection, the initial (unloaded) geometry is slightly altered. The lower most element along CD is now slightly inclined, so that the node at *D* is placed 2 mm to the left of its previous location. Linear buckling analysis and nonlinear static analysis are requested.

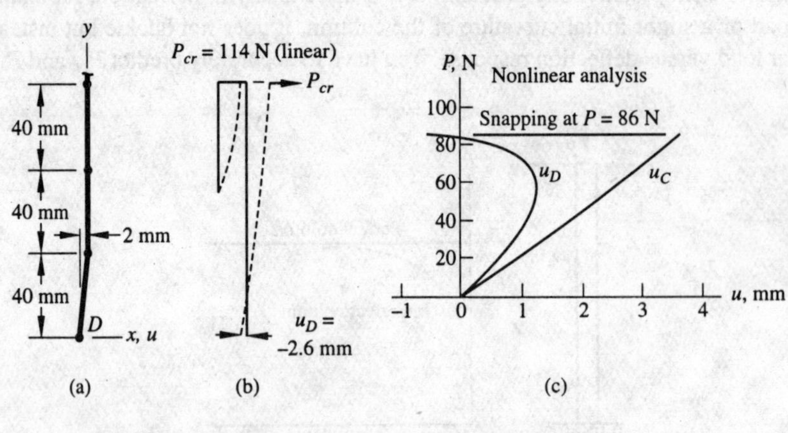

Fig. 6.22 *Sketch for problem 8*

6.9 A shallow pinned arch is subjected to a vertical load at its crown as shown in the sketch. Compare the results of linear static analysis and nonlinear static analysis and comment on the structural behaviour.

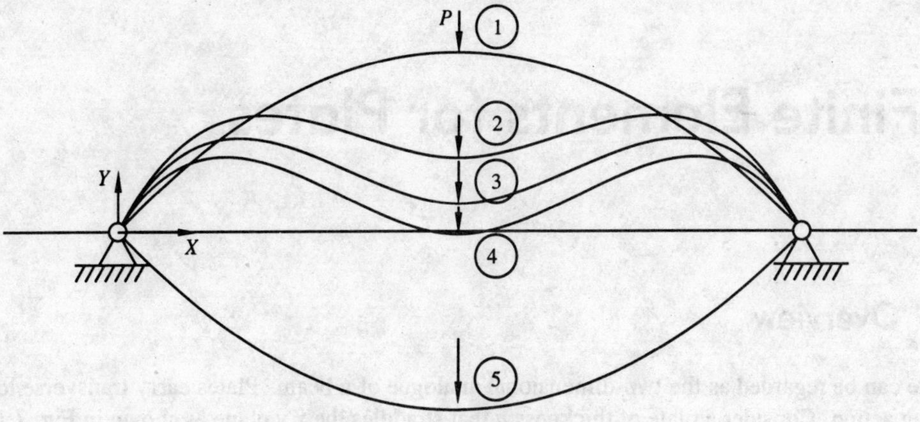

Fig. 6.23 *Sketch for problem 9*

6.10 A frame consisting of a vertical and horizontal member and loaded by a vertical point load, *P*, is shown in the sketch. Elastic material behaviour is assumed. Initial large deformations under increasing load, followed by a snap buckling at constant load are expected. Perform a nonlinear static analysis and verify this behaviour.

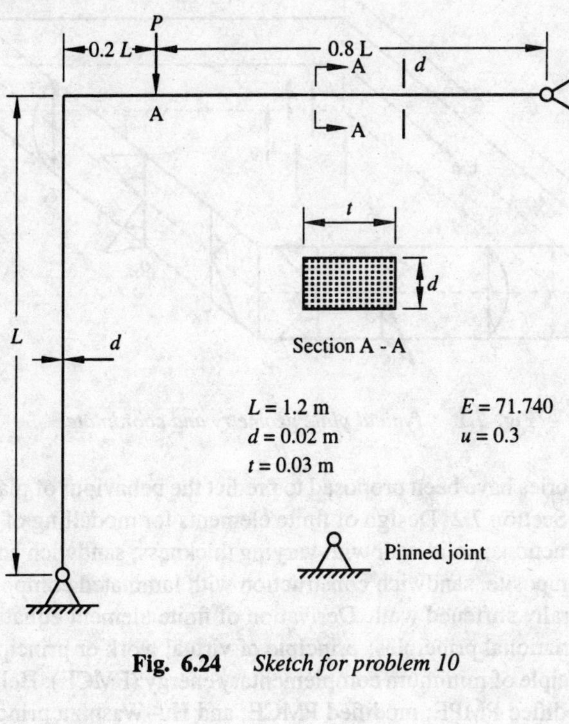

Fig. 6.24 *Sketch for problem 10*

7 Finite Elements for Plates

7.1 Overview

A plate can be regarded as the two-dimensional analogue of a beam. Plates carry transverse loads by bending action. Consider a plate of thickness h that straddles the $x\,y$ plane as shown in Fig. 7.1. Plate surfaces are at $Z = \pm h/2$ and the plate mid surface is at $Z = 0$. The plate can have different plan forms (rectangular, circular, annular, rhombic, etc.,) and a range of thickness 'h' in comparison with the plan form dimension 'a'. In practice (h/a) can vary from 0.1 (thick plate) to 0.001 (thin plate).

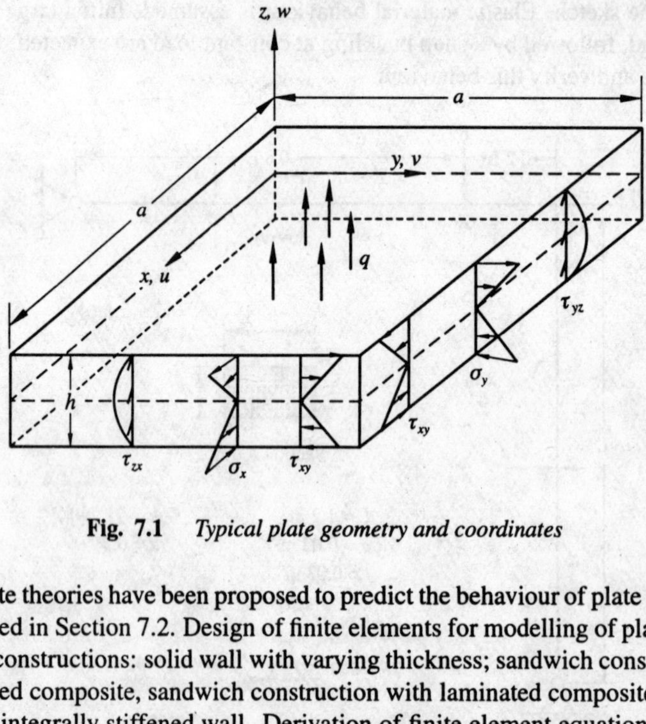

Fig. 7.1 *Typical plate geometry and coordinates*

A number of plate theories have been proposed to predict the behaviour of plate structures and some of these are presented in Section 7.2. Design of finite elements for modelling of plates should consider the following wall constructions: solid wall with varying thickness; sandwich construction with metallic facings; laminated composite, sandwich construction with laminated composite facings; discretely stiffened wall; and integrally stiffened wall. Derivation of finite element equations can be based on a number of alternative variational principles: principle of virtual work or principle of minimum potential energy (PMPE); principle of minimum complementary energy (PMCE); Hellinger–Reissner mixed variational principle; modified PMPE; modified PMCE; and Hu–Washizu principle. Plate behaviours to be studied are: bending; buckling; vibration, and transient response, both linear and nonlinear.

A library of plate elements of different shapes (triangular, quadrilateral) and increasing orders (linear, quadratic, and cubic) widely used in practice are presented in Sections 7.3 and 7.4. The shear-locking phenomenon associated with shear deformable plate elements is given explicit attention. A set of standard test problems to assess element accuracy is included. The presentation identifies some topics for advanced study.

7.2 Plate Theories

A rational basis to distinguish between different plate theories is the a priori assumptions made about the displacement field. The classical plate theory proposed by Kirchoff [7.1] is limited to thin plates. The shear deformation theory proposed by Mindlin [7.2] is applicable to thick and thin plates. An improved shear deformation theory due to Reddy [7.4] also restores the parabolic distribution of transverse shear stresses. Higher order theories have been proposed and they are useful in some advanced applications such as impact dynamics.

7.2.1 Classical plate theory

This theory is based on the displacement field (see Fig. 7.1).

$$u\,(xyz) = -Z\frac{\partial w}{\partial x},$$

$$v\,(xyz) = -Z\frac{\partial w}{\partial y},$$

$$w\,(xyz) = w\,(xy)\,. \tag{7.1}$$

The corresponding linear strains are

$$\{\boldsymbol{\varepsilon}\} = -Z\,\{\chi_b\}\,, \tag{7.2}$$

where

$$\{\boldsymbol{\varepsilon}\}^T = (\varepsilon_{xx}, \varepsilon_{yy}, \gamma_{xy})\,,$$

$$\{\chi_b\}^T = \left(\frac{\partial^2 w}{\partial x^2}, \frac{\partial^2 w}{\partial y^2}, 2\frac{\partial^2 w}{\partial x \partial y}\right) \tag{7.3}$$

and $\varepsilon_{zz} = \gamma_{yz} = \gamma_{xz} = 0$.

The normal stress σ_{zz} is considered negligible in a thin plate theory. The stresses σ_{xx}, σ_{yy}, and τ_{xy} in the plate, however, vary linearly with distance from the mid-surface; transverse shear stresses τ_{yz} and τ_{xz} are also present and vary quadratically through the thickness. The stresses give rise to bending moments M_x, M_y and twisting moment M_{xy}. Accordingly,

$$M_x = \int_{-h/2}^{+h/2} \sigma_{xx} z dz.$$

$$M_y = \int_{-h/2}^{+h/2} \sigma_{yy} z dz.$$

$$M_{xy} = \int_{-h/2}^{+h/2} \tau_{xy} z dz. \tag{7.4}$$

These stress resultants are related to the curvatures as

$$\{\mathbf{M}\} = [\mathbf{D}]\{\chi_b\}, \tag{7.5}$$

where $\{\mathbf{M}\}^T = (M_x M_y M_{xy})$.

Assuming a homogeneous isotropic linear elastic material model, the constitutive matrix $\mathbf{D}$ is

$$[\mathbf{D}] = \begin{bmatrix} D_{11} & D_{12} & D_{13} \\ D_{21} & D_{22} & D_{23} \\ D_{31} & D_{32} & D_{33} \end{bmatrix} \equiv \frac{Eh^3}{12(1-\nu^2)} \begin{bmatrix} 1 & \nu & 0 \\ \nu & 1 & 0 \\ 0 & 0 & \frac{(1-\nu)}{2} \end{bmatrix}. \tag{7.6}$$

The potential energy functional π_p is

$$\pi_p = \frac{1}{2} \int_{\text{Area}} \{\mathbf{M}\}^T \{\chi_b\} dxdy - \int_{\text{Area}} P_z W dxdy - \sum_i P_i W_i. \tag{7.7}$$

In Eqn. (7.7), P_z denotes transverse distributed load and P_{zi} denotes transverse point load and there is only one variable $w(xy)$.

The Kirchoff plate theory neglects transverse shear deformation, which can be significant for thick plates, sandwich plates, and laminated composite plates. A large number of finite elements for modelling of thin plates have been developed and these are called Kirchoff plate elements. In fact, a catalog of plate bending finite elements is readily available [7.5]. Some of them are presented in Section 7.3.

7.2.2 Shear deformation theory

This theory is based on the displacement field (see Fig. 7.1)

$$U(xyz) = u(xy) + z\beta_x(xy).$$

$$V(xyz) = v(xy) + z\beta_y(xy).$$

$$W(xyz) = w(xy). \tag{7.8}$$

In Eqn. (7.8), (u, v, w) are the displacements of a point $(x, y, 0)$, and β_x and β_y, are the rotations of a line initially normal to the mid-plane about the Y- and X-axes, respectively.

The corresponding membrane strains $\boldsymbol{\varepsilon}_m$, bending curvatures χ_b, and transverse shear strains γ_s are evaluated as

$$\{\boldsymbol{\varepsilon}_m\}^T = \left(\frac{\partial u}{\partial x}, \frac{\partial v}{\partial y}, \left(\frac{\partial u}{\partial y} + \frac{\partial v}{\partial x}\right)\right). \tag{7.9}$$

$$\{\chi_b\}^T = \left(\frac{\partial \beta_x}{\partial x}, \frac{\partial \beta_y}{\partial y}, \left(\frac{\partial \beta_y}{\partial x} + \frac{\partial \beta_x}{\partial y}\right)\right). \tag{7.10}$$

$$\{\gamma_s\}^T = \left(\left(\frac{\partial w}{\partial x} - \beta_x\right), \left(\frac{\partial w}{\partial y} - \beta_y\right)\right). \tag{7.11}$$

The effect of the lateral normal stress σ_{zz} is neglected and the membrane stress resultants **N**, bending stress resultants **M** and transverse shear stress resultants **Q** are then evaluated using

$$\begin{bmatrix} \{\mathbf{N}\} \\ \{\mathbf{M}\} \\ \{\mathbf{Q}\} \end{bmatrix} = \begin{bmatrix} [A] & [B] & [0] \\ [B] & [D] & [0] \\ [0] & [0] & [E] \end{bmatrix} \begin{bmatrix} \{\boldsymbol{\varepsilon}_m\} \\ \{\chi_b\} \\ \{\gamma_s\} \end{bmatrix}. \tag{7.12}$$

In Eqn. (7.12)

$\{\mathbf{N}\}^T = (N_x, N_y, N_{xy})$ are the membrane forces;
$\{\mathbf{M}\}^T = (M_x, M_y, M_{xy})$ are the bending and twisting moments;
$\{\mathbf{Q}\}^T = (Q_{yz}, Q_{xz})$ are the transverse shear forces;

$$N_x = \int_{-h/2}^{+h/2} \sigma_{xx} dz.$$

$$N_y = \int_{-h/2}^{+h/2} \sigma_{yy} dz.$$

$$N_{xy} = \int_{-h/2}^{+h/2} \tau_{xy} dz.$$

$$Q_{yz} = \int_{-h/2}^{+h/2} \tau_{yz} dz.$$

$$Q_{xz} = \int_{-h/2}^{+h/2} \tau_{xz} dz.$$

Assuming a homogeneous isotropic linear elastic material model, the constitutive matrices **A, B, D,** and **E** are

$$[\mathbf{A}] = \frac{Eh}{1-\nu^2} \begin{bmatrix} 1 & \nu & 0 \\ \nu & 1 & 0 \\ 0 & 0 & \dfrac{(1-\nu)}{2} \end{bmatrix}$$

$$[\mathbf{B}] = [0].$$

$$[\mathbf{D}] = \frac{Eh^3}{12(1-\nu^2)}\begin{bmatrix} 1 & \nu & 0 \\ \nu & 1 & 0 \\ 0 & 0 & \dfrac{(1-\nu)}{2} \end{bmatrix},$$

$$[\mathbf{E}] = GhS\begin{bmatrix} 1 & 0 \\ 0 & 1 \end{bmatrix},$$

with $S = \pi^2/12$ or 5/6 (shear correction factor)

The potential energy functional π_p is

$$\pi_p\,(uvw\beta_x\beta_y) = \frac{1}{2}\int\limits_{\text{Area}} \left([\mathbf{N}]^T\{\boldsymbol{\varepsilon}_m\} + \{\mathbf{M}\}^T\{\chi_b\} + \{\mathbf{Q}\}^T\{\gamma_s\}\right)dxdy - \int\limits_{\text{Area}} P_z W dxdy - \sum P_{Zi}W_i. \tag{7.13}$$

In Eqn. (7.13), there are five independent variables; P_z denotes transverse distributed load and P_{zi} denotes transverse point load.

The Mindlin plate theory includes shear deformation and therefore is applicable to thick plates, sandwich plates, laminated composite plates, etc. A large number of shear flexible plate finite elements based on this theory have been proposed [7.7 to 7.10]. Mindlin plate elements, implemented in commercial FEM systems are therefore presented in Section 7.4.

7.2.3 Improved shear deformation theory

This theory eliminates the discrepancy between the distribution of transverse shear stresses of the Mindlin theory and the actual distribution thereby avoids the use of shear correction factor.

This theory is based on the displacement field (see Fig. 7.1).

$$\begin{aligned} U(xyz) &= u(xy) + z\beta_x(xy) + z^3\bar{\Phi}_x(xy), \\ V(xyz) &= V(xy) + z\beta_y(xy) + z^3\bar{\Phi}_y(xy), \\ W(xyz) &= w(xy), \end{aligned} \tag{7.14}$$

with

$$\begin{aligned} \Phi_x &= -\frac{4}{3h^2}\left(\beta_x + \frac{\partial w}{\partial x}\right), \\ \Phi_y &= -\frac{4}{3h^2}\left(\beta_y + \frac{\partial w}{\partial y}\right). \end{aligned} \tag{7.15}$$

Eqns (7.15) ensure vanishing transverse shear stresses at the top and bottom faces of the plate. Therefore in Eqns. (7.14), (u, v, w) are the displacements of a point $(x, y, 0)$, and β_x and β_y are the rotations of a line initially normal to the mid-plane about the Y and X axes, respectively.

The potential energy functional π_p based on this theory contains first-order derivatives of u, v, β_x, β_y and derivatives up to second order of W. Design of plate bending elements based on this theory therefore demand C^1 continuous interpolation functions for W while C^0 continuous interpolation functions are sufficient for u, v, β_x, β_y. Even for a simple three-noded triangular plate element, if the nodal degrees of freedom are restricted to be u_i v_i w_i θ_{xi} θ_{yi}, finding interpolation functions for W remains a challenge. On the other hand, if known C^0 continuous interpolation functions are used to approximate W also over the element domain, then there is no assurance on the element performance. In view of these observations, no specific element formulation based on the improved shear deformation theory is presented here. However, this is certainly a topic identified for research. Design of a four-noded quadrilateral plate element is equally challenging.

In summary, the deformation of a plate can be described by classical plate theory, or by first-order shear deformation theory, or by third-order shear deformation theory. A review of these plate theories and other refined theories can be found in Reddy [7.3].

7.3 Kirchoff Plate Elements

7.3.1 Non-conforming elements

The first finite elements designed for bending analysis of thin plates were based on the Kirchoff plate theory. The four-noded rectangular plate finite element shown in Fig. 7.2 is the simplest. At each node i the permitted degrees of freedom are w_i, $\theta_{xi}\left(or \dfrac{\partial w}{\partial y}\right)$, $\theta_{yi}\left(or -\dfrac{\partial w}{\partial x}\right)$. The nodal degrees of freedom vector $\mathbf{q}$ is

$$\{\mathbf{q}\}^T = (w_1\theta_{x1}\theta_{y1}w_2\theta_{x2}\theta_{y2}w_3\theta_{x3}\theta_{y3}w_4\theta_{x4}\theta_{y4}).$$

The displacement W is approximated over the element domain using the polynomial

$$W = \alpha_1 + \alpha_2 x + \alpha_3 y + \alpha_4 x^2 + \alpha_5 xy + \alpha_6 y^2 + \alpha_7 x^3 + \alpha_8 x^2 y + \alpha_9 xy^2 + \alpha_{10} y^3 + \alpha_{11} x^3 y + \alpha_{12} xy^3$$

or

$$W = \lfloor \mathbf{P} \rfloor \{\boldsymbol{\alpha}\}. \tag{7.16}$$

It is possible to relate the polynomial coefficients $\boldsymbol{\alpha}$ to the nodal degrees of freedom $\mathbf{q}$ as

$$[\mathbf{A}]\{\boldsymbol{\alpha}\} = \{\mathbf{q}\}$$

$$\text{and } \{\boldsymbol{\alpha}\} = [\mathbf{A}]^{-1}\{\mathbf{q}\} \tag{7.17}$$

Combining Eqns. (7.16) and (7.17), we can write

$$W = \lfloor \mathbf{P} \rfloor [\mathbf{A}]^{-1}\{\mathbf{q}\} = [\mathbf{N}]\{\mathbf{q}\}. \tag{7.18}$$

In Eqn. (7.18), $\mathbf{N}$ denotes the matrix of shape functions.

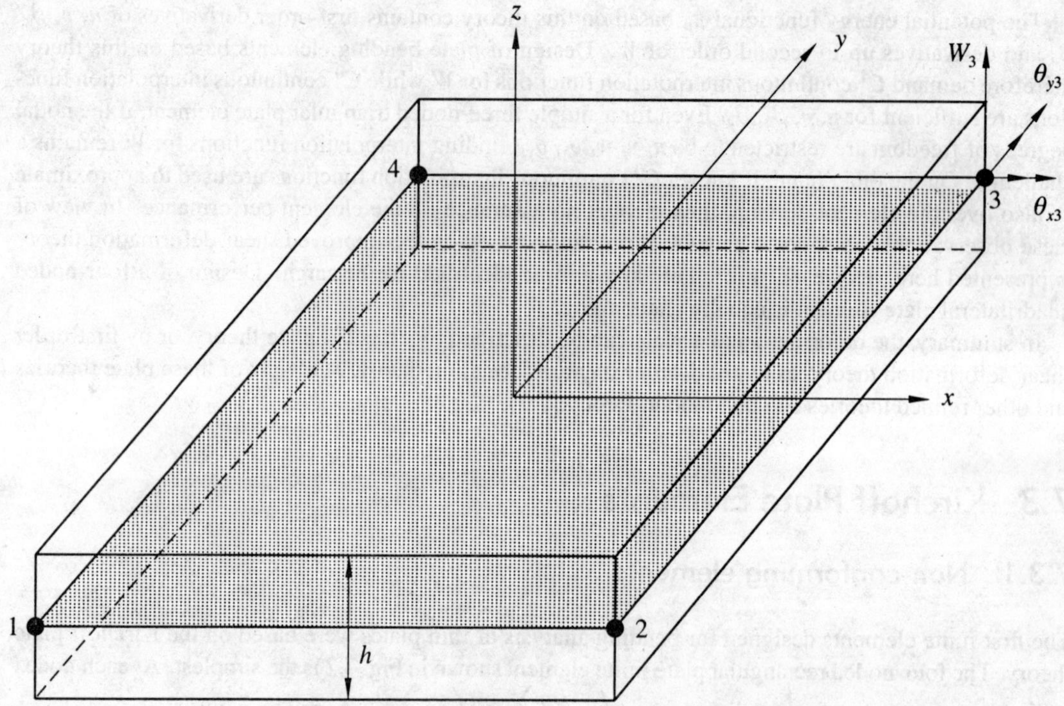

Fig. 7.2 *Four-node rectangular plate element*

Consequently, we can write

$$\{\chi_b\} = [\mathbf{B}_b]\{\mathbf{q}\}. \tag{7.19}$$

$$\{\mathbf{M}\} = [\mathbf{D}][\mathbf{B}_b]\{\mathbf{q}\}. \tag{7.20}$$

$$\pi_p(\mathbf{q}) = \frac{1}{2}\{\mathbf{q}\}^T[\mathbf{k}]\{\mathbf{q}\} - \{\mathbf{q}\}^T\{\mathbf{f}^p\} - \{\mathbf{q}\}^T\{\mathbf{f}^c\}. \tag{7.21}$$

$$[\mathbf{k}] = \int_{A^{(e)}} [\mathbf{B}_b]^T[\mathbf{D}][\mathbf{B}_b]dxdy. \tag{7.22}$$

$$\{\mathbf{f}^p\} = \int_{A^{(e)}} [\mathbf{N}]^T P_z dxdy. \tag{7.23}$$

It is therefore possible to compute the element stiffness matrix, mass matrix, load vectors, etc.

Eqn. (7.18) ensures continuity of w along inter-element boundaries, whereas the normal slope continuity along inter-element boundaries is not satisfied. This element therefore is non-conforming (or incompatible). It is complete because it can represent states of constant curvature and twist. Accordingly, this element cannot guarantee a lower bound on computed deflections.

Restriction to rectangular shape is unacceptable for practical applications. However, generalization to an arbitrary quadrilateral shape is rather difficult.

It is surprisingly difficult to design a simple three-noded triangular plate finite element shown in Fig. 7.3 based on the Kirchoff plate theory that can represent states of constant curvature and twist (completeness) and gives good results when subjected to standard tests. Consequently, the discrete Kirchoff theory (DKT) plate elements are in common use. The essential feature of a DKT plate element formulation [7.15] is that transverse shear strain is set to zero at a finite number of points in the element, rather than at every point. The result is an incompatible thin plate element, which usually converges rapidly with mesh refinement.

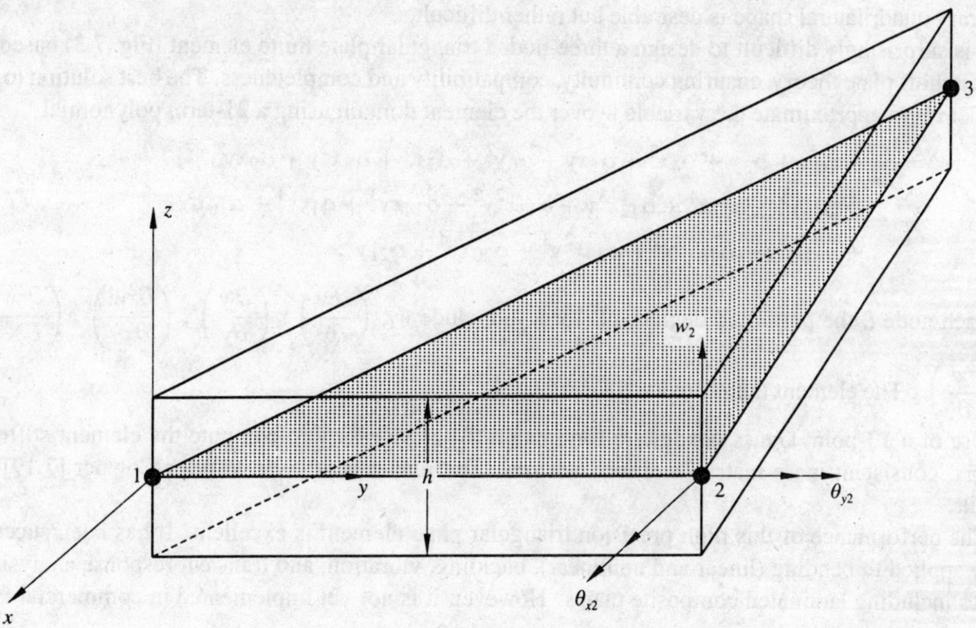

Fig. 7.3 *Three-node triangular plate element*

After so many manipulations it is not apparent how a DKT plate element will behave. The user should perform numerical experiments to learn about element behaviour.

Semiloof plate elements, both triangular and quadrilateral in shape, developed by Irons and reported in [7.16] are the best among DKT plate elements. However, they are not yet implemented in commercial FEM systems.

7.3.2 Conforming elements

The four-noded rectangular plate element shown in Fig. 7.2 can be made fully conforming by permitting each node i to have w_i, $\left(\dfrac{\partial w}{\partial x}\right)_i$, $\left(\dfrac{\partial w}{\partial y}\right)_i$, $\left(\dfrac{\partial^2 w}{\partial x \partial y}\right)_i$ as the nodal degrees of freedom. The displacement w is approximated over the element domain using the polynomial

$$
\begin{aligned}
W = \alpha_1 + \alpha_2 x + \alpha_3 y + \alpha_4 x^2 + \alpha_5 xy + \alpha_6 y^2 + \alpha_7 x^3 + \alpha_8 x^2 y + \alpha_9 xy^2 \\
+ \alpha_{10} y^3 + \alpha_{11} x^3 y + \alpha_{12} x^2 y^2 + \alpha_{13} xy^3 + \alpha_{14} x^3 y^2 + \alpha_{15} x^2 y^3 + \alpha_{16} x^3 y^3 .
\end{aligned}
\tag{7.24}
$$

It is possible to relate the polynomial coefficients $\boldsymbol{\alpha}$ to the nodal degrees of freedom $\mathbf{q}$ and hence write Eqn. (7.24) as

$$W = [\mathbf{N}]\{\mathbf{q}\}. \tag{7.25}$$

The shape function matrix $\mathbf{N}$ for this element is available in Reddy [7.17].

The computation of element stiffness matrix, mass matrix, and load vectors then follows the standard procedure.

Restriction to a rectangular shape is unacceptable for practical applications. Generalization to an arbitrary quadrilateral shape is desirable but rather difficult.

It is surprisingly difficult to design a three-noded triangular plate finite element (Fig. 7.3) based on the Kirchoff plate theory, ensuring continuity, compatibility and completeness. The best solution to the problem is to approximate the variable w over the element domain using a 21-term polynomial

$$\begin{aligned} W = {} & \alpha_1 + \alpha_2 x + \alpha_3 y + \alpha_4 x^2 + \alpha_5 xy + \alpha_6 y^2 + \alpha_7 x^3 + \alpha_8 x^2 y + \alpha_9 xy^2 \\ & + \alpha_{10} y^3 + \alpha_{11} x^4 + \alpha_{12} x^3 y + \alpha_{13} x^2 y^2 + \alpha_{14} xy^3 + \alpha_{15} y^4 + \alpha_{16} x^5 \\ & + \alpha_{17} x^4 y + \alpha_{18} x^3 y^2 + \alpha_{19} x^2 y^3 + \alpha_{20} xy^4 + \alpha_{21} y^5 . \end{aligned} \tag{7.26}$$

At each node i, the permitted degrees of freedom include w_i, $\left(\frac{\partial w}{\partial x}\right)_i$, $\left(\frac{\partial w}{\partial y}\right)_i$, $\left(\frac{\partial^2 w}{\partial x^2}\right)_i$, $\left(\frac{\partial^2 w}{\partial y^2}\right)_i$, $\left(\frac{\partial^2 w}{\partial x \partial y}\right)_i$. The element therefore has 18 d.o.f.

Use of a 13-point Gauss quadrature formula [7.18] is necessary to compute the element stiffness matrix, consistent mass matrix, and load vectors. The reader is advised to refer Cowper [7.19] for details.

The performance of this high precision triangular plate element is excellent. It has been successfully applied to bending (linear and nonlinear), buckling, vibration, and transient response analysis of plates including laminated composite plates. However, it is not yet implemented in commercial FEA software.

7.4 Mindlin Plate Elements

7.4.1 The four-node quadrilateral element

Figure 7.4 shows the geometry and coordinate systems used for design of a general quadrilateral plate element based on the Mindlin plate theory. The geometry is modelled using well-known bi-linear interpolation functions $\mathbf{N}_i(\xi\eta)$ and Cartesian coordinates $(X_i Y_i 0)$ of nodes:

$$\begin{Bmatrix} X \\ Y \end{Bmatrix} = \sum_{i=1}^{4} N_i(\xi\eta) \begin{Bmatrix} X_i \\ Y_i \end{Bmatrix}. \tag{7.27}$$

The thickness h at any point within the element is also interpolated in terms of the nodal thicknesses $h_i (i = 1, 4)$

$$h = \sum_{i=1}^{4} N_i(\xi\eta) h_i, \tag{7.28}$$

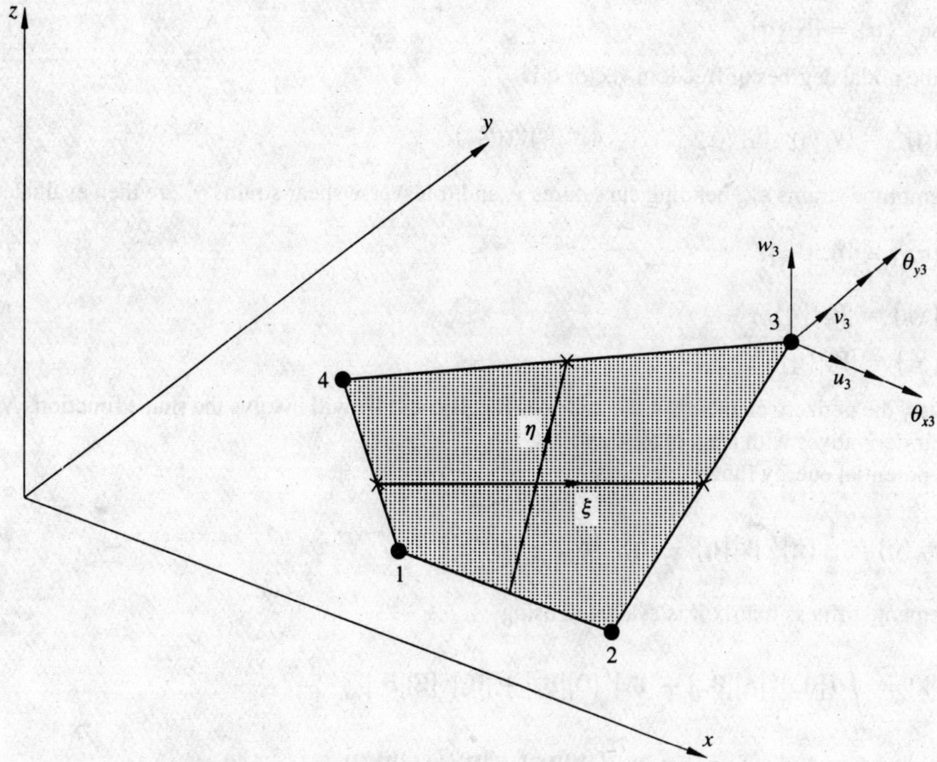

Fig. 7.4 *Four-node quadrilateral Mindlin plate element*

where

$$N_1 = \frac{1}{4}(1-\xi)(1-\eta);$$
$$N_2 = \frac{1}{4}(1+\xi)(1-\eta);$$
$$N_3 = \frac{1}{4}(1+\xi)(1+\eta);$$
$$N_4 = \frac{1}{4}(1-\xi)(1+\eta). \tag{7.29}$$

In order to predict the coupled stretching and bending behaviour of plates, Mindlin's theory uses five independent variables u, v, w, β_x, and β_y. These variables are interpolated over the element domain using the same functions $N_i(\xi\eta)$ leading to an isoparametric formulation

$$\begin{Bmatrix} u \\ v \\ w \\ \beta_x \\ \beta_y \end{Bmatrix} = \sum_{i=1}^{4} N_i(\xi\eta) \begin{Bmatrix} u_i \\ v_i \\ w_i \\ \theta_{yi} \\ -\theta_{xi} \end{Bmatrix},$$

$$\text{or} \quad \{\mathbf{u}\} = [\mathbf{N}]\{\mathbf{q}\}, \tag{7.30}$$

where the nodal degrees of freedom vector $\mathbf{q}$ is

$$\{\mathbf{q}\}^T = (u_1 v_1 w_1 \theta_{x1} \theta_{y1}, \ldots, \ldots, u_4 v_4 w_4 \theta_{x4} \theta_{y4}).$$

The membrane strains $\boldsymbol{\varepsilon}_m$, bending curvatures $\boldsymbol{\chi}_b$and transverse shear strains $\boldsymbol{\gamma}_s$ are then evaluated as

$$\{\boldsymbol{\varepsilon}_m\} = [\mathbf{B}_m]\{\mathbf{q}\}. \tag{7.31}$$

$$\{\boldsymbol{\chi}_b\} = [\mathbf{B}_b]\{\mathbf{q}\}. \tag{7.32}$$

$$\{\boldsymbol{\gamma}_s\} = [\mathbf{B}_s]\{\mathbf{q}\}. \tag{7.33}$$

Obviously the nonzero elements of the matrices $\mathbf{B}_m$, $\mathbf{B}_b$, and $\mathbf{B}_s$ will involve the shape functions $N_i(\xi\eta)$ and their derivatives with respect to X and Y.

The potential energy functional π_p is

$$\pi_p(\mathbf{q}) = \frac{1}{2}\{\mathbf{q}\}^T[\mathbf{k}]\{\mathbf{q}\} - \{\mathbf{q}\}^T\{\mathbf{f}^p\} - \{\mathbf{q}\}^T\{\mathbf{f}^c\}. \tag{7.34}$$

The element stiffness matrix $\mathbf{k}$ is evaluated using

$$\begin{aligned}[\mathbf{k}] = \int_{A^{(e)}} \big([\mathbf{B}_m]^T[\mathbf{A}][\mathbf{B}_m] + [\mathbf{B}_b]^T[\mathbf{D}][\mathbf{B}_b] + [\mathbf{B}_s]^T[\mathbf{E}][\mathbf{B}_s] \\ + [\mathbf{B}_m]^T[\mathbf{B}][\mathbf{B}_b] + [\mathbf{B}_b]^T[\mathbf{A}][\mathbf{B}_m]\big)\,dxdy.\end{aligned} \tag{7.35}$$

The consistent load vector $\mathbf{f}^p$ due to laterally distributed load P_z is

$$\{\mathbf{f}^p\} = \int_{A^{(e)}} [\bar{\mathbf{N}}]^T \left(\sum_{i=1}^{4} N_i(\xi\eta) P_{zi}\right) dxdy. \tag{7.36}$$

In Eqn. (7.36), $\bar{\mathbf{N}}$ is the row corresponding to the variable W in the shape function matrix $\mathbf{N}$.

The integrals in Eqns. (7.35) and (7.36) are evaluated using Gaussian quadrature formulae

$$[\mathbf{k}] = \sum_{i=1}^{\text{NGP}} \sum_{j=1}^{\text{NGP}} W_i W_j \left[\Phi(\xi_i \eta_j)\right],$$

$$\{\mathbf{f}^p\} = \sum_{i=1}^{\text{NGP}} \sum_{j=1}^{\text{NGP}} W_i W_j \left[\psi(\xi_i \eta_j)\right],$$

where

$$\begin{aligned}\Phi(\xi\eta) = \big([\mathbf{B}_m]^T[A][\mathbf{B}_m] + [\mathbf{B}_b]^T[\mathbf{D}][\mathbf{B}_b] + [\mathbf{B}_s]^T[E][\mathbf{B}_s] \\ + [\mathbf{B}_m]^T[\mathbf{B}][\mathbf{B}_b] + [\mathbf{B}_b]^T[A][\mathbf{B}_m]\big)\,|\mathbf{J}|;\end{aligned}$$

$$\psi(\xi\eta) = [\bar{\mathbf{N}}]^T \left(\sum_{i=1}^{4} \mathbf{N}_i(\xi\eta) P_{zi} \right) |\mathbf{J}|;$$

$|\mathbf{J}|$ is the determinant of the Jacobian matrix $\mathbf{J}$.

$$[\mathbf{J}] = \sum_{i=1}^{4} \begin{bmatrix} \frac{\partial N_i}{\partial \xi} X_i & \frac{\partial N_i}{\partial \xi} Y_i \\ \frac{\partial N_i}{\partial \eta} X_i & \frac{\partial N_i}{\partial \eta} Y_i \end{bmatrix} = \begin{bmatrix} J_{11} & J_{12} \\ J_{21} & J_{22} \end{bmatrix}.$$

Choosing the number of Gauss points NPG = 2, the sampling point locations and weighting coefficients are given in Table 3.2.

It has been observed that violent shear stress oscillations are predicted within the elements and the spurious shear stresses result in a strong artificial stiffening of the elements when applied to thin plates. This effect of *shear-locking* is more pronounced when the elements are geometrically distorted and used for stress predictions.

To design efficient and reliable Mindlin plate elements, there is a need to alleviate the shear-locking effects. A number of researchers have advocated the application of reduced integration techniques. Another successful approach is to use a mixed interpolation of the displacements, rotations, and transverse shear strains. Some researchers have resorted to the assumed strain method. The reader is advised to refer [7.7 to 7.10] for details.

7.4.2 Higher order elements

A family of Mindlin plate elements whose interpolation functions are quadratic and cubic in order, that have a good mathematical basis and are reliable and efficient has also been proposed. The quadrilateral elements are named as QUADn, where n refers to the number of nodes and $n = 8, 12$ (for serendipity family) and $n = 9, 16$ (for Lagrange family). Similarly the triangular elements are named as TRIAn with $n = 3, 6$ and 10 for linear, quadratic and cubic order elements.

The design of all these elements is similar to that of the four-noded quadrilateral element presented here. The required shape functions and Gauss quadrature formulae for all the elements proposed are available.

A notable feature of the higher order elements is that they can be distorted to have curved edges. With so many options available for individual element formulation, element behaviour cannot be deduced solely from the assumed displacement field. In particular, the shear-locking effect may still persist!

One expects that shear deformable plate elements in commercial software will be free from shear-locking effects and spurious zero energy modes, and will be able to represent states of constant curvature and twist. The analyst should study how a plate element behaves by using it in standard test cases. In particular, one may wish to know if accuracy is lost when the element becomes very thin (due to shear-locking), as aspect ratio increases, and as quadrilateral shapes become distorted from a rectangle. In summary, the analyst should perform numerical experiments to learn about element behaviour using benchmark tests.

7.5 Test Problems

7.5.1 Patch-tests

A plane element must be able to display states of constant σ_{xx}, σ_{yy}, and τ_{xy} if it has to pass patch-tests. A plate element must be able to display these states in each z = constant layer, which means that a valid plate element must pass patch-tests for states of constant M_x, M_y and M_{xy}.

The most frequently used patch test is a configuration of five arbitrary quadrilateral elements as shown in Fig. 7.5 The loading for bending, shear and twist cases is also shown in the figure. A patch-test for constant curvature (or for constant M_x), for constant shear (or for constant Q_{xz}), and for constant twist (or for constant M_{xy}) is identified in Fig. 7.5. Verify whether the plate elements in the FEA software you use pass these tests.

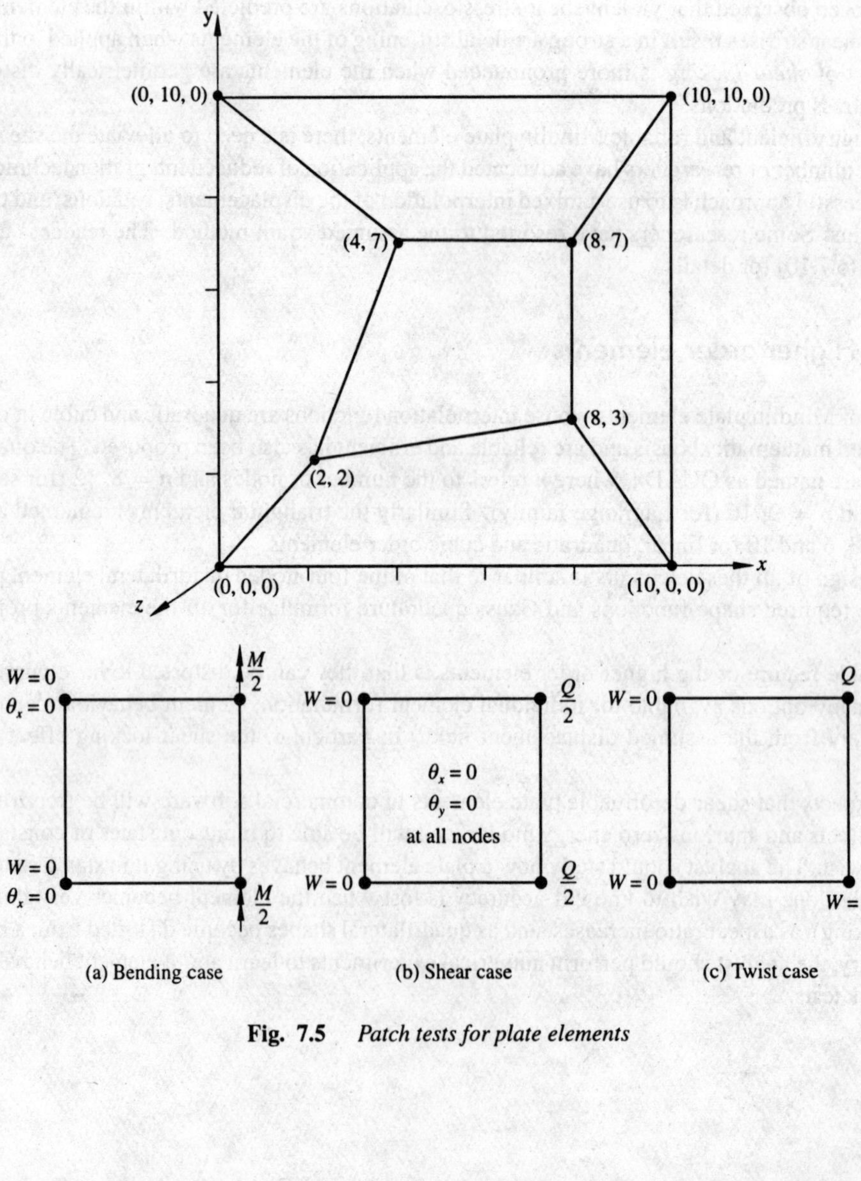

Fig. 7.5 *Patch tests for plate elements*

7.5.2 Square plate bending tests

A square plate with side length 'a' and thickness 'h' has all edges either simply supported or clamped. The plate is laterally loaded by either a uniformly distributed load q or a concentrated centre force P. Centre deflections and bending moments for these cases are reported in [7.1]. For finite element analysis, only one quadrant need be discretized. Typically, a uniform mesh is used and one examines the convergence of solutions as the mesh is refined and compares the converged solutions with exact solutions. These tests have to be performed for $(h/a) = 0.001, 0.01$ and 0.1. Carefully observe if the element under test exhibits shear-locking behaviour.

7.5.3 Rhombic plate bending test

Figure 7.6 shows a 30° skew plate, with simply-supported edges, under uniformly distributed load q. The plate has thickness h and side length a. An exact solution is available in Morley [7.6] for the central deflection W_c. Report converged solutions for $(h/a) = 0.001, 0.01$, and 0.1 and compare the same with the exact solution, observe if the solutions converge rather slowly due to the singularity in the bending moments at the obtuse corners.

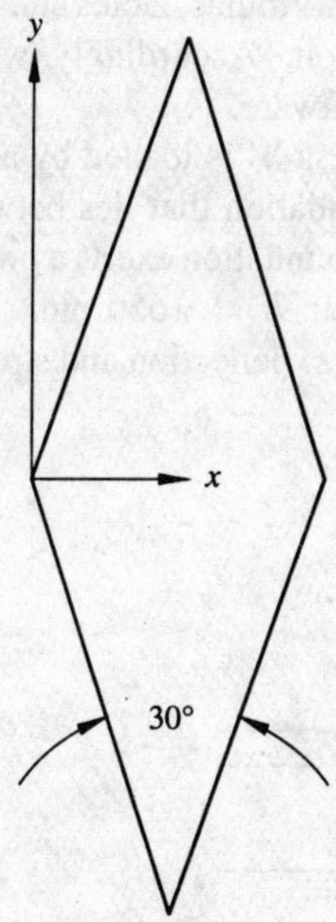

Fig. 7.6 *Rhombic plate*

7.5.4 Circular plate bending test

As an example, where the element geometry is distorted, a circular plate of radius a, thickness h with a clamped edge is considered. The plate is laterally loaded by either a uniformly distributed load q or a concentrated centre force P. Exploiting symmetry, the first quadrant is discretized. Typically the converged solution is compared with the exact solution if available. These tests have to be performed for $h/a = 0.001, 0.01$ and 0.1. Carefully observe if there are oscillations in the prediction of transverse shear stresses.

7.5.5 Bending of a corner-supported square plate

This is a test case where the finite element model introduces very few global constraints so that spurious zero energy modes in the plate element under test will show up. Here a square plate of side length 'a' and thickness 'h' is supported at the four corners (i.e., $w = 0$) and is acted upon by a uniformly distributed lateral load of intensity q. Symmetry allows a quarter of the plate to be modelled. A convergence study gives absurd results if singularity arises from spurious zero energy mechanisms.

7.6 Practical Applications

7.6.1 Plate on elastic foundation

Textbooks and handbooks provide useful formulas, tables and charts for beams on an elastic foundation but not for plates on an elastic foundation. Accordingly, we propose finite element modelling and analysis using a general purpose FEA software.

A large steel plate of constant thickness 't' is loaded by a concentrated force P. The plate is supported entirely by a Winkler elastic foundation that lies between the plate and a flat base that is assumed to be rigid (Fig. 7.7). A Winkler foundation exerts a pressure k_0 on the plate for each millimeter of z-direction deflection. Given the data: $L = 1650$ mm, $t = 30$ mm, $E = 200$ Gpa, $\nu = 0.3$, $k_0 = 0.2$ MPa/mm, $P = 1.0$ N, the load vs. deflection and stress distribution in the plate are required.

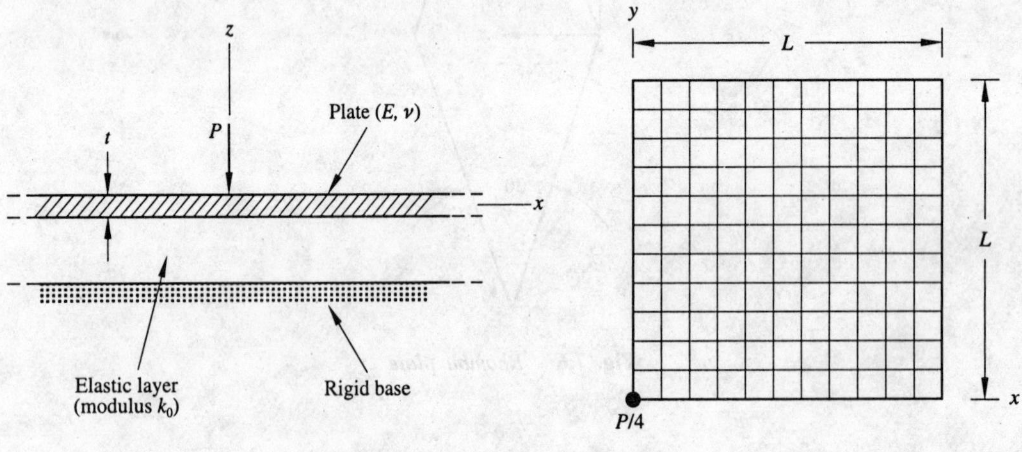

Fig. 7.7 *(a) Cross-section of a plate on an elastic foundation (b) FE mesh on a quadrant with z-direction force $P/4$ at $x = y = 0$*

Analysis can be confined to the neighbourhood of the load. Thus there is no need to extend the finite element model whose outer portion would be essentially undeflected and unstressed. The chosen L meets these criteria.

Figure 7.7(b) shows a 11 by 11 mesh of four-noded quadrilateral plate elements. A uniform mesh is shown although a graded mesh should have greatest refinement around the load point. A load of $P/4$ is applied at the node located at $x = 0$, $y = 0$. Symmetry is imposed about the x- and y-axes; that is $\theta_x = 0$ along the x-axis and $\theta_y = 0$ along the y-axis. Displacement W is unrestrained at all nodes.

Some FEA software allows an elastic foundation option for plate elements. The user might instead connect each node of the FE model to the base by a linear spring. This would be a tedious procedure with a graded mesh, for which the appropriate spring stiffnesses might differ from one another. A better option would be to place the plate on a layer of 3-D solid elements, whose material model is anisotropic with only E_z nonzero if a Winkler type elastic foundation is to be simulated.

The reader is advised to refer Cook [7.12] for typical results. Instead, another analysis is requested with a finer mesh around the load point. The computed results can then be correlated with those in Cook [7.12].

For large loads, the plate may lift off its foundation demanding a nonlinear analysis because the zone of lift off is unknown at the outset.

7.6.2 Bending stress concentration around holes in plates

Handbooks and textbooks lack information on stress concentration factors for holes in plates undergoing bending. Accordingly, we propose the application of FEM/GPP to such problems.

A rectangular plate of length L, width 'b' and thickness 'h' with a central circular cutout of radius 'a' is loaded in a four-point bending. Accurate stress analysis of the plate with and without the hole is requested leading to the definition and quantification of bending stress concentration factor for different (a/b) ratios. Report converged accurate solutions for $h/b = 0.001, 0.01$ and 0.1 and provide critical assessment of the same.

Holes of different shapes such as elliptical, rectangular with rounded corners with different orientations may induce more severe stress concentrations.

7.6.3 Simulation of static indentation tests of laminated composite circular plates

Studies have shown that static indentation tests and low velocity lateral impact tests produce the same damage. Therefore, numerical simulation of static indentation tests is preferred to lateral impact tests.

The geometry, loading, support and ply-orientations/stacking sequences used in the simulation are shown in Fig. 7.8. Finite element discretization of one quarter of the plate is also shown. The orthotropic elastic properties of unidirectional graphite/epoxy composite materials used in the computation are: $E_1 = 131$ GPa, $E_2 = 13.0$ GPa, $G_{12} = 6.4$ GPa, $\nu_{12} = 0.34$.

The calculated load versus load point displacement and spatial variation of the lateral displacement w are also presented in Fig. 7.8. In general, the results predicted using nonlinear static analysis (Procedure #2) are in good agreement with the test data. Linear static analysis (Procedure #1) is totally unacceptable. A comparison of stress distributions, nature and extent of damage induced lead to similar conclusions. The reader is advised to refer Lakshminarayana et al., [7.13] for details.

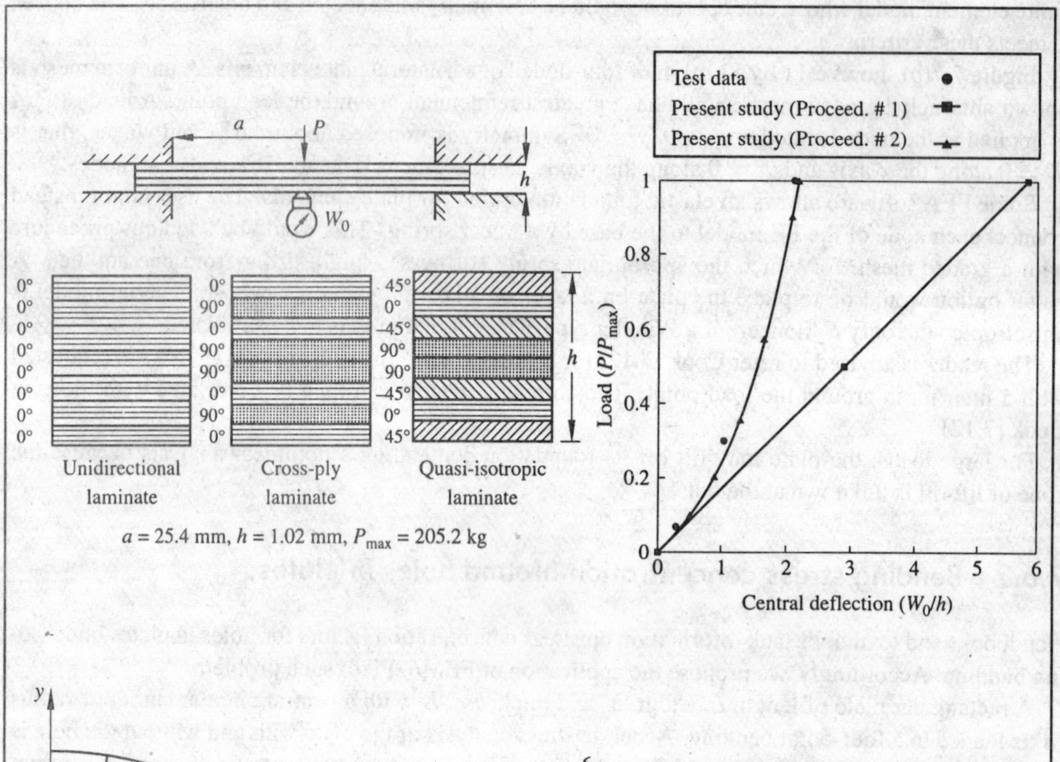

Fig. 7.8 *Finite element simulation of static indentation tests on graphite/epoxy composite circular plates*

7.6.4 Lateral impact response analysis of laminated composite plates

For reliable prediction of dynamic behaviour of composite structures subjected to lateral impact (to simulate bird strike), the finite element method in general and a general purpose FEA program in particular offer a powerful tool.

The test shown as insert in Fig. 7.9 consists of a laminated graphite/epoxy composite, square plate with all edges simply-supported and subjected to lateral impact at its centre. This test is simulated by the contact force history $P(t) = P_0 \sin(\pi t/t_0)$. The basic problem is to calculate the displacement history w (x, y, t). It is then possible to determine the strain history, spacial and temporal variation of ply stresses (including inter laminar stresses) across the laminate thickness, and the nature and extent of damage.

For impact by a 12.7 mm diameter steel ball with a velocity $V_0 = 3$ m/s, $P_0 = 266.9$ N and $t_0 = 0.2$ ms. Figure 7.9 shows the computed central displacement time history using linear transient response analysis. The reader is advised to refer Lakshminarayana et all [7.14] for details.

7.7 Closure

Kirchoff plate theory and Mindlin plate theory provide the mathematical models for the design of finite elements for plate type structures. Assumed the displacement functions, the minimum potential energy functional approach is preferred to derive element equations. Gauss quadrature formulae for triangular and quadrilateral domains are used to compute element stiffness matrix and load vectors. Plate elements presented are suitable for implementation in commercial FEM systems and can model solid walls with varying thickness, sandwich construction with metallic facings, laminated composites, sandwich construction with laminated composite facings, discretely stiffened wall and integrally stiffened walls. Restriction to linear bending analysis is relaxed. More advanced applications to buckling, vibration, and transient response analysis – both linear and nonlinear – are therefore included as computational problems. These problems can be solved using a general purpose FEA software. The reader can refer Crisfield [7.20, 7.21] for details of element design and solution procedures.

A set of standard test problems to assess element performance is included. Mindlin plate elements when applied to predict the behaviour of thin plates exhibit the shear-locking phenomenon. Elimination of shear-locking remains a topic for research.

A lateral pressure p is carried mostly by bending action in a plate and there is an almost linear relation between lateral deflection W and lateral pressure p until the maximum deflection is roughly half the thickness h. With greater deflection, membrane stresses support an increasing portion of the lateral pressure load. Thus the stiffness of the plate appears to increase, To model large deflections of a plate, we need shell elements which account for both membrane strains and bending curvatures, and a nonlinear analysis procedure. Design of shell elements is addressed in the next chapter. Nonlinear finite element analysis is addressed in the last chapter.

In summary, analysts must both anticipate structural behaviour and understand element behaviour in order to make a suitable choice of finite element, mesh, and analysis procedure.

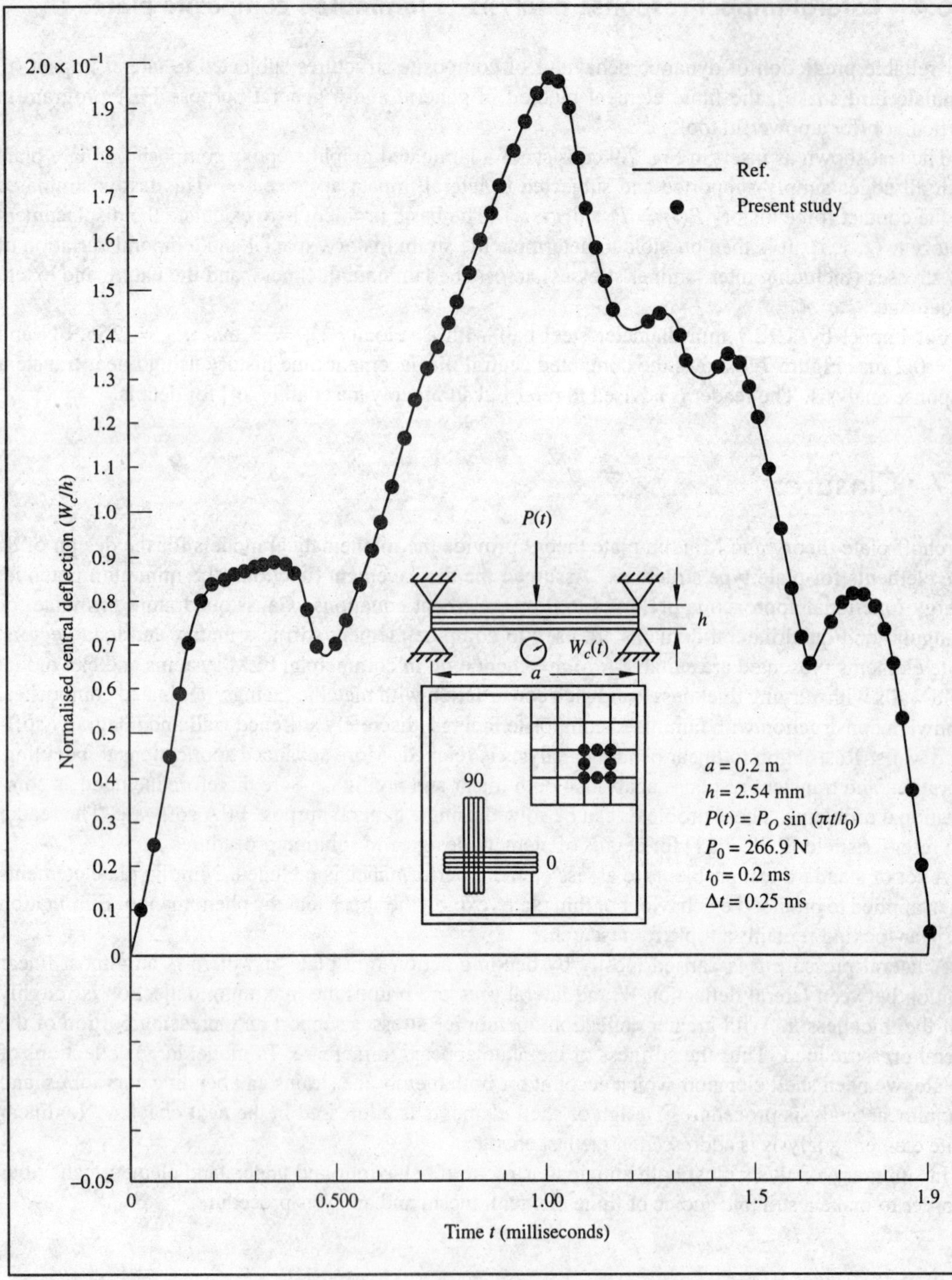

Fig. 7.9 *Lateral impact response of laminar composite plates*

7.8 References

[7.1] Timoshenko, S. and Woinowsky-Krieger, S. *Theory of plates and shells*. McGraw Hill, New York, 2nd ed., 1959.

[7.2] Mindlin, R.D. Influence of rotatory inertia and shear on flexural motions of isotropic elastic plates. *J. Applied Mechanics*, 18:31–38, 1951.

[7.3] Reddy, J.N. A review of refined theories of laminated composite plates. *Shock and Vibration Digest*, 22:3–17, 1990.

[7.4] Reddy, J.N. A general non-linear third order theory of plates with moderate thicknesses. *J. Nonlinear Mechanics*, 25:677–686, 1990.

[7.5] Hrabok, M.M and Hrudey, T.M. A review and catalog of plate bending finite elements. *Computers and Structures*, 19:4979–495, 1984.

[7.6] Morley, L.S.D. *Skew plates and structures*. Pergamon Press, Oxford, 1963.

[7.7] Hughes, T.J.R. and Tezduyer, T.E. Finite elements based upon mindlin plate theory, with particular reference to the four-node bilinear isoparametric element. *J. appl. Mech.*, 48:587–596, 1981.

[7.8] Pugh, E.D.L., Hinton, E., and Zienkiewiez, O.C. A study of quadrilateral plate bending elements with reduced integration. *Int. J. Numer. Meths. Engrg.*, 12:1059–1079, 1978.

[7.9] Bathe, K.J. and Dvorkin, E.N. A four-node plate bending element based on mindlin/reissner plate theory and a mixed interpolation. *Int. J. Numer. Meths. Engrg.*, 21:367–383, 1986.

[7.10] Hinton, E. and Huang, H.C. A family of quadrilateral mindlin plate elements with substitute shear strain fields. *Computers and Structures*, 23:409–431, 1986.

[7.11] Rossow, M.P. Efficient c° finite element solutions of simply—supported plates of polygonal shapes. *ASME Journal of Applied Mechanics*, 44:347–349, 1977.

[7.12] Cook, R.D. *Finite Element Modeling For Stress Analysis*. John Wiley, New York, 1995.

[7.13] Lakshminarayana, H.V., Boukhili, R., and Gauvin, R.. Finite element simulation of impact tests of laminated composite plates. *Composite Structures*, 28:47–59, 1994.

[7.14] Lakshminarayana, H.V., Boukhili, R., and Gauvin, R.. Impact response analysis of laminated composite plates, using finite element method: A parametric study. *Journal of the institution of Engineers, India*, 75:87–91, 1994.

[7.15] Batoz, J.L. An explicit formulation for an efficient triangular plate bending element. *Int. J. Numer Methods in Engng*, 18:1077–1089, 1982.

[7.16] Irons, B. and Ahmad, S. *Techniques of Finite Elements*. Ellis Horwood, Chichester, 1980.

[7.17] Reddy, J.N. *An Introduction to the Finite Element Method*. McGraw Hill, New York, 2nd ed., 1993.

[7.18] Cowper, G.R. Gaussian quadrature formulas for triangles. *Int. J. Numer Methods in Engng*, 7:405–408, 1973.

[7.19] Cowper, G.R., Kosko, E., Lindberg, G.M., and Olson, M.D. A high precision triangular plate bending element. Technical report, Aeronautical Report LR–514, National Research Council of Canada, Ottawa, 1968.

[7.20] Crisfield, M.A. Finite elements and solution procedures for structural analysis. *ESSENTIALS*, 1, 1986.

[7.21] Crisfield, M.A. *Non linear Finite Element Analysis of Solids and Structures*. John Wiley & Sons, Chichester, U.K., 1991.

7.9 Computational Problems

When specific instructions are not stated, report significant values of stress resultants, stresses, or displacements, as appropriate. Exploit symmetry if permitted. Choose convenient numbers and consistent units for material properties, dimensions, and loads. Unless directed otherwise, assume that thicknesses are uniform and the material model is homogeneous, isotropic.

7.1 Perform patch-test on the plate elements in the FEA software you use. You may wish to adapt the Fig. 7.5 to cases of pure twist and pure bending.

7.2 Do plate elements in the FEA software you use take transverse shear deformation into account? Find out by computational experiments of a cantilever plate. Plan carefully the support conditions, loading, mesh, material properties, and how the numerical results will be interpreted in order to answer the question.

7.3 For a square plate of thickness h and side length a, compute the center deflections, center bending moments, and the Von-Mises equivalent stress distribution at the top and bottom surfaces. Consider the following cases:

(a) Simply-supported edges, uniformly distributed load.
(b) Simply-supported edges; central concentrated force.
(c) Clamped edges; uniformly distributed load.
(d) Clamped edges; concentrated center force.

Report converged results for $(h/a) = 0.001, 0.01$, and 0.1.

7.4 Problem 7.3 can be repeated, using thermal load instead of mechanical load. A simple thermal load is $\Delta T = \left(\frac{2Z}{h}\right) T_0$, where T_0 is the surface temperature relative to the initial stress free temperature.

7.5 Two circular plates, one solid and another annular, are shown in cross section with some possible cases of axisymmetric mechanical load. For thermal load case, the temperature field is $\Delta T = (2Z/t)\, T_0$, where T_0 is the surface temperature relative to the initial stress free temperature. The outer edge $r = a$ may be simply-supported or clamped. Report converged results for $(t/a) = 0.001, 0.01$, and 0.1.

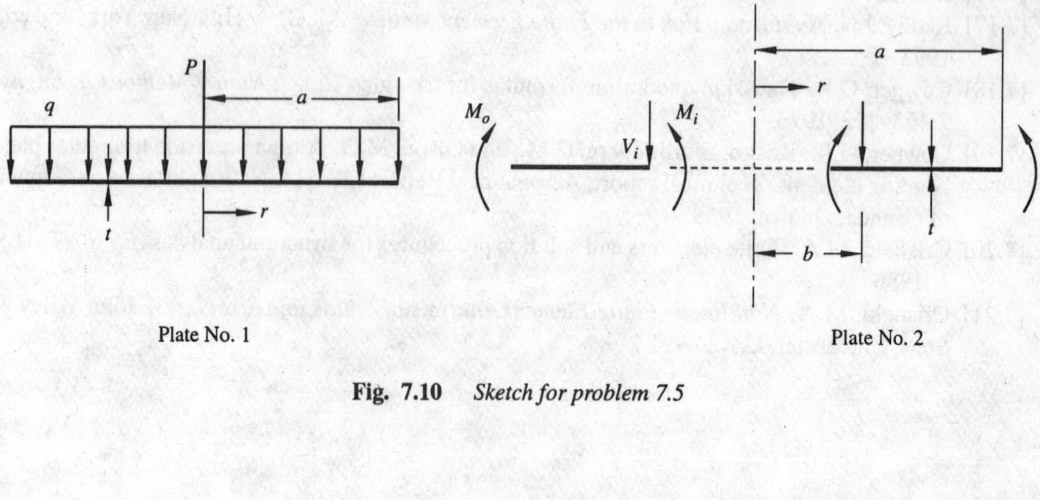

Fig. 7.10 *Sketch for problem 7.5*

7.6 Investigate the natural frequencies and modes of lateral vibration of a flat square plate (side length a and thickness h) with a central circular hole of diameter $0.2a$.

(a) Let the plate be unsupported.
(b) Let the plate be simply supported on all outer edges.
(c) Let the plate be clamped on all outer edges.
(d) Let the plate be clamped on one outer edge.

Report converged solutions for $(h/a) = 0.001, 0.01, \& \ 0.1$.

7.7 Harmonic response analysis can be applied to Problem 7.3. Great variety is afforded by options in the number of forcing functions, their directions and points of application, their frequency, and their phase angles. In each case report converged results for $(h/a) = 0.001, 0.01, 0.1$.

7.8 Dynamic response analysis can be applied to problem 7.3. Great variety is afforded by options in the number of forcing functions, their directions and points of application, and their manner of time variation. Some possible load time histories are shown in the sketch. In each case report converged results for $(h/a) = 0.001, 0.01, 0.1$.

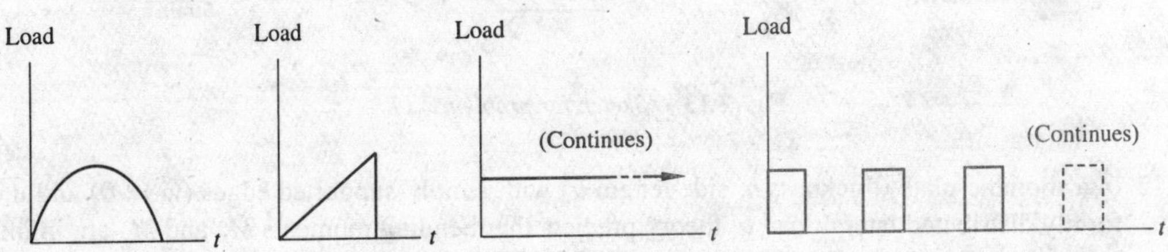

Fig. 7.11 *Sketch for problem 7.8*

7.9 Consider a square plate (side length a and thickness h) loaded by a force P at its center. The plate rests on elastic foundation that can push against the plate but cannot pull on it. Determine the extent of contact and the intensity of the contact pressure between the plate and the foundation as a function of load P. Report converged results for $(h/a) = 0.001, 0.01$, and 0.1.

7.10 The sketch shows a large circular cutout in a flat plate under compression. The ligament dimension h may be very small in comparison with width H. Investigate the magnitude of load required to produce buckling or collapse. Consider the supports on the plate sides just adequate to prevent rigid body motion. Report results for $t/H = 0.001, 0.01, 0.1$, where t denotes plate thickness.

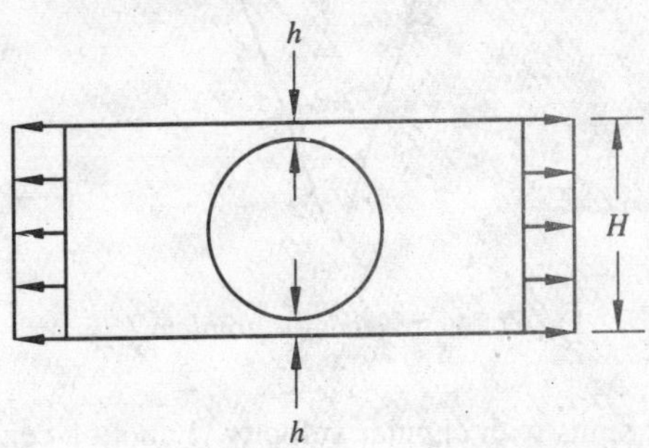

Fig. 7.12 *Sketch for problem 7.10*

7.11 The curved cantilever plate is subjected to an out-of-plane load F. Use a stress strain relation line in the sketch below. Apply a load sufficient to produce yielding. Determine the extent of yielding and the residual stresses upon unloading.

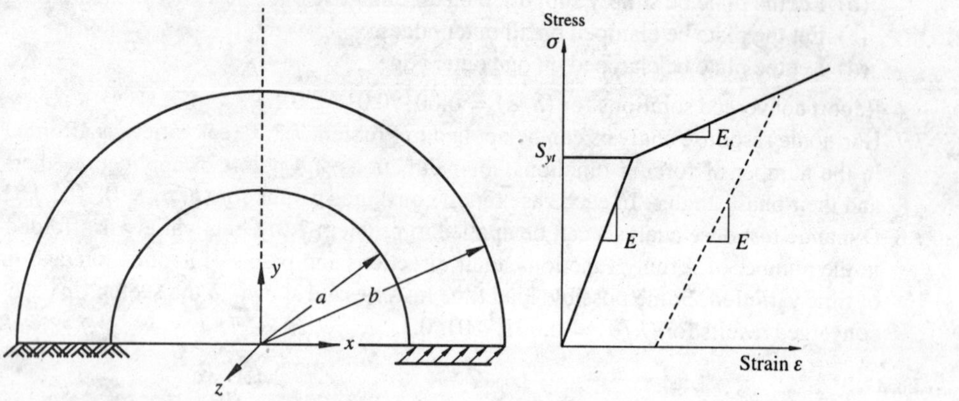

Fig. 7.13 *Sketch for problem 7.11*

7.12 In a rhombic plate (thickness h, side length a) with simply-supported edges ($w = 0$) and uniformly distributed lateral load q, theory predicts that bending moments M_x and M_y are infinite but of opposite sign at an obtuse corner such as the one at $x = y = 0$ [7.11]. Considerable mesh refinement may be needed to approximate this behaviour. Report converged and accurate solutions for $h/a = 0.001, 0.01, 0.1$ and for $\alpha = 30°, 45°$, and $60°$.

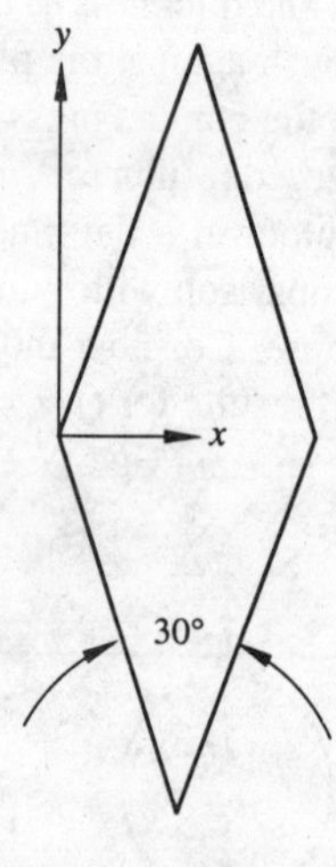

Fig. 7.14 *Sketch for problem 7.12*

7.13 A thin, flat circular disk spins with angular velocity Ω about a central axes normal to the plane of the disk. Investigate the effect of Ω on the fundamental frequency of vibration. Determine the temperature T_0 that causes buckling if the temperature T varies linearly with the radial coordinate, from zero at the centre to T_0 at the outside edge.

7.14 Rectangular plates are welded together to form the T shaped body shown. It rotates at angular velocity Ω about an axis normal to the plane of the paper at A. Investigate the effect of Ω, stress stiffening and spin softening on the natural frequencies of vibration. Clearly specify the support conditions you decide to use at A.

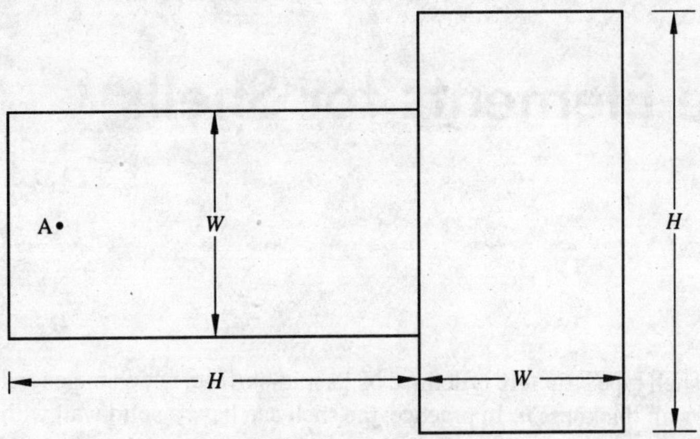

Fig. 7.15 *Sketch for problem 7.14*

7.15 The cantilever plate shown is initially at rest. The sketch includes a formula for the period of vibration T_n of mode n, in which $b_1 = 3.52$, $b_2 = 22.0$, and $b_n = (2n-1)^2\pi^2/4$ for $n > 2$. Starting at time $t = 0$, load P increases linearly to P_0 and then drops suddenly to zero. Construct a finite element model of four-noded quadrilateral plate elements and compute the deflection versus time plot for the right-hand end of the plate. Consider: (a) $t_0 = T_1/4$. (b) $t_0 = T_2/4$. (c) $t_0 = T_3/4$. for $(L/h) = 10, 100$, and 1000, and $(W/h) = 10$, where W is the width of the plate.

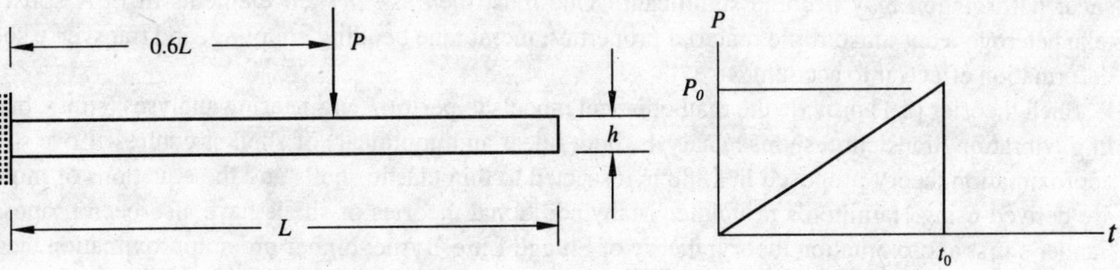

Fig. 7.16 *Sketch for problem 7.15*

8 Finite Elements for Shells

8.1 Overview

The geometry of a shell type structure is defined by its mid-surface, which in general is a curved surface in space, and has a wall thickness h. In practice, the shell can have a solid wall with varying thickness; sandwich construction with bonded metallic facings; laminated composite; sandwich construction with laminated composite facings; discretely stiffened; or integrally stiffened wall. A shell has two principal radii of curvature at every point: R_1 and R_2 ($R_1 > R_2$). A thin-walled shell ($h/R_2 \leq 0.01$) can be very strong if membrane action dominates. However, no shell structure is completely free on bending stresses. They appear at or near cutouts, cracks, point loads, line loads, reinforcements, junctions, changes of curvature, and supports. In short, any concentration of load or geometric discontinuity can be expected to induce bending stresses, often much larger than membrane stresses, but usually quite localized near the load or discontinuity.

Finite element models for engineering analysis of shell structures should account for both membrane strains and bending curvatures. If the shell wall thickness h is large ($h/R_2 \leq 0.1$), transverse shear deformation may become significant. One must then ask if shell elements in FEA software take heterogeneous anisotropic material properties, membrane bending coupling, and transverse shear deformation effects into account.

Shell theories [8.1] provide the mathematical models to perform engineering analyses (stress, buckling, vibration, transient response analysis, both linear and nonlinear) of shell structures. Love's first approximation theory proposed in 1888 is restricted to thin elastic shells and the equations of motion are derived using Hamilton's principle. Many additional theories of shells have also been proposed: Sander's first approximation theory; theory of Flugge-Lure-Byrne; higher order approximation theory incorporating transverse shear and normal stress effects; and nonlinear theory of thin elastic shells. For the derivation of the equations of motion based on each one of these theories, the reader can refer Kraus [8.1]. However, these equations are very complicated, so the numerical solutions for engineering purposes must usually be obtained by finite element analysis.

A library of finite elements for shells has been proposed [8.2, 8.3]: flat shell elements; isoparametric shell elements; and axisymmetric shell elements. Defect-free shell elements are perhaps the most difficult to design, and shell behaviour is often difficult to anticipate. Users may wish to test shell elements in commercial FEA software using a set of standard test problems for which results are known in advance. Elimination of membrane and shear-locking effects in isoparametric shell elements continues to be a research goal. The presentation identifies some topics for advanced study.

8.2 Flat Shell Elements: Facet Approximation

The earliest of finite element modeling approach—the facet approximation, is based on replacing the shell surface with plane triangular or quadrilateral elements.

8.2.1 Three-node triangular shell element

The most direct way to design the shell element shown in Fig. 8.1 is to superimpose a plane-stress element and a plate-bending element described in earlier chapters. The user defines the element geometry in a global coordinate system using $X_iY_iZ_i(i = 1,3)$ and $h_i(i = 1,3)$. Software typically generates matrices and vectors for an element in an element local Cartesian coordinate system $\bar{X}\bar{Y}\bar{Z}$ shown in Fig. 8.1. The $\bar{X}$-axis is defined by the coordinates of nodes 1 and 2. The element lies in the $\bar{X}\bar{Y}$ plane, which contains the nodes 1, 2 and 3. The $\bar{Z}$-axis is normal to the $\bar{X}\bar{Y}$ plane. The nodal d.o.f in the local coordinate system $\bar{X}\bar{Y}\bar{Z}$ and in the global coordinate system XYZ are also shown in Fig. 8.1. The software automatically transforms the element matrices and vectors to the global system for assembly of elements.

If the Allman triangle is chosen as the plane stress element, the shell normal rotation d.o.f $\theta_{\bar{z}i}$ at node i is present. If a Mindlin plate element is chosen as the plate-bending element, then transverse shear deformation effect is included. The assembled stiffness matrix will not be singular even if all elements that share node i happen to be flat and coplanar at node i. There is no need to include a stabilization matrix and one expects that commercial FEA software will do so.

The membrane strains $\boldsymbol{\varepsilon}_m$, bending curvature χ_b and transverse shear strains γ_s, membrane stress resultants **N**, bending stress resultants **M** and transverse shear stress resultants **Q** are all recovered in the element local coordinate system $\bar{X}\bar{Y}\bar{Z}$ only. The software outputs the stresses $\sigma_{\bar{x}\bar{x}}$, $\sigma_{\bar{y}\bar{y}}$, $\tau_{\bar{x}\bar{y}}$ at the top, bottom and middle surfaces for each element.

8.2.2 Four-node quadrilateral shell element

The element shown in Fig. 8.2 can be produced by combining quadrilateral plane-stress and plate-bending elements designed in earlier chapters. The user defines the element geometry in the global coordinate system using $X_iY_iZ_i(i = 1,4)$ and the nodal thicknesses $h_i(i = 1,4)$. The software typically generates matrices and vectors in an element local coordinate system $\bar{X}\bar{Y}\bar{Z}$ as shown in Fig. 8.2. The $\bar{X}$-axis is defined by the coordinates of nodes 1 and 2. The element lies in the $\bar{X}\bar{Y}$ plane, which contains the nodes 1, 2 and 4. The $\bar{Z}$-axis is normal to the $\bar{X}\bar{Y}$ plane. The nodal d.o.f in the two systems is also shown in Fig. 8.2. The software automatically transforms the element matrices and vectors to the global system for assembly of elements.

If the MacNeal quadrilateral is chosen as the plane stress element, the shell normal rotation d.o.f $\theta_{\bar{z}i}$ at node i is nonzero. If a Mindlin plate element is chosen as the plate bending element, then transverse shear deformation effects are included. The assembled stiffness matrix will not be singular even if all elements that share node i happen to be flat and coplanar at node i. There is no need to include stabilization matrix and one expects that commercial FEA software will do so.

However, a four-node quadrilateral is in general a warped element because its nodes are not all coplanar. A modest amount of warping can seriously degrade the performance of an element. Commercial FEA software may allow only a very small amount of warping. Much greater warping is

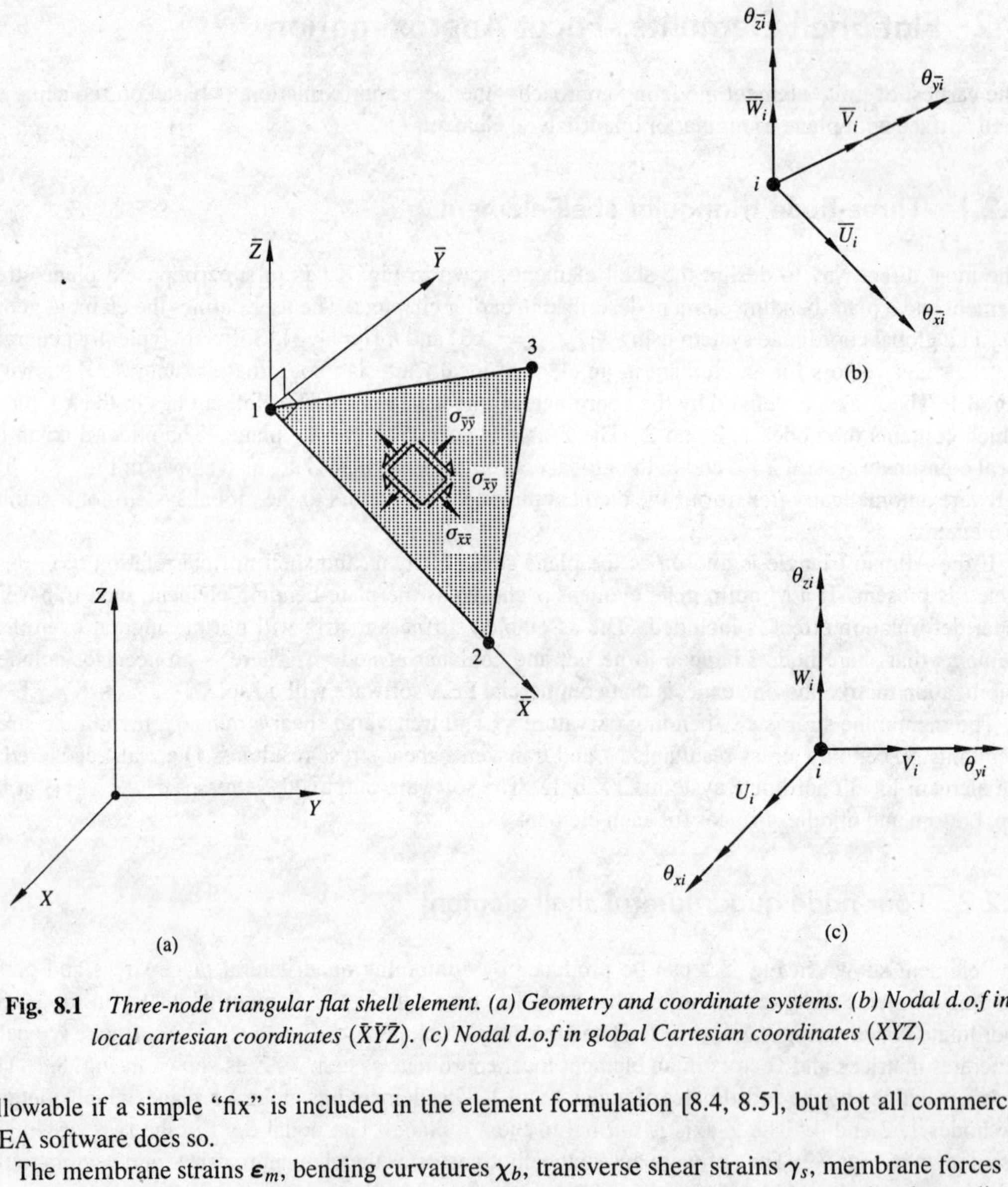

Fig. 8.1 *Three-node triangular flat shell element. (a) Geometry and coordinate systems. (b) Nodal d.o.f in local cartesian coordinates ($\bar{X}\bar{Y}\bar{Z}$). (c) Nodal d.o.f in global Cartesian coordinates (XYZ)*

allowable if a simple "fix" is included in the element formulation [8.4, 8.5], but not all commercial FEA software does so.

The membrane strains $\boldsymbol{\varepsilon}_m$, bending curvatures χ_b, transverse shear strains γ_s, membrane forces **N**, bending moments **M** and transverse shear forces **Q** are all recovered in the element local coordinate system $\bar{X}\bar{Y}\bar{Z}$ only. The software computes the stress components $\sigma_{\bar{x}\bar{x}}$, $\sigma_{\bar{y}\bar{y}}$, $\tau_{\bar{x}\bar{y}}$ at the top, bottom, and middle surfaces for each element.

Advantages of a flat shell element include simplicity of element formulation, simplicity in the description of element geometry, and the element's ability to represent rigid body motion without strain.

Disadvantages of a flat shell element include the approximation of a smoothly curved shell surface by flat or slightly warped facets, introducing fold lines where the elements meet. There is discretization error associated with the lack of coupling between membrane and bending actions within individual elements. Discretization error can of course be reduced by using smaller elements.

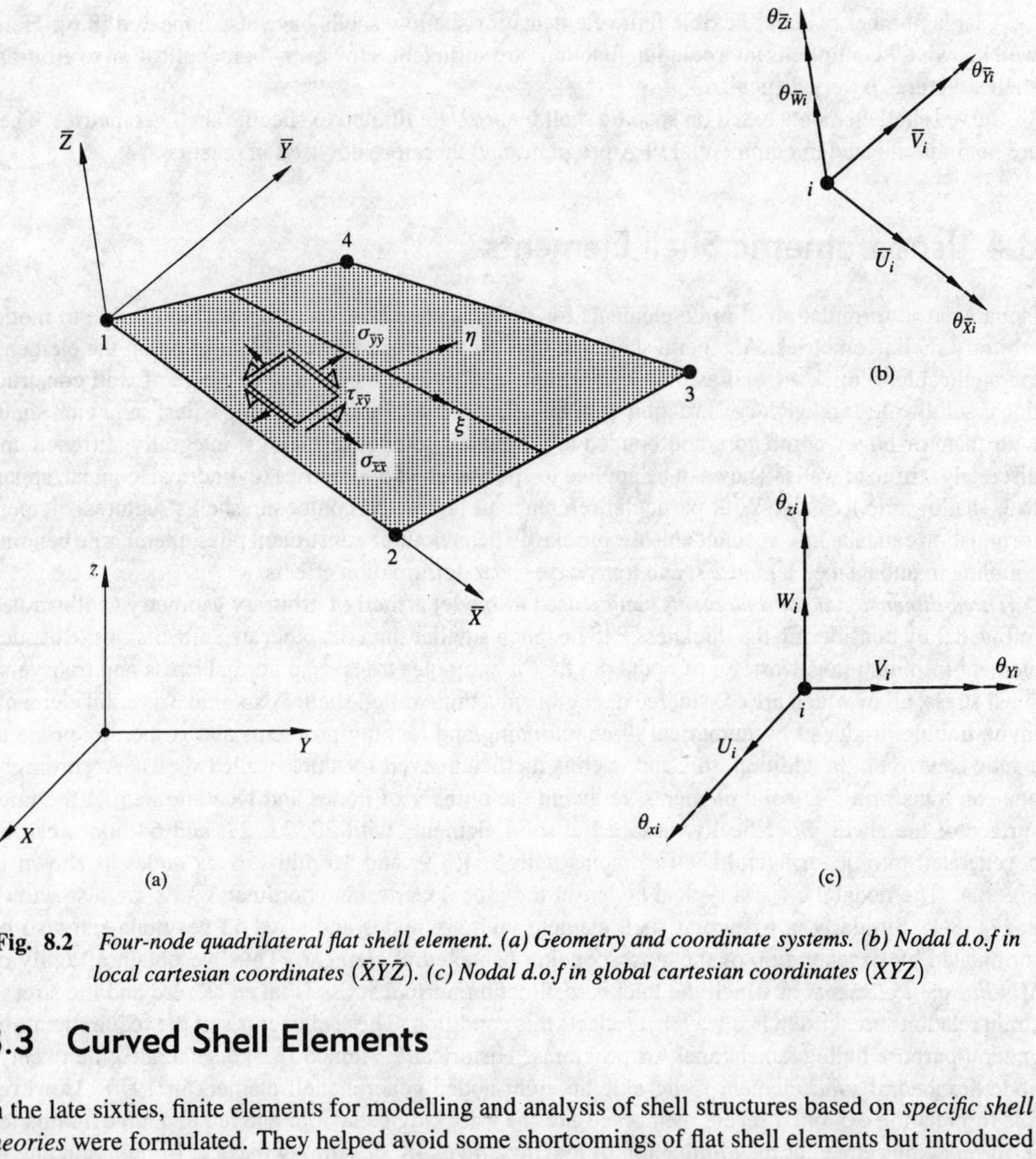

Fig. 8.2 *Four-node quadrilateral flat shell element. (a) Geometry and coordinate systems. (b) Nodal d.o.f in local cartesian coordinates* $(\bar{X}\bar{Y}\bar{Z})$. *(c) Nodal d.o.f in global cartesian coordinates* (XYZ)

8.3 Curved Shell Elements

In the late sixties, finite elements for modelling and analysis of shell structures based on *specific shell theories* were formulated. They helped avoid some shortcomings of flat shell elements but introduced other difficulties. Membrane and bending actions are coupled within the element. More user input data are needed to specify the element geometry. Element formulation is complicated, as it involves a shell theory, of which there are many.

The design of triangular and quadrilateral curved shell elements based on specific shell theories has proved to be a challenging task [8.2]. For example, a three-noded triangular shallow shell element based on Kirchoff–Love theory needs C^1 continuous interpolation functions and this means that higher-order polynomials and nodal degrees of freedom up to second-order derivatives of the displacement W normal to the mid-surface have to be retained. These are usually based on curvilinear coordinates and their application to arbitrary shaped shell structures becomes very difficult. So, majority of available formulations are limited to *specific shell geometries* (cylindrical, conical, spherical, shallow, etc).

A large number of shear flexible finite elements for shallow shells have also appeared [8.6]. Here, well known C^0 continuous interpolation functions are sufficient. However, their application to arbitrary shell structures is very difficult.

Curved shell elements based on specific shell theories are limited to specific shell geometries. They are not implemented in commercial FEA programs and therefore not used in practice.

8.4 Isoparametric Shell Elements

Isoparametric formulation of finite elements for shells is attractive, effective and applicable to model arbitrary shell geometries. A general shell theory is employed in the formulation so that the elements are applicable to thick as well as thin-walled shells. They can model a wide range of wall constructions: solid wall; sandwich construction with metallic facings; laminated composites; sandwich shells with foam or honey-comb core and bonded to laminated composite facings; integrally stiffened and discretely stiffened walls. They can be applied to specific shell geometries (cylindrical, conical, spherical, shallow, etc.) easily. With particular reference to laminated composite shell structures, element formulation can take into account anisotropic elastic behaviour of constituent plies, membrane bending coupling in unbalanced laminates, and transverse shear deformation effects.

Three-dimensional solid elements can be used to model a shell of arbitrary geometry as illustrated in Fig. 8.3 by considering the thickness h to be much smaller than the other two dimensions. But such element formulation is wasteful of nodal d.o.f, as it computes transverse normal stress and transverse shear stress, all of which are considered negligible in a thin-walled shell. Also, thin 3-D solid elements invite trouble produced by numerical ill-conditioning and locking problems and is thereby prone to erratic behaviour. In addition, this approach is inefficient even for thick-walled shells. Accordingly, one can transform 3-D solid elements, reducing the number of nodes and locating them at the mid-surface of the shell. Specifically, hexahedral solid elements with 20, 32, 27, and 64 nodes can be degenerated into quadrilateral shell elements with 8, 12, 9, and 16 mid-surface nodes as shown in Fig. 8.4. The nodal d.o.f at a typical node i in the global Cartesian coordinates XYZ are also shown in Fig. 8.4. Similarly, a triangular shell element with six nodes and six d.o.f per node can also be formulated by degeneration of the corresponding pentahedral element. Thus we obtain a family of *Mindlin shell elements* in which the thickness direction normal stress is taken as zero and the stress–strain relations are written in a way that reflects this condition. These elements suit the requirements of general-purpose finite element analysis programs. Historically, Ahmad [8.7] degenerated the twenty-node hexahedral solid element to develop an eight-noded general shell element in 1970. However this formulation exhibited severe membrane and shear-locking behaviour and led to a large number of re-formulations aimed at the elimination of locking effects [8.3]. Among these, a formulation due to Lakshminarayana [8.8] is presented in the next section.

8.4.1 Eight-node quadrilateral shell element

Geometry and coordinate systems

A curved shell element of quadrilateral shape is shown in Fig. 8.5(a). The outer and inner surfaces of the element are curved, whereas the sections across the thickness are generated by straight lines. The geometric description of the element requires the specification of two vectors at each of the eight

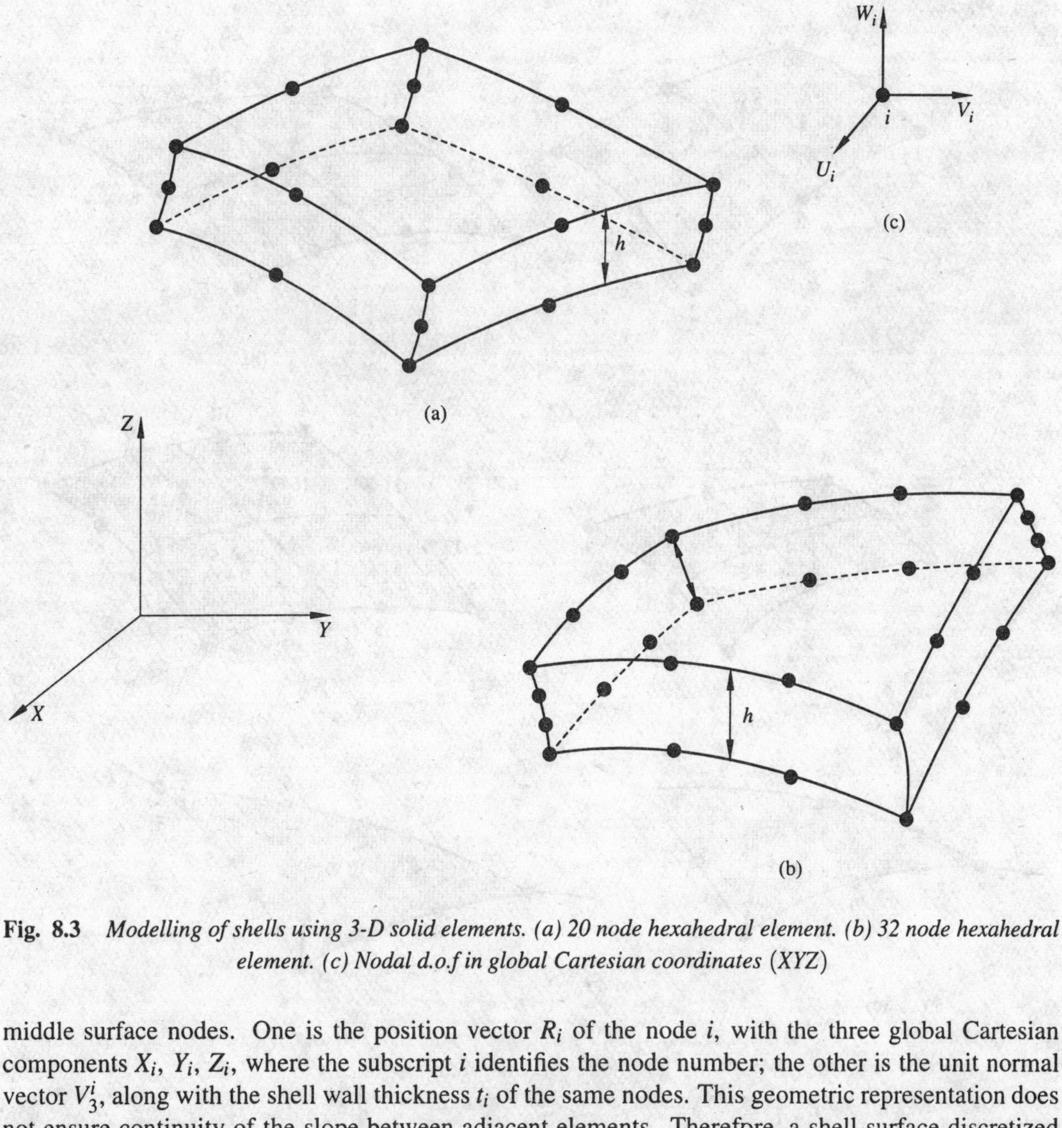

Fig. 8.3 *Modelling of shells using 3-D solid elements. (a) 20 node hexahedral element. (b) 32 node hexahedral element. (c) Nodal d.o.f in global Cartesian coordinates (XYZ)*

middle surface nodes. One is the position vector R_i of the node i, with the three global Cartesian components X_i, Y_i, Z_i, where the subscript i identifies the node number; the other is the unit normal vector V_3^i, along with the shell wall thickness t_i of the same nodes. This geometric representation does not ensure continuity of the slope between adjacent elements. Therefore, a shell surface discretized with these elements cannot be exact. However, the accuracy of the approximation increases as the number of elements in the mesh is increased. The element formulated for the analysis of shells with continuous as well as discontinuous slopes and curvatures carries six engineering degrees of freedom (three translations and three rotations) at each of its eight mid-surface nodes. The nodal d.o.f in the global Cartesian coordinates *XYZ* are illustrated in Fig. 8.5(b). Stiffness for rotation about the normal to the element middle surface is zero at all the nodes. Therefore, a small fictitious rotational spring is introduced at each node corresponding to this d.o.f.

Referring to Fig. 8.5(b), let ξ, η be the two curvilinear coordinates in the element's middle surface and τ a linear coordinate in the thickness direction. If we assume that ξ, η, τ vary between -1 and $+1$ on the respective faces of the element, we can write a relationship between the global Cartesian

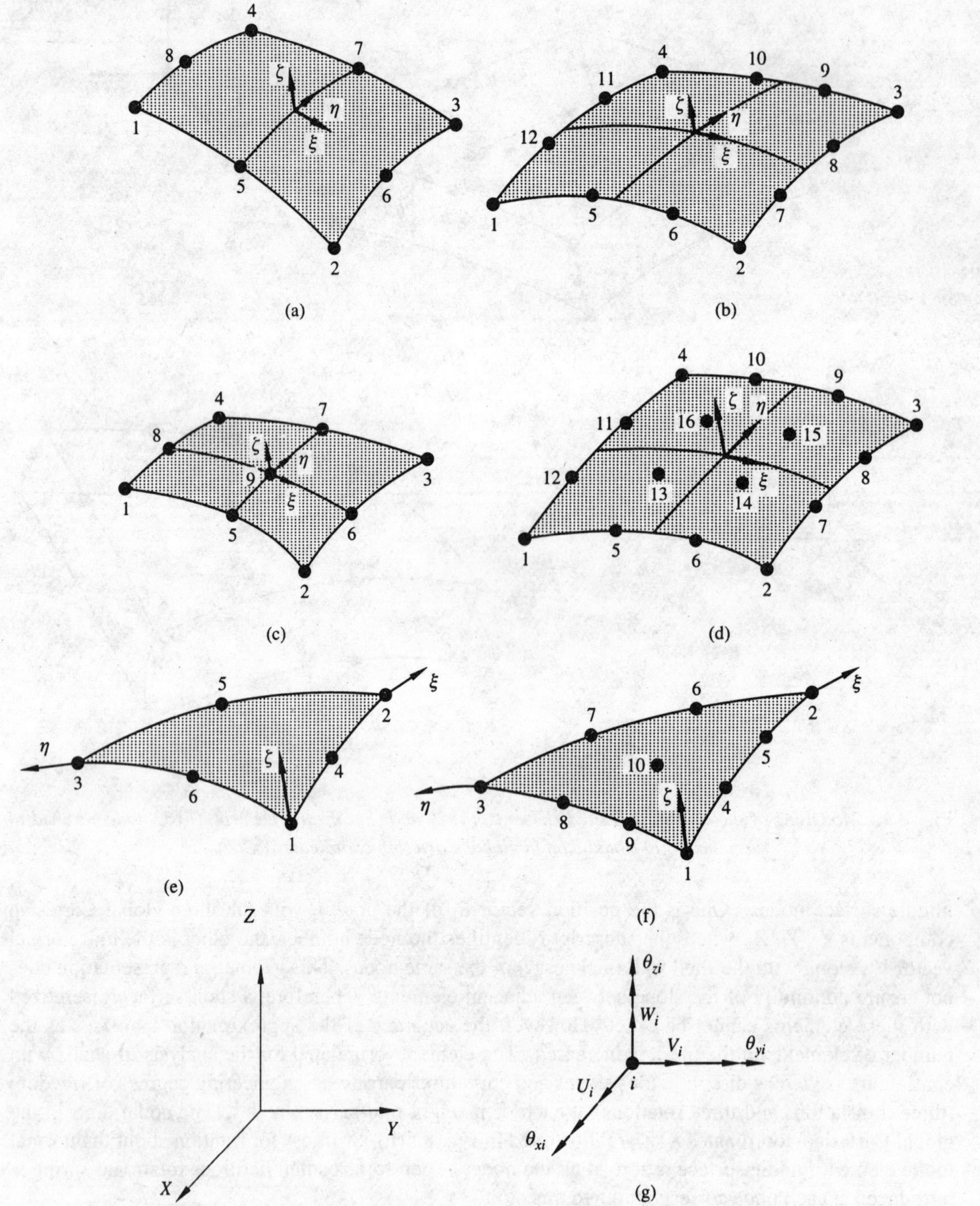

Fig. 8.4 *Isoparametric shell elements (a) QUAD8 (b) QUAD12 (c) QUAD16 (e) TRIA6 (f) TRIA10. (g) Nodal d.o.f in global Cartesian coordinates (XYZ)*

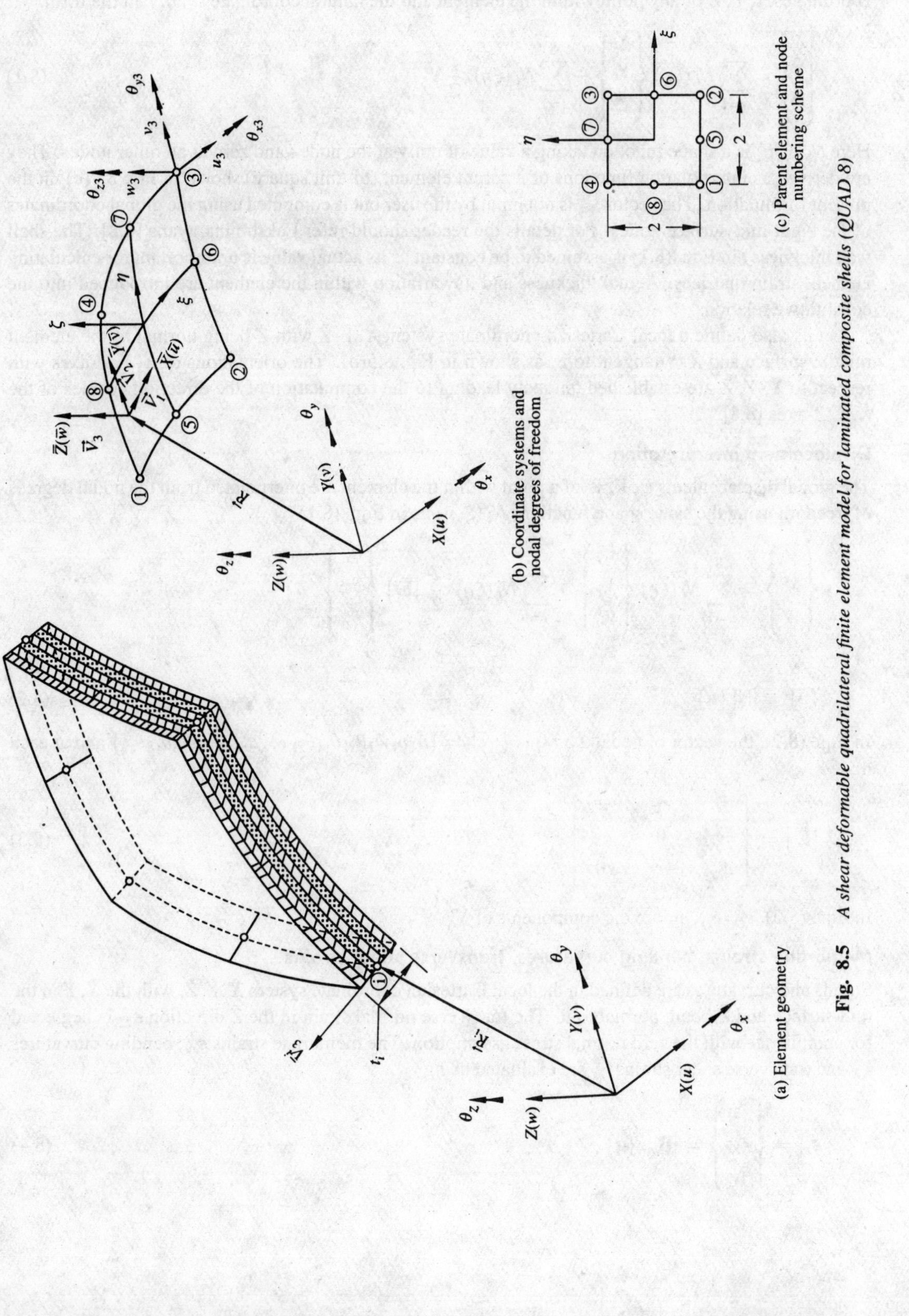

Fig. 8.5 *A shear deformable quadrilateral finite element model for laminated composite shells (QUAD 8)*

coordinates X, Y, Z of any point within the element and the natural coordinates ξ, η, τ in the form.

$$\begin{bmatrix} X \\ Y \\ Z \end{bmatrix} = \sum_{i=1}^{8} N_i(\xi\eta) \begin{Bmatrix} X_i \\ Y_i \\ Z_i \end{Bmatrix} + \sum_{i=1}^{8} N_i(\xi\eta)\tau \frac{t}{2} \mathbf{V}_3^i \tag{8.1}$$

Here $N_i(\xi,\eta)$ is a shape function taking a value of unity at the node i and zero at all other nodes. They are derived as interpolation functions of a parent element (bi-unit square) shown in Fig. 8.7(c) in the present formulation. The vector $\mathbf{V}_3^i$ is not input by the user but is computed using the global coordinates of the eight mid-surface nodes. For details the reader should refer Lakshminarayana [8.8]. The shell wall thickness t in Eqn. (8.1) is assumed to be constant as its actual value is unimportant for calculating element strain matrices. Actual thickness and its variation within the element are introduced into the constitutive relations.

Let us also define a local Cartesian coordinate system $\bar{X}, \bar{Y}, \bar{Z}$ with $\bar{Z}$ being normal to the element middle surface and $\bar{X}, \bar{Y}$ tangent to it, as shown in Fig. 8.5(b). The orientations of $\bar{X}, \bar{Y}, \bar{Z}$-axes with respect to X, Y, Z are established uniquely leading to the computation of the direction cosines of the $\bar{X}, \bar{Y}, \bar{Z}$-axes [8.8].

Displacement interpretation

The global displacements U, V, W of a point within the element are interpolated from the nodal degrees of freedom using the same shape functions $N_i(\xi,\eta)$ as in Eqn. (8.1)

$$\begin{Bmatrix} U \\ V \\ W \end{Bmatrix} = \sum_{i=1}^{8} N_i(\xi\eta) \begin{Bmatrix} U_i \\ V_i \\ W_i \end{Bmatrix} + \sum_{i=1}^{8} N_i(\xi\eta)\tau \frac{t}{2} [\bar{\mathbf{E}}^i] \begin{Bmatrix} \theta_{xi} \\ \theta_{yi} \\ \theta_{zi} \end{Bmatrix}$$

or

$$\{\mathbf{u}\} = [\mathbf{N}]\{\mathbf{q}\}. \tag{8.2}$$

In Eqn. (8.2), the vector of nodal d.o.f $\mathbf{q}$ is $\{\mathbf{q}\}^T = (u_1 v_1 w_1 \theta_{x1} \theta_{y1} \theta_{z1} \ldots u_8 v_8 w_8 \theta_{x8} \theta_{y8} \theta_{z8})$ and at each node i

$$[\bar{\mathbf{E}}^i] = \begin{bmatrix} 0 & e_{\bar{z}z} & -e_{\bar{z}y} \\ -e_{\bar{z}z} & 0 & e_{\bar{z}x} \\ e_{\bar{z}y} & -e_{\bar{z}x} & 0 \end{bmatrix}. \tag{8.3}$$

In Eqn. (8.3), $e_{\bar{z}x}, e_{\bar{z}y}, e_{\bar{z}z}$ are the components of $\bar{\mathbf{V}}_3^i$.

Membrane strains, bending curvatures, transverse shear strains

Strains and curvatures are defined in the local Cartesian coordinate system $\bar{X}, \bar{Y}, \bar{Z}$, with the $\bar{X}, \bar{Y}$ in the mid-surface and $\bar{Z}$ being normal to it. The transverse normal strain in the $\bar{Z}$ direction $\boldsymbol{\varepsilon}_{\bar{z}\bar{z}}$ is neglected for compliance with the zero normal stress assumption. The membrane strains $\boldsymbol{\varepsilon}_m$, bending curvatures χ_b and transverse shear strains γ_s are evaluated using

$$\boldsymbol{\varepsilon}_m = \begin{Bmatrix} \varepsilon_{\bar{x}\bar{x}} \\ \varepsilon_{\bar{y}\bar{y}} \\ \gamma_{\bar{x}\bar{y}} \end{Bmatrix} = [\mathbf{B}_m]\{\mathbf{q}\}, \tag{8.4}$$

$$\chi_b = \begin{Bmatrix} \chi_{\bar{x}\bar{x}} \\ \chi_{\bar{y}\bar{y}} \\ \chi_{\bar{x}\bar{y}} \end{Bmatrix} = [\mathbf{B}_b]\{\mathbf{q}\}, \tag{8.5}$$

$$\gamma_s = \begin{Bmatrix} \gamma_{\bar{x}\bar{z}} \\ \gamma_{\bar{y}\bar{z}} \end{Bmatrix} = [\mathbf{B}_s]\{\mathbf{q}\}. \tag{8.6}$$

The formulae to compute the matrices $\mathbf{B}_m$, $\mathbf{B}_b$, and $\mathbf{B}_s$ are derived in Ref. [8.8].

Stress resultants and constitutive relations

The stress resultants corresponding to the strains and curvatures are defined and related by constitutive relations in the usual manner. The general form of the constitutive relation used in the present formation is

$$\begin{Bmatrix} \mathbf{N} \\ \mathbf{M} \\ \mathbf{Q} \end{Bmatrix} = \begin{bmatrix} [\mathrm{A}] & [\mathrm{B}] & [0] \\ [\mathrm{B}] & [\mathrm{D}] & [0] \\ [0] & [0] & [\mathrm{E}] \end{bmatrix} \begin{Bmatrix} \boldsymbol{\varepsilon}_m \\ \chi_b \\ \gamma_s \end{Bmatrix}, \tag{8.7}$$

where $\{\mathbf{N}\}^T = (N_{\bar{x}\bar{x}}, N_{\bar{y}\bar{y}}, N_{\bar{x}\bar{y}})$ are the membrane forces per unit length;
$\{\mathbf{M}\}^T = (M_{\bar{x}\bar{x}}, M_{\bar{y}\bar{y}}, M_{\bar{x}\bar{y}})$ are the bending and twisting moments per unit length; and
$\{\mathbf{Q}\}^T = (Q_{\bar{x}\bar{z}}, Q_{\bar{y}\bar{z}})$ are the transverse shear forces per unit length.

We note that the element stresses have been defined in the local Cartesian coordinate system $\bar{X}, \bar{Y}, \bar{Z}$. Therefore it is also necessary to specify the elasticity matrices **A**, **B**, **D** and **E** in the same coordinate system. $A_{ij}, B_{ij}, D_{ij} (i,j = 1,2,3)$ and $E_{ij} (i,j = 1,2)$ are the so-called membranes, bending-membrane coupling, bending, and transverse-shear stiffness coefficients. The general form of Eqn. (8.7) is included in this element formulation to enable the analyst to model laminated anisotropic materials and sandwich wall construction in addition to homogeneous isotropic materials and solid wall construction. Actual thickness t and its variation within the element are accounted for in the calculation of **A**, **B**, **D** and **E** matrices. It is also a standard practice to include shear correction factors in evaluating the **E** matrix. When thermal loads are considered, Eqn. (8.7) has to be modified to include thermal forces and thermal moments.

For a homogeneous isotropic material model and restricting to linear elastic behavior, A_{ij} is proportional to Et; $B_{ij} = 0$; D_{ij} is proportional to $Et^3/12(1-\nu^2)$, and E_{ij} is proportional to GtS; where E is the Young's modulus of elasticity, ν is Poisson's ratio, G is the shear modulus, S is the shear correction factor, and t is the thickness.

Stiffness matrix

The formula to compute the element stiffness matrix **K** is derived from the strain energy (SE) of the element

$$\mathrm{SE} = \frac{1}{2} \int_{A^{(e)}} (\boldsymbol{\varepsilon}_m^T \mathbf{A} \boldsymbol{\varepsilon}_m + \boldsymbol{\varepsilon}_m^T \mathbf{B} \chi_b + \chi_b^T \mathbf{B} \boldsymbol{\varepsilon}_m + \chi_b^T \mathbf{D} \chi_b + \gamma_S^T \mathbf{E} \gamma_S)\, d(\text{area}), \tag{8.8}$$

where the integration is taken over the element mid-surface area $A^{(e)}$. Using Eqns. (8.4) to (8.6), we get

$$\mathrm{SE} = \frac{1}{2}\{\mathbf{q}\}^T[\mathbf{K}]\{\mathbf{q}\}. \tag{8.9}$$

with

$$[\mathbf{K}] = \int_{A^{(e)}} \{[\mathbf{B}_m]^T[\mathbf{A}][\mathbf{B}_m] + [\mathbf{B}_m]^T[\mathbf{B}][\mathbf{B}_b] + [\mathbf{B}_b]^T[\mathbf{B}][\mathbf{B}_m] + [\mathbf{B}_b]^T[\mathbf{D}][\mathbf{B}_b] + [\mathbf{B}_s]^T[\mathbf{E}][\mathbf{B}_s]\}d(\text{area}). \tag{8.10}$$

When numerical integration technique using Gauss quadrature formula is applied to Eqn. (8.10), the equation for computation of element stiffness matrix **K** becomes

$$[\mathbf{K}] = \sum_g W_g[\Phi(\xi\eta)]_g, \tag{8.11}$$

where

$$[\Phi(\xi\eta)] = \{[\mathbf{B}_m]^T[\mathbf{A}][\mathbf{B}_m] + [\mathbf{B}_m]^T[\mathbf{B}][\mathbf{B}_b] + [\mathbf{B}_b]^T[\mathbf{B}][\mathbf{B}_m] + [\mathbf{B}_b]^T[\mathbf{D}][\mathbf{B}_b] + [\mathbf{B}_s]^T[\mathbf{E}][\mathbf{B}_s]\}_g \mathbf{J}_g^*. \tag{8.12}$$

In Eqn. (8.12), $\mathbf{J}_g^*$ appears because d (area) in Eqn. (8.10) was replaced by $\mathbf{J}^* d\xi d\eta$ so that the integration now can be performed over the parent element domain. It is also implied that the strain matrices $\mathbf{B}_m$, $\mathbf{B}_b$, and $\mathbf{B}_s$ have to be evaluated at each Gauss point g and the elasticity matrices **A**, **B**, **D** and **E** have to be specified with reference to the local Cartesian coordinates $\bar{X}\bar{Y}\bar{Z}$ again at each Gauss point. $\mathbf{J}^*$ is also evaluated at each Gauss point.

Three Gauss points are required in the ξ, η directions ($g = 9$) for the exact integration of parabolic shape functions used in the present element. The weighting factors W_g and Gauss point locations$(\xi\eta)_g$ are tabulated in Ref. [8.8].

Pressure load

The work equivalent nodal forces in the global coordinate directions X, Y, Z resulting from normal pressure P acting in the $\bar{Z}$ direction over the element's mid-surface is evaluated using:

$$\{\mathbf{f}_i^P\} = \sum_{g=1}^{9} W_g J_g^* P_g N_i(\xi_g, \eta_g) \begin{Bmatrix} e_{\bar{z}x} \\ e_{\bar{z}y} \\ e_{\bar{z}z} \end{Bmatrix}. \tag{8.13}$$

In Eqn. (8.13), the subscripts g and i refer to Gauss point and node number respectively. We note that nodal point moments are null. The normal pressure P may be uniform or may vary over the element mid-surface. In the latter case, the normal pressure at the Gauss point P_g is interpolated from corner node values specified by the user as

$$P_g = \sum_{i=1}^{4} h_i(\xi_g\eta_g)P_i, \tag{8.14}$$

where $h_i(\xi\eta)$ are the bi-linear interpolation functions

$$h_i(\xi\eta) = \frac{1}{4}(1 + \xi\xi_i)(1 + \eta\eta_i). \tag{8.15}$$

Stress recovery

Element stresses will be recovered at the Gauss points. Stress resultants will be computed in the local Cartesian coordinate system $\bar{X}, \bar{Y}, \bar{Z}$ in planes parallel to the mid-surface. The stress resultants are calculated using Eqn. (8.7). At each Gauss point, global coordinates, orientations of the local Cartesian coordinate axes, and shell wall thickness are also printed out along with the values of **N**, **M**, and **Q**. Using them one can compute shell surface (mid, top, bottom) stresses $\sigma_{\bar{x}\bar{x}}$, $\sigma_{\bar{y}\bar{y}}$, $\tau_{\bar{x}\bar{y}}$, $\tau_{\bar{x}\bar{z}}$, $\tau_{\bar{y}\bar{z}}$. We note that the transverse shear stresses $\tau_{\bar{x}\bar{z}}$ and $\tau_{\bar{y}\bar{z}}$ are in fact zero at the top and bottom surfaces. However, the values reported are the average values across the thickness. Their maximum values occur on the mid-surface and are equal to 1.5 times the average value. If nodal point stresses are to be reported, there is a need for a rational and consistent procedure to ensure that the stresses at the common nodes of adjoining elements are computed in the same direction before averaging.

Alleviation of membrane and shear-locking effects

The present formulation with the standard approach of using exact integration (3×3 mesh of Gauss points) designated as QUAD8 exhibits both shear-locking and membrane-locking phenomenon.

Membrane and bending actions are coupled within the element so that it is harder to avoid membrane locking, that is great over-stiffness in bending because details of element formulation cause spurious membrane strains to appear in association with bending curvatures, and membrane stiffness is far greater than bending stiffness if the shell is thin-walled.

A phenomenon known as shear-locking is also observed when the eight-nodes Mindlin shell element is applied to thin-walled shells. The interpolation functions used for **u** over the element domain should admit that for thin-walled shells, the transverse shear strains $\boldsymbol{\gamma}_s$ should be zero. However, if by virtue of the shape functions used on **u** the transverse shear strains $\boldsymbol{\gamma}_s$ may not be zero everywhere, then an erroneous shear strain energy, which can be much larger compared with the bending stain energy is included in the finite element analysis of shells. Hence, the resulting finite element model is much too stiff and predicts much smaller deflections than the exact solution based on Kirchoff-Love thin shell theory.

Use of reduced integration (2×2 network of Gauss points) in the valuation of Eqn. (8.9) generally results in improved performance, and the resulting shell element is designated QUAD8*. Numerical experiments with QUAD8* on a set of standard test problems proposed by MacNeal and Harder [8.9] revealed persistence of locking effects. Even worse it revealed spurious zero energy modes in addition to the rigid body modes.

The task now is to modify the QUAD8 element to alleviate the locking effects using the assumed strain method. The details of the approach are well documented in Ref. [8.8] and are not repeated here. In practice, programming and computer implementation of the assumed strain method involves replacing the original membrane strain matrix $\mathbf{B}_m$ and transverse shear strain matrix $\mathbf{B}_s$ by the modified matrices denoted as $\bar{\mathbf{B}}_m$ and $\bar{\mathbf{B}}_s$ everywhere. This will also enable us to employ exact integration (3×3 network of Gauss points) in the computation of the stiffness matrix, thereby avoiding spurious zero energy modes. The new element is designated QUAD8**.

The QUAD8** element was critically evaluated using a standard set of test problems [8.9]. The element is free of membrane and shear- locking problems, has no spurious zero energy modes and yields numerical results which are quite insensitive to mesh distortion. Good accuracy and fast convergence to theoretical solutions were observed for both displacements and stress resultants in a number of example problems.

A family of Mindlin shell elements illustrated in Fig. 8.4 can be designed following the approach for the QUAD8 element. Some of them are implemented in commercial FEM systems and widely used in practice.

8.5 Axisymmetric Shell Elements

Finite elements for modelling and analysis of shells of revolution shown in Fig. 8.6 are called axisymmetric shell elements. In cross-section, an element for a shell of revolution resembles a 2-D beam element. Like an element for a solid of revolution, an element for a shell of revolution has nodal circles rather than nodal points. Typically there are two, three and four nodal circles for linear quadratic and cubic order axisymmetric shell elements.

Specifically, the formulation of a linear element resembles the 2-D beam element, in that it uses a cubic polynomial for the lateral displacement $\bar{W}$, and linear polynomial for meridional displacement $\bar{u}$ as shown in Fig. 8.6(a). Each nodal circle has two translations (one radial, the other axial) and one rotation as d.o.f. This conical shell element has advantages and disadvantages like those of other flat shell elements.

The quadratic and cubic order elements have 3 and 4 nodal circles and 3 d.o.f at each node as shown in Fig. 8.6(b) and (c).

Like axisymmetric solids, axisymmetric shells under non-axisymmetric loading can be analyzed using Fourier series. Thus the load is represented as the sum of its Fourier components, an analysis is performed for each component, and results are superposed.

It is possible to include transverse shear deformation effects in the formulation of axisymmetric shell elements also. The meridional displacement $\bar{u}$, lateral displacement $\bar{W}$ and meridional rotation $\bar{\beta}$ are independent variables. They are interpolated over the element domain in terms of their values at each nodal circle using well-known C^0 continuous shape functions. The element strain energy now is due to membrane strains $\boldsymbol{\epsilon}_m^T = (\boldsymbol{\epsilon}_s, \boldsymbol{\epsilon}_\theta)$, bending curvatures $\chi_b = \chi_s$, and transverse shear strain γ_s. Accordingly it is possible to compute meridional stress σ_s and hoop stress $\sigma_{\theta\theta}$ at the top, mid, and bottom surfaces of each element.

8.6 Test Problems

MacNeal and Harder [8.9] have proposed a standard set of problems to test finite element accuracy. A description of each of these problems in terms of geometry, boundary conditions, applied loads, material properties, and finite element meshes is documented in [8.9]. Theoretical results for all problems are also tabulated and these have to be used to present test results in a normalized form, i.e., the reported value is the ratio of the value obtained to the theoretical value. The test results are also reduced to a letter grade, using the following rules:

Rule	Grade
2% > error > 0	A
10% > error > 2%	B
20% > error > 10%	C
50% > error > 20%	D
error > 50%	F

For designers of new finite elements, these test problems help to find and correct element errors and weaknesses. Almost every problem is capable of evoking results ranging from excellent to poor. Therefore a low grade in one or more problems should not disqualify the element under test for practical applications.

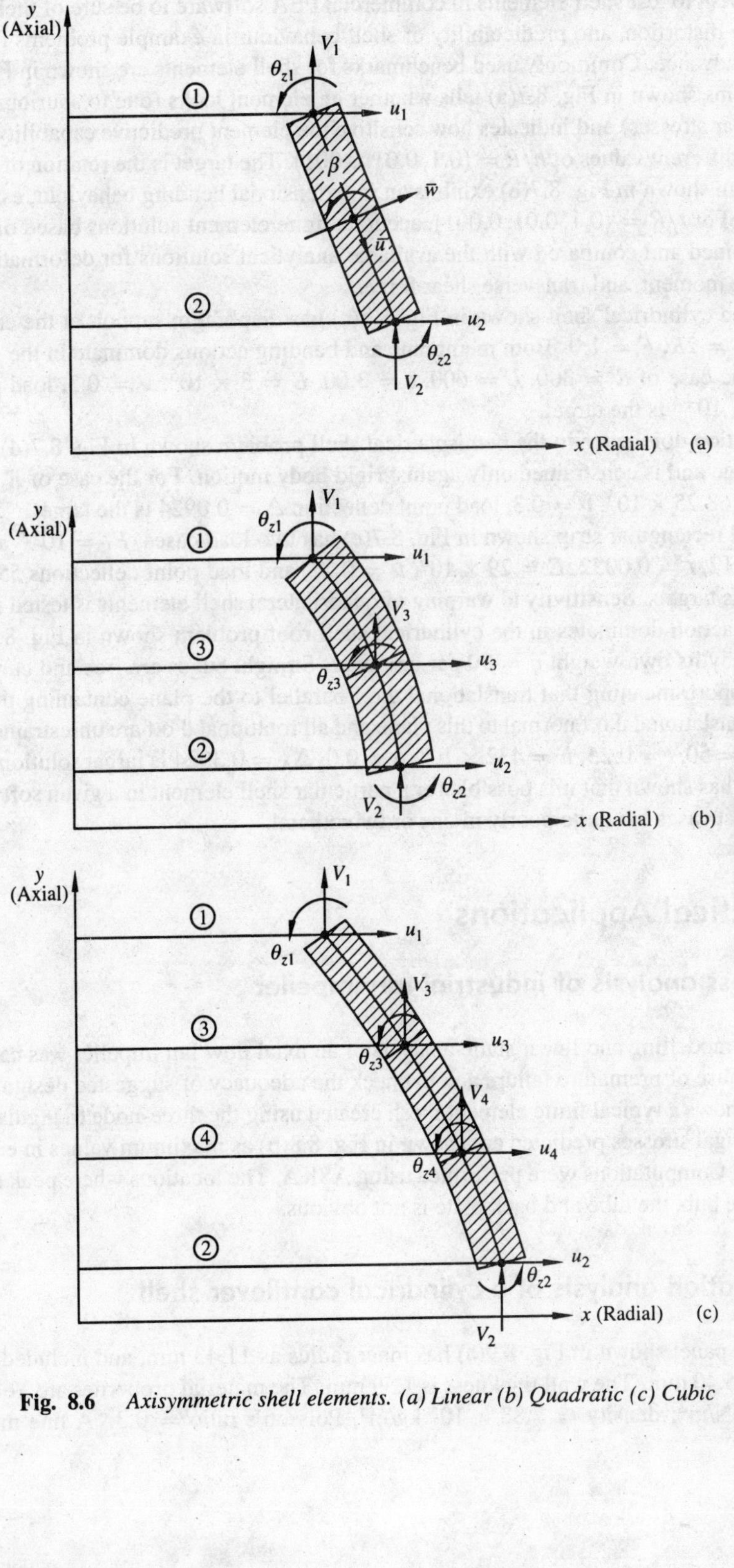

Fig. 8.6 *Axisymmetric shell elements. (a) Linear (b) Quadratic (c) Cubic*

One may wish to test shell elements in commercial FEA software to be sure of their validity, sensitivity to shape distortion, and predictability of shell behaviour in example problems for which results are known in advance. Commonly used benchmarks for shell elements are shown in Fig. 8.7.

The problems shown in Fig. 8.7(a) tells whether an element locks (due to spurious membrane and transverse shear stresses) and indicates how sensitive the element predictive capability is due to shape distortion for different values of $h/R = (0.1, 0.01, 0.001)$. The target is the rotation of the loaded face.

The problem shown in Fig. 8.7(b) exhibits an in-extensional bending behaviour, especially for thin walled shells. For $t/R = (0.1, 0.01, 0.001)$, accurate finite element solutions based on a convergence study are obtained and compared with the available analytical solutions for deformations, membrane force, bending moment, and transverse shear force.

The pinched cylindrical shell shown in Fig. 8.7(c) has diaphragm support at the ends; $t/R = 0.1$, $0.01, 0.001$; $L = 2R$; $F = 1.0$. Both membrane and bending actions dominate in the shell behaviour. For the specific case of $R = 300$, $L = 600$, $t = 3.00$, $E = 3 \times 10^6$, $\nu = 0.3$, load point deflection $\Delta = 0.1825 \times 10^{-4}$ is the target.

Bending action dominates in the hemispherical shell problem shown in Fig. 8.7(d). The shell features a free edge and is constrained only against rigid body motion. For the case of $F = 2.0$, $R = 10$, $t = 0.04$, $E = 68.25 \times 10^6$, $\nu = 0.3$, load point deflection $\Delta = 0.0924$ is the target.

The twisted rectangular strip shown in Fig. 8.7(e) has two load cases; $F_1 = 10^{-6}$ and $F_2 = 10^{-6}$; $b = 1.1$, $L = 12$, $t = 0.0032$, $E = 29 \times 10^6$, $\nu = 0.22$, and load point deflections 5556×10^{-6} and 1294×10^{-6} as targets. Sensitivity to warping of quadrilateral shell elements is tested in this problem.

Membrane action dominates in the cylindrical shell roof problem shown in Fig. 8.7(f). The shell roof is loaded by its own weight $q = 90$ per unit area. Straight edges are free and curved edges have diaphragm support, meaning that translational d.o.f parallel to the plane containing the curve are restrained but translational d.o.f normal to this plane and all rotational d.o.f are unrestrained. Specifically for $R = 25$, $L = 50$, $t = 0.25$, $E = 432 \times 10^6$, $\nu = 0.0$, $\Delta_A = 0.3024$ is target solution.

Experience has shown that it is possible for a particular shell element in a given software to do very well in most test cases but quite poorly in one or two others!

8.7 Practical Applications

8.7.1 Stress analysis of industrial fan impeller

Finite element modelling and linear static analysis of an axial flow fan impeller was carried out to investigate the cause of premature failure and to check the adequacy of suggested design modifications. Figure 8.8(a) shows a typical finite element mesh created using the three-node triangular flat shell elements. Centrifugal stresses predicted are shown in Fig. 8.8(b) as maximum values in each component of the impeller. Computations were performed using ASKA. The locations where peak stresses appear in the blade, the hub, the tube and back plate is not obvious.

8.7.2 Vibration analysis of a cylindrical cantilever shell

The cylindrical panel shown in Fig. 8.9(a) has inner radius as 11.43 mm, and included angle of 142° and length of 25.40 mm. The wall thickness is 1.59 mm. The material properties are Young's modulus $= 2.07 \times 10^7$ N/m^2; density $= 7.83 \times 10^3$ kg/m^3, Poisson's ratio $= 0.3$. A fine mesh (25 × 25)

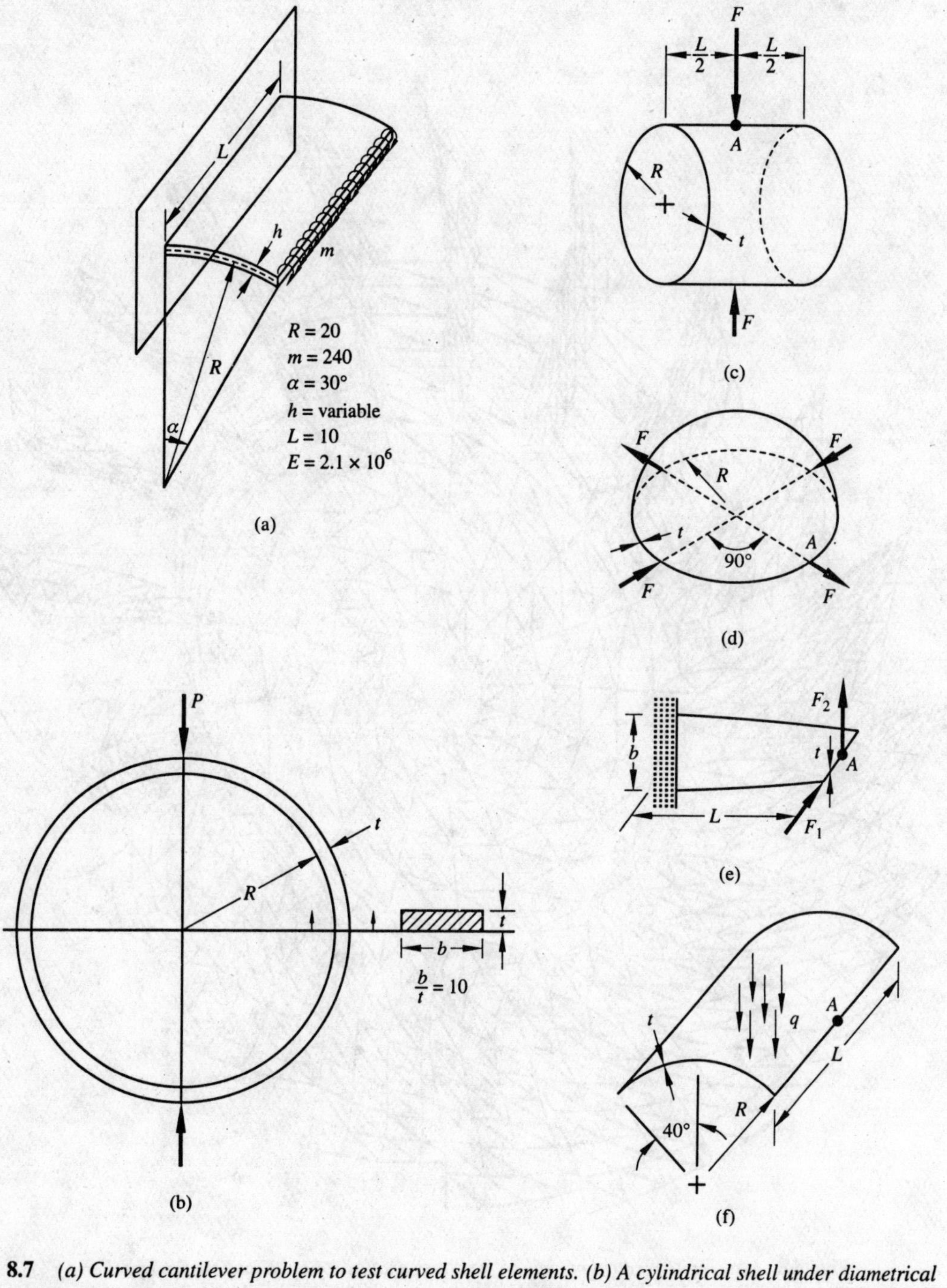

Fig. 8.7 *(a) Curved cantilever problem to test curved shell elements. (b) A cylindrical shell under diametrical compression. (c) Pinched cylinder. (d) Hemispherical shell with four uniformly spaced radial forces in the equatorial plane. (e) Rectangular strip with a 90° twist from end to end. (f) Cylindrical shell roof loaded by its own weight*

Fig. 8.8 *(a) Finite element modelling for stress analysis of an axial flow fan impeller*

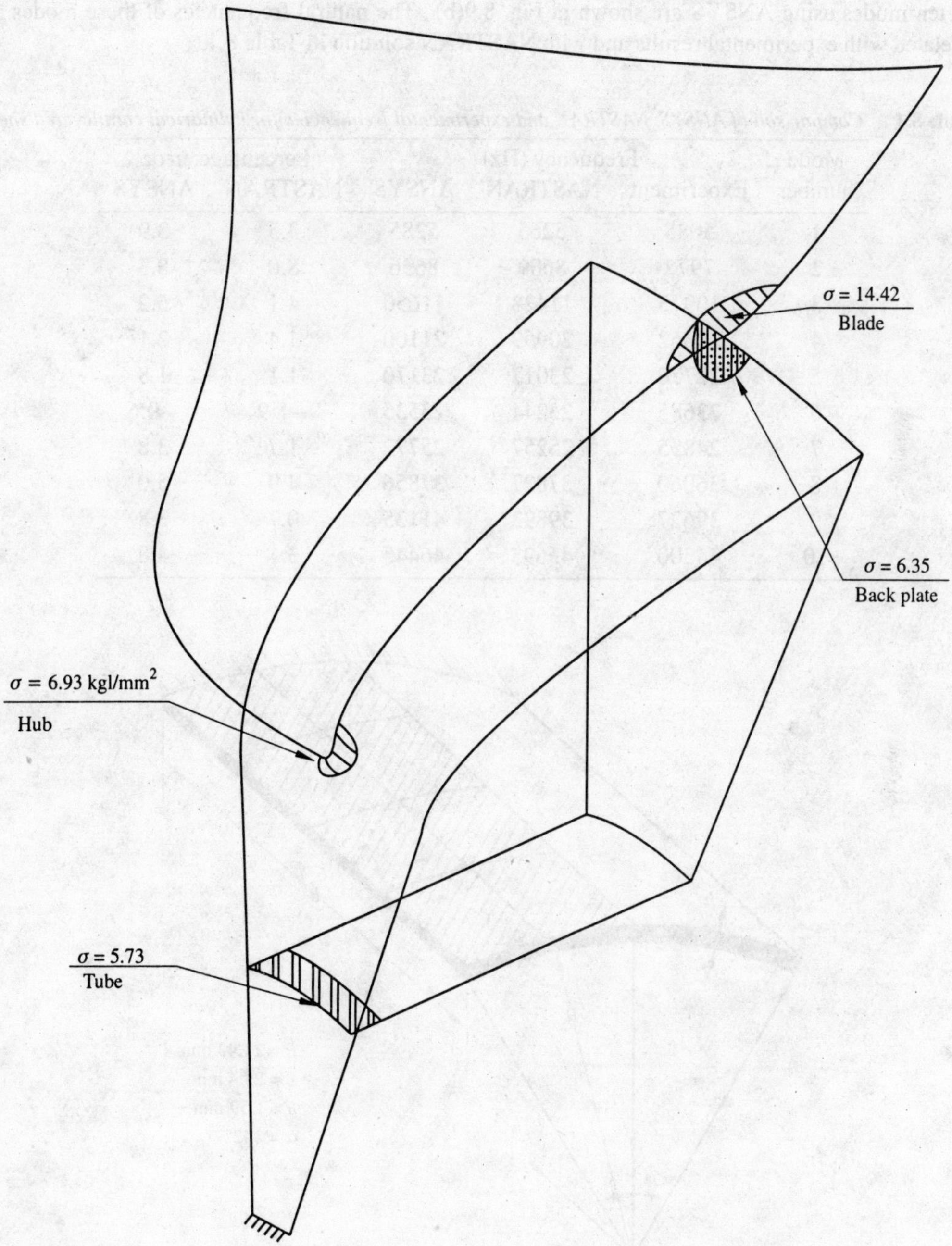

Fig. 8.8 *(b) Peak stresses in parts of an axial flow fan impeller*

of eight-node quadrilateral shell elements was used. Predicted natural frequencies and modes for the first ten modes using ANSYS are shown in Fig. 8.9(b). The natural frequencies of these modes are correlated with experimental results and with NASTRAN solution in Table 8.1.

Table 8.1 *Comparison of ANSYS, NASTRAN and experimental frequencies for cylindrical cantilevered shell*

Mode	Frequency (Hz)			Percentage error	
number	Experiment	NASTRAN	ANSYS	NASTRAN	ANSYS
1	5088	5265	5285	3.5	3.9
2	7972	8609	8636	8.0	8.3
3	10973	11428	11650	4.1	6.2
4	20662	20959	21100	1.4	2.1
5	22770	23017	23170	1.1	1.8
6	23685	23244	23535	−1.9	−0.6
7	24835	25257	25772	1.7	3.8
8	36060	37827	37856	4.9	5.0
9	39627	39892	41135	0.7	3.8
10	44300	45693	46445	3.1	4.8

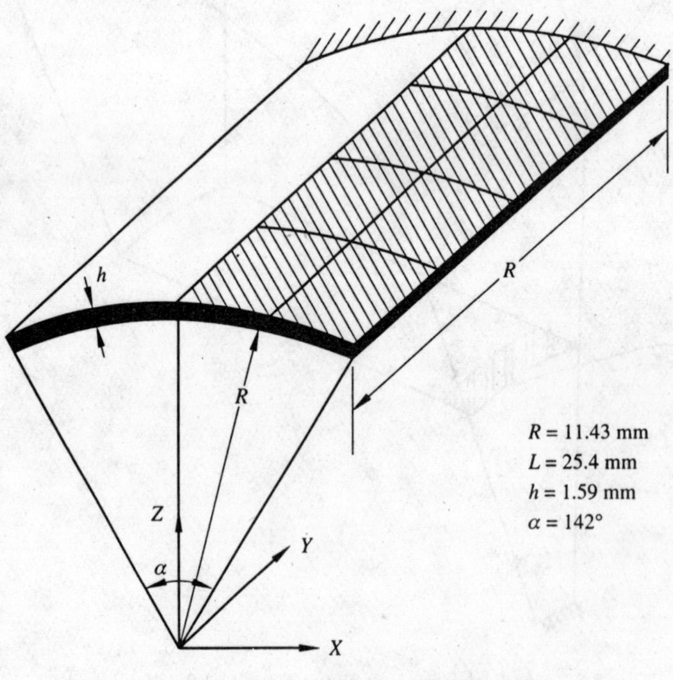

Fig. 8.9 *(a) Open cylindrical cantilevered shell*

Fig. 8.9 *(b) Resonant mode shapes form ANSYS solution for cylindrical cantilevered shell in the form of contour plots of the out-of-plane displacements*

8.7.3 Buckling analysis of a blade stiffened panel

Geometry and loading of a blade stiffened panel is shown in Fig. 8.10(a). Figure 8.10(b) shows a typical finite element mesh created using four-node quadrilateral flat shell elements. Linear buckling analysis was carried out using NASTRAN. Predicted buckling modes for longitudinal compression and shear loading are illustrated in Fig. 8.10(c) and 8.10(d) respectively.

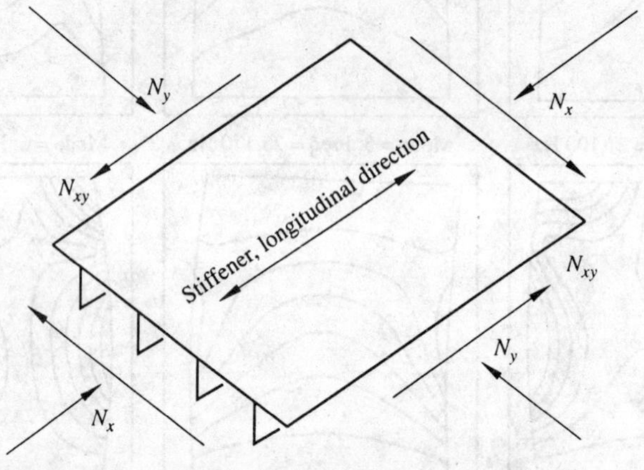

(a) Composite blade stiffened panel

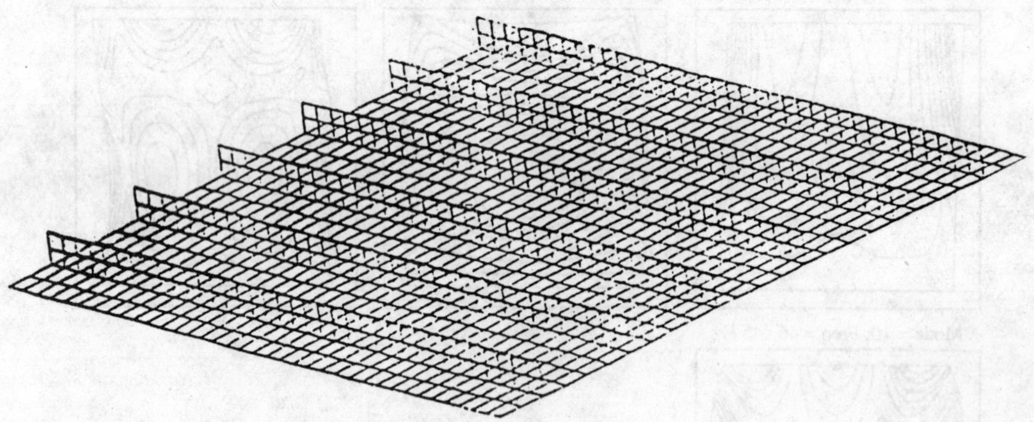

(b) Finite element model

Fig. 8.10 *Composite blade stiffened panel—buckling analysis (continues)*

8.7.4 Nonlinear behaviour of a curved panel

The geometry, loading, support conditions for a curved composite panel are shown in Fig. 8.11. The structure was modelled using eight-node quadrilateral shell elements in ABAQUS. A fine enough mesh was used to assure converged results. The predicted load vs. load point deflection curve is also shown

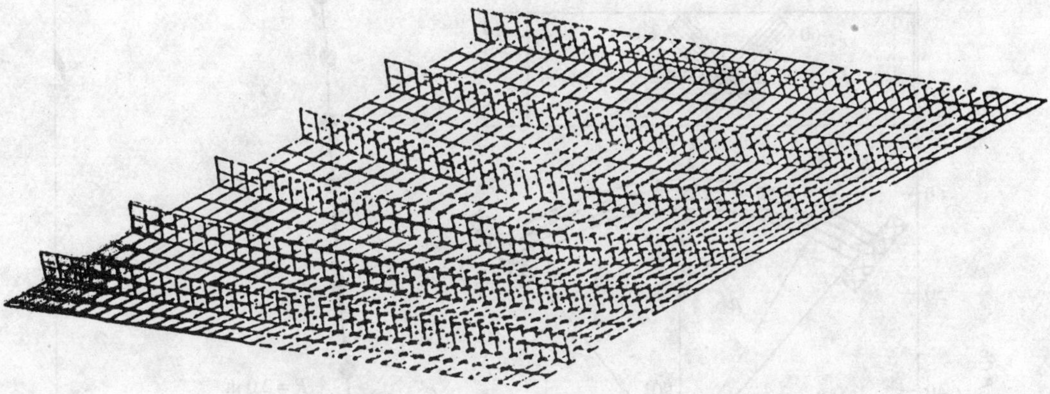

(c) Buckling mode shape for longitudinal compression

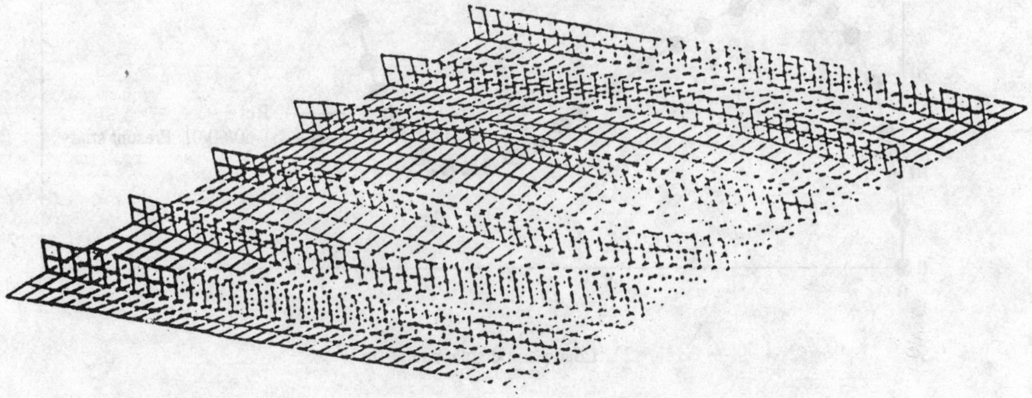

(d) Buckling mode shape for pure shear

Fig. 8.10 *Composite blade stiffened panel—buckling analysis (concluded)*

in Fig. 8.11. Evidently, the shell exhibits highly nonlinear, snap buckling behaviour. Linear static analysis is inadequate for this behaviour.

8.8 Closure

Design of simple-to-use and reliable shell elements has been and continues to be an active area of finite element research. Shell elements currently implemented in commercial FEA software may or may not represent the state-of-the-art. Shell elements are perhaps the most difficult finite elements to formulate, and the structural behaviour of shells is often difficult to anticipate. Therefore, one must anticipate the shell behaviour in order to make a suitable choice of elements, meshes, and analysis procedures.

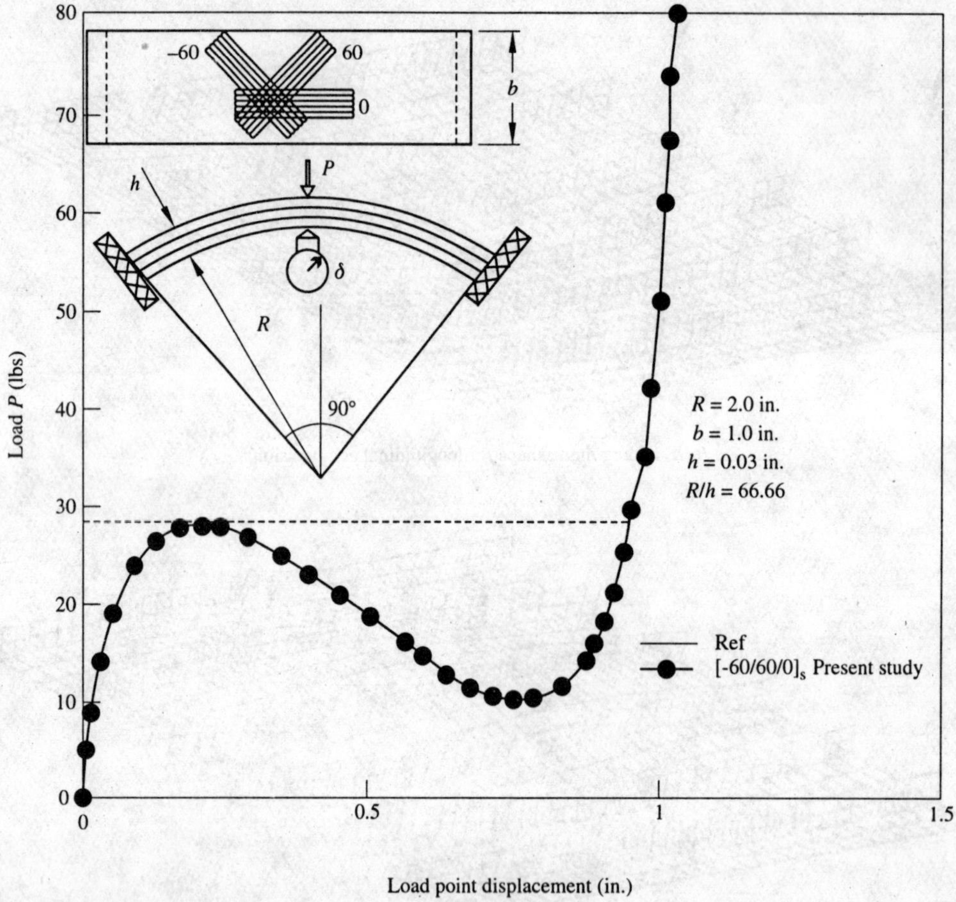

Fig. 8.11 *Geometry, loading, support conditions for a curved composite panel*

It is informative to check if the triangular and quadrilateral flat shell element formulations in a commercial FEA program include the drilling d.o.f In addition, there is a need to verify if the warping effect has been accounted for the quadrilateral shape.

Simple and effective curved shell elements for thin-walled shells are often designed using the discrete Kirchoff theory (DKT) approach. Semiloof shell elements, triangular and quadrilateral in shape developed by Irons and presented in Ref. [8.2] represent the state-of-the-art. However, they are not yet implemented in commercial FEA software and therefore are not used in practice.

Mindlin shell elements—six-noded triangular; eight-noded quadrilateral; and twelve-noded quadrilateral are perhaps the state of art general shell elements. It is still not clear whether the last word on the topic of membrane and shear-locking effects in these elements has been said. Elimination of locking phenomenon continues to be a focus of current research.

Shells of revolution featuring junctions, with continuous or discontinuous slopes and curvatures can be efficiently modelled and analyzed using axisymmetric shell elements. Provision is made to compute meridional and hoop stresses on the top, mid, and bottom surfaces of each element.

Critical assessment of the performance of shell elements in commercial FEA software is required before they are applied in practice. The standard set of test problems and example problems presented in this chapter are aimed at this requirement.

Shell structures in practice are complex and their behaviour is equally complex. However, shell elements in general and general-purpose programs in particular are capable of handling these with flying colours. This is demonstrated in the practical applications. Shell elements designed for implementation in a commercial FEA program should be able to model a solid wall with varying thickness; sandwich construction with bonded metallic facings; multi-layered fiber reinforced plastic composites; discretely stiffened walls and integrally stiffened walls. The self-imposed restriction to linear static analysis in the presentation is relaxed in the proposed computational problems. Commercial FEA programs do have the capabilities for stress, buckling, vibration and transient response analyses. The behaviour can be linear or nonlinear. Each computational problem demands formulation of additional matrices and vectors at the element level and use of advanced numerical methods for the solution of global matrix equations.

8.9 References

[8.1] Kraus, H. *Thin Elastic shells.* John Wiley and Sons, New York, 1967.

[8.2] Ashwell, D.G. and Gallagher, R.H, editors. *Finite Elements for Thin Shells and Curved Members.* Wiley, London, 1976.

[8.3] Gilewski, W and Radwanske, M. A survey of finite element models for the analyses of moderately thick shells. *Finite Element in Analysis and Design*, 9:1–21, 1991.

[8.4] MacNeal, R.H. *Finite Elements: Their Design and Performance.* Marcel Dekker, New York, 1994.

[8.5] Naganarayana, B.P. and Prathap, G. Force and moment corrections for the warped four node quadrilateral plane shell element. *Computers & Structures*, 33:1107–1115, 1989.

[8.6] Noor, A.K. and Mathers, M.D. Shear flexible finite element models of laminated composite plates and shells. Technical report, NASA TND–8044, 1975.

[8.7] Ahmad, S., Irons, B.M, and Zienkiwicz, O.C. Analysis of thick and thin shell structures by curved finite elements. *Int. J. Numer. Meths. Engng*, 2:419–451, 1970.

[8.8] Lakshminarayana, H.V. and Kailash, K. A shear deformable curved shell element of quadrilateral shape. *Computers & Structures*, 33:987–1001, 1989.

[8.9] Mac Neal, R.H. and Harder, R.L. A proposed standard set of problems to test finite element accuracy. *Finite Elements in Analysis and Design*, 1:3–20, 1985.

[8.10] Lakshminarayana, H.V. and Viswanath, S. A high precision triangular laminated anisotropic cylindrical shell finite element. *Computers and Structures*, 8:633–640, 1978.

8.10 Computational Problems

Report converged finite element solutions. Exploit symmetry if possible. Choose convenient numbers and consistent units for material properties, dimensions, and loads. Unless directed otherwise, assume that thicknesses are uniform and the material is isotropic.

8.1 A typical assembly of a hemispherical shell and a cylindrical shell as encountered in a missile component is shown in the sketch. The upper chamber is subjected to an axial thrust load where the lower chamber is pressurized. A rigorous stress analysis is requested.

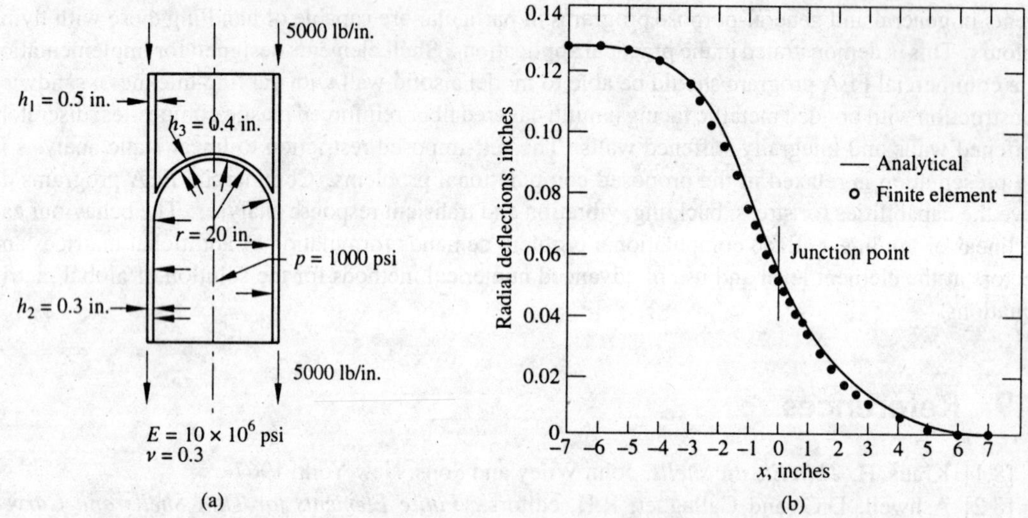

Fig. 8.12 *Sketch for problem 8.1*

8.2 A common steam power plant component is the omega expansion joint shown in the sketch. Predict and plot the meridional stress and hoop stress at the inner and outer surfaces of the joint.

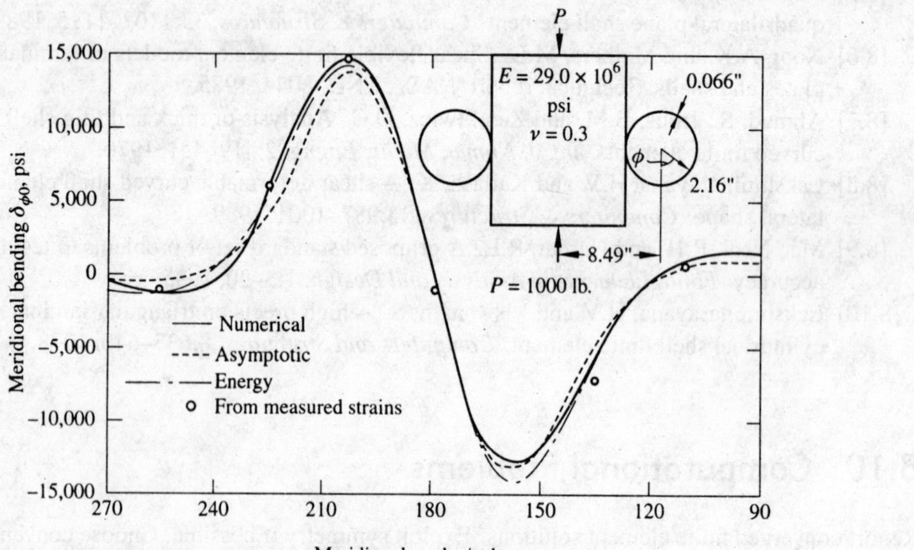

Fig. 8.13 *Sketch for problem 8.2*

8.3 Sketch shows a hemispherical cap of constant thickness h and radius R that is clamped at its edge and loaded by an internal pressure P. Compute and plot the meridional and hoop stresses at inner and outer surfaces for $R/h = 10, 100, 1000$. Normalize the results with $\sigma_{\text{ref}} = PR/h$. Comparison solutions are reported in [8.1].

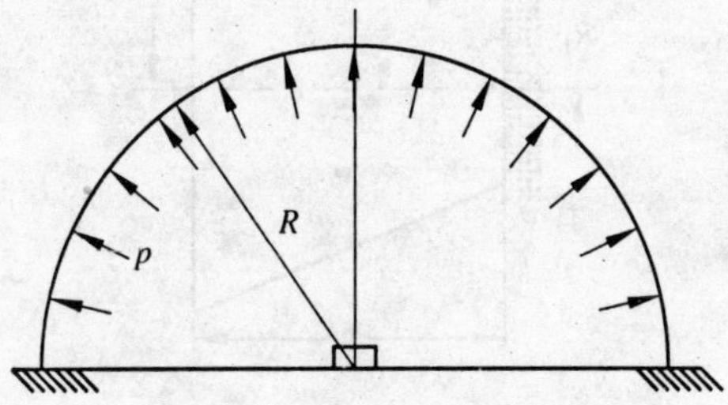

Fig. 8.14 *Sketch for problem 8.3*

8.4 A thin shell of revolution in the shape of a truncated cone is simply supported around its base, as shown in the sketch. A uniformly distributed line load q is applied to the top. If the cone is rather flat, it is called "Belleville Spring" and has a highly nonlinear load versus deflection relation and exhibits snap buckling behaviour. You are directed to predict this behaviour with the right choice of element, mesh, and analysis procedure.

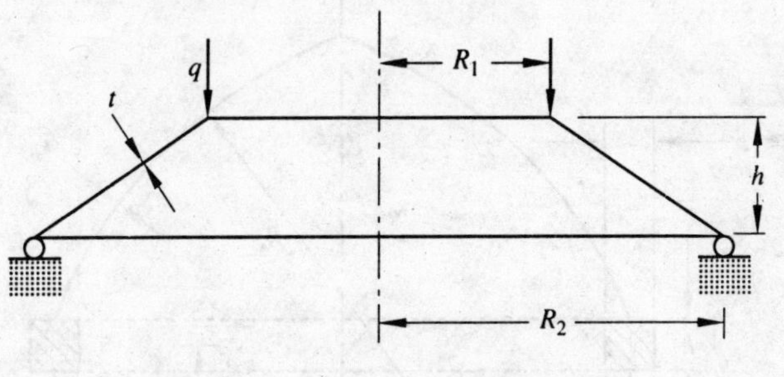

Fig. 8.15 *Sketch for problem 8.4*

8.5 The sketch depicts two channel sections, welded together to create an angle Φ between them and supported as shown. Investigate the natural frequencies and modes for $\Phi = 90°$ and $120°$.

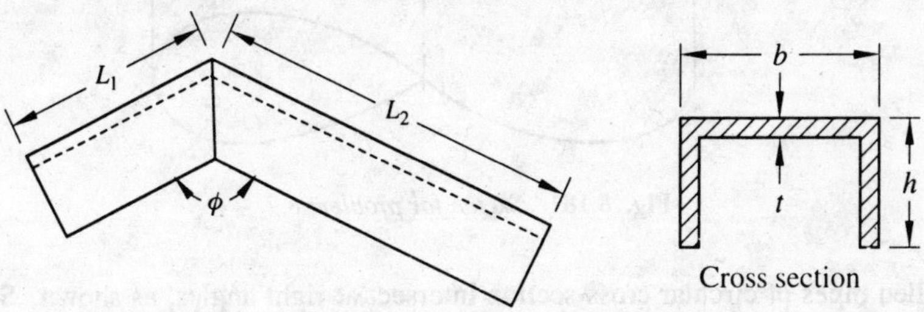

Fig. 8.16 *Sketch for problem 8.5*

8.6 Investigate the natural frequencies and modes of the conical shell shown.

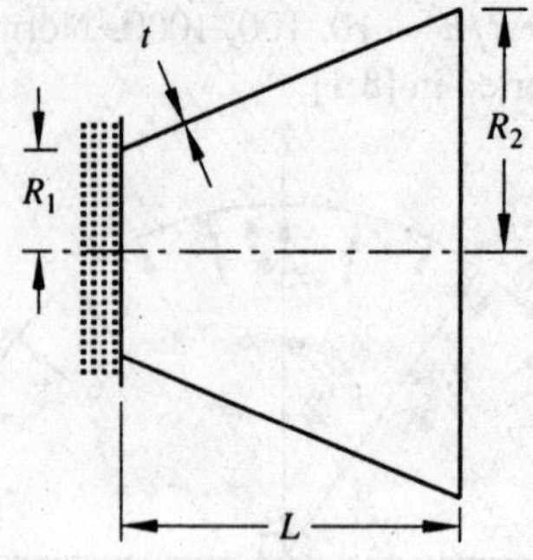

Fig. 8.17 *Sketch for problem 8.6*

8.7 A thin-walled shell of revolution consists of a cylinder of radius R_c capped by a closure whose meridian is a circular arc of radius R_s, as shown in the sketch. The base of the cylinder may be considered fixed. Internal pressure P is applied. Investigate the junction stresses. Predict and plot the meridional and hoop stresses at the inner and outer surfaces. Normalize these stresses w.r.t. $\sigma_0 = PR_C/t_C$.

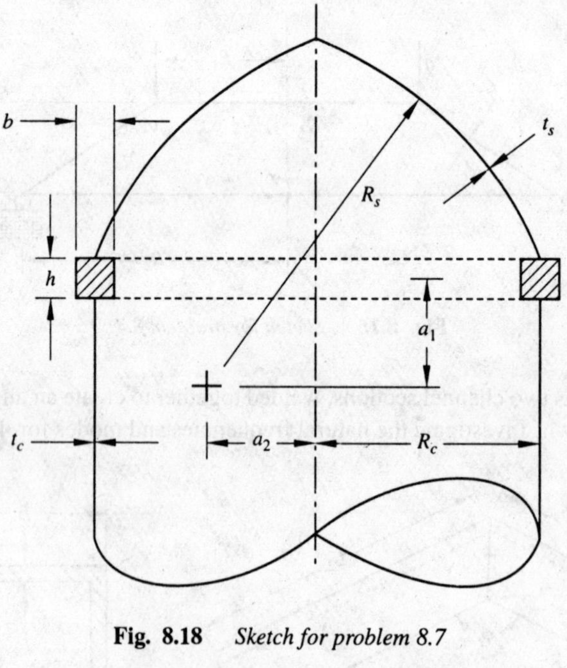

Fig. 8.18 *Sketch for problem 8.7*

8.8 Thin-walled pipes of circular cross-section intersect at right angles, as shown. Stresses at and near the intersection are of interest. Six load cases are shown. For convenience loads may be applied to a circular disk on the end of the joint.

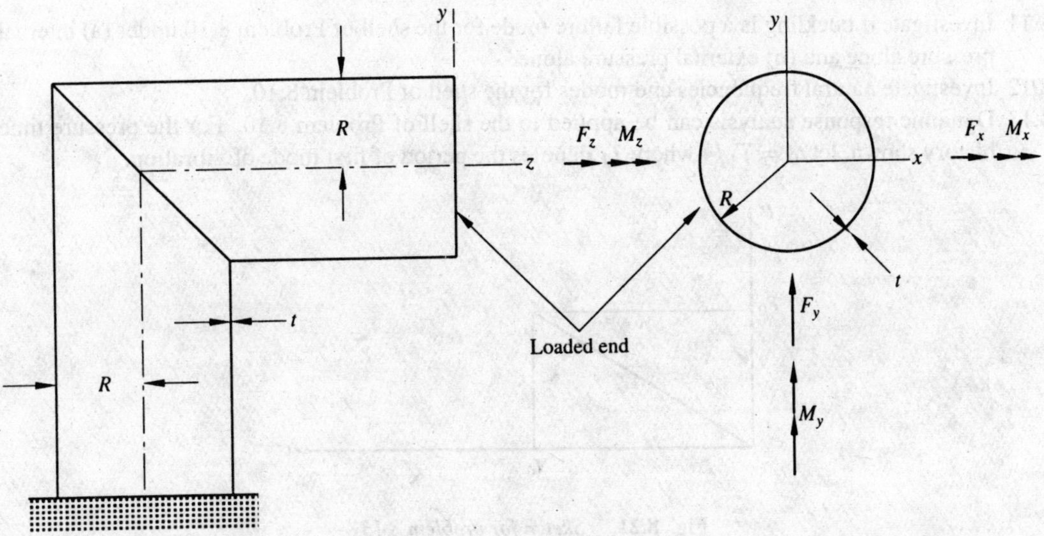

Fig. 8.19 *Sketch for problem 8.8*

8.9 For the intersecting cylindrical shells of Problem 8.8, predict the natural frequencies and modes.

8.10 The sketch shows an end closure on a cylindrical shell of radius R_c and thickness t_c. The closure consists of a toroidal knuckle of meridional radius R_t and thickness t_t and a spherical cap of radius R_s and thickness t_s. Investigate the stress distribution due to (a) internal pressure, and (b) temperature that varies linearly through the wall thickness. The base of the cylinder may be considered fixed.

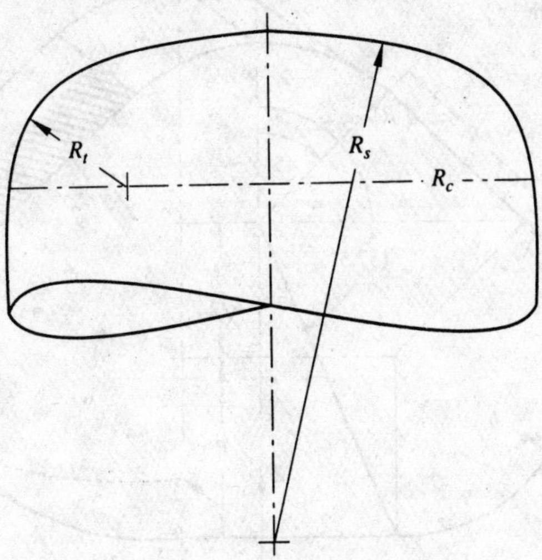

Fig. 8.20 *Sketch for problem 8.10*

8.11 Investigate if buckling is a possible failure mode for the shell of Problem 8.10 under (a) internal pressure alone and (b) external pressure alone.

8.12 Investigate natural frequencies and modes for the shell of Problem 8.10.

8.13 Dynamic response analysis can be applied to the shell of Problem 8.10. For the pressure time history shown, let $t_0 = T_1/4$ where T_1 denotes the period of first mode of vibration.

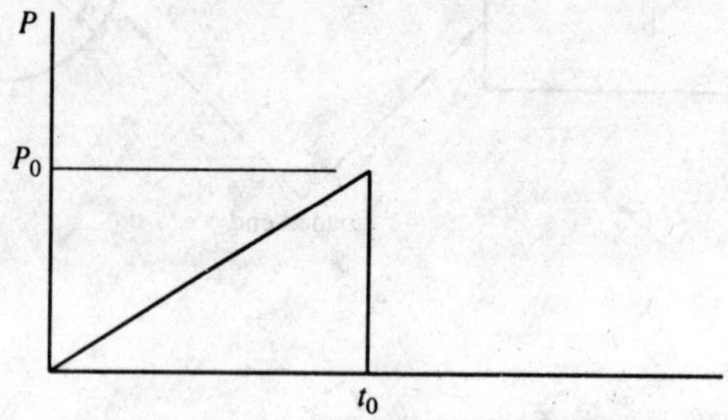

Fig. 8.21 *Sketch for problem 8.13*

8.14 Stress, buckling, vibration analysis of a pear-shaped cylinder of various lengths is requested. One end is fully clamped while the other is free. This prismatic shell features varying and discontinuous curvature. Use internal pressure for stress analysis and external pressure for buckling analysis. Include pre-stress effects in vibration analysis. Use $(L/R) = 2, 4, 16$.

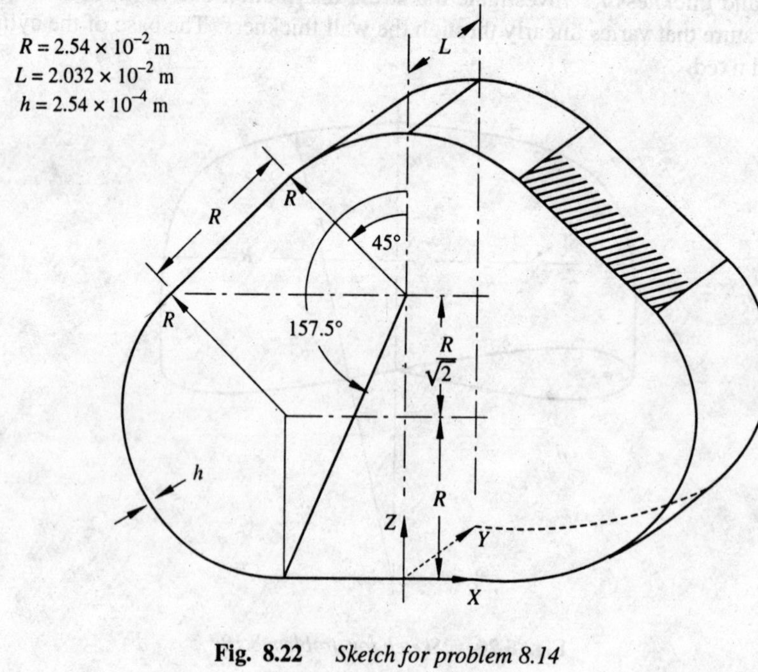

Fig. 8.22 *Sketch for problem 8.14*

8.15 A thin-walled cylindrical shell with a circular cutout under axial tension is shown in the sketch. Predict and plot the stress distribution in the vicinity of the cutout. Comparison solutions are available in Lakshminarayana [8.10].

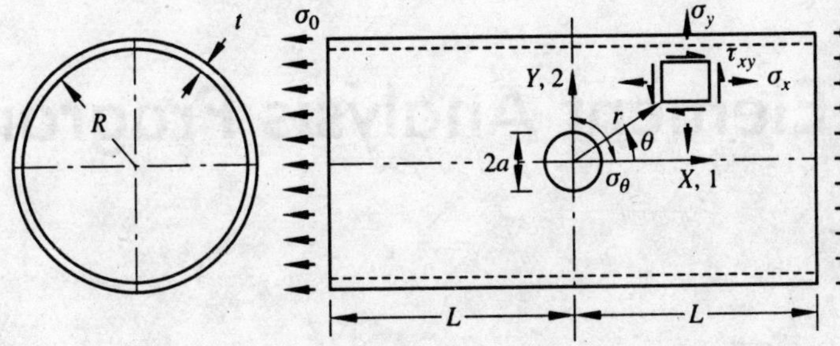

Geometric Parameters	Material Constants		
		Isotropic (Aluminum)	Orthotropic Gloss-Epoxy Composite
$\frac{R}{t} = 100$	E_{11}	100×10^6 psi	6.13×10^6 psi
$\frac{a}{t} = 10$	E_{22}	10.0×10^6 psi	1.42×10^6 psi
$\frac{L}{R} = 3$	G_{12}	3.84×10^6 psi	0.53×10^6 psi
$a = 1.0$ in.	ν_{12}	0.3	0.27
	ν_{21}	0.3	0.0625

Fig. 8.23 *Sketch for problem 8.15*

9 Finite Element Analysis Programs

9.1 Overview

Computer implementation of finite elements and solution procedures for engineering analysis is addressed. The end product is a general-purpose finite element analysis program. For such software to be used as an effective CAE tool, the programming should be hardware independent. The chosen finite elements and numerical methods must be accurate and reliable. The program should be executable on a given platform of choice – single processor, multi-processor, parallel processor, etc.

A general purpose FEA program consists of three modules: a pre-processor, a solver, and a post-processor. Commercial FEA programs can handle very large number of nodes and nodal degrees of freedom provided a powerful hardware is made available. User's manual, theoretical manual, and verification problems manual, document a commercial FEA program.

Surveys of general-purpose programs for finite element analysis have been published [9.1]. At present FEA programs are used rather than written. Understanding of the organization, capabilities, and limitations of commercial FEA programs is generally more important than an ability to develop or even modify a FEA code. The emphasis on programming the FEM which was a major preoccupation in many recent textbooks [9.2 to 9.4] is therefore absent in this book.

The purpose of this chapter is to describe the organization and desirable capabilities of a general-purpose FEA program. A brief description of widely distributed and extensively used commercial FEA codes is included so that the reader is aware of their current capabilities.

Benchmark constitutes a standard set of test problems devised to assess the performance of FEA codes.

The practical issue of developing a viable FEA program and its implementation in the PC environment is a much larger challenge. Typically, it involves hundreds of human year's effort.

9.2 FEA Program: Organization

The four components shown in Fig. 9.1 are common to virtually all general-purpose FEA programs. The INPUT phase enables the user to provide information relating to geometric representation, finite element discretization, support conditions, applied loads, and material properties. The more sophisticated commercial FEM systems facilitate automated generation of nodes and elements and provide access to a material property database. Plotting of the finite element model is also possible so that errors if any, in the input phase, may be detected and corrected prior to performing computations.

The finite element library comprises the element matrix generation modules. Herein resides the coded formulative process for the individual finite elements. Ideally, the element library is open-ended

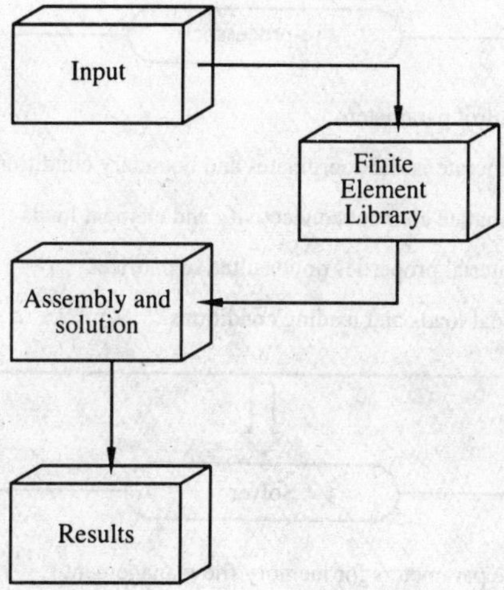

Fig. 9.1 *Components of a general purpose finite element analysis program*

and capable of accommodating new elements to any degree of complexity. This phase generates the required element matrices and vectors.

The assembly module includes all matrix operations necessary to position the element matrices for connection to neighbouring elements and the connection process itself. The latter operation thereby produces the global matrix equation of the finite element model.

The solution phase operates on the governing matrix equation of the problem derived in the previous phase. In the case of a linear static analysis, this may mean no more than the solution of a set of linear algebraic equations for a known right-hand side. In the case of linear vibration and buckling analysis, this may mean the extraction of eigen values and eigen vectors. Transient response analysis will require computations over a time history of applied load.

Finally, the results phase provides the analyst with a record of the solution. The record is commonly a printed list of nodal d.o.f, element strains and stresses, reaction forces corresponding to constrained degrees of freedom and a host of other requested information. As in input phase, there is a trend toward graphical output of results such as plots of displacement and stress contours, modes of vibration and buckling, etc.

A commercial FEM system therefore consists of three basic modules: pre-processor; solver; and post-processor. These modules and their functions are illustrated in Fig. 9.2. The pre-processor allows the user to create geometry or input CAD geometry, and provides the tools for meshing the geometry. The solver takes the finite element model provided by the pre-processor and computes the required response. The post-processor takes the data from the solver and presents it in a form that the user can understand.

9.3 FEA Program: Capabilities

The desirable features of a general-purpose FEA program are a large number of material models; a good library of finite elements; a good number of analysis procedures; and ability to manage the associated data. A brief discussion on these follows.

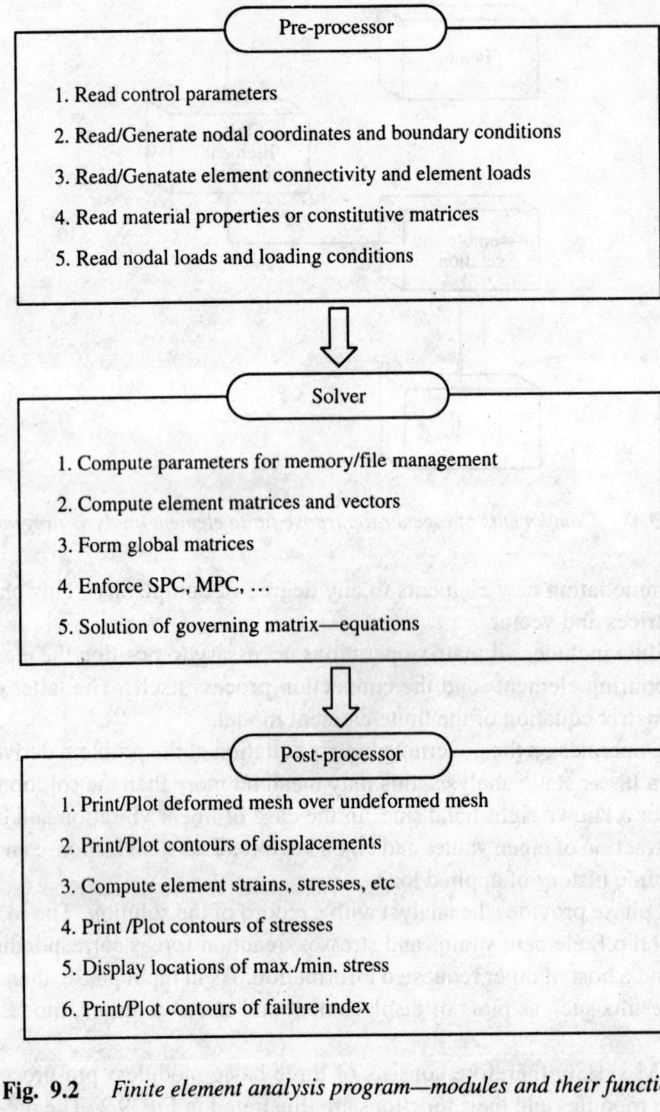

Fig. 9.2 *Finite element analysis program—modules and their functions*

9.3.1 Material models

To cover a large number of metallic and non-metallic materials and a wide range of their behaviour, a general-purpose FEA program should provide a library of material models.

- Homogeneous, isotropic, linear, elastic
- Orthotropic
- Anisotropic
- Laminated composite
- Nonlinear elastic
- Elastic plastic

- Viscoelastic
- Viscoplastic
- Hyperelastic
- Temperature-dependent material properties

9.3.2 Element library

The available elements are for solid, structural, thermal and fluid flow analysis. They can be classified as follows:

- One-dimensional elements
  - 1-D, 2-D, 3-D bar elements
  - Linear/quadratic/cubic in order
- Two-dimensional elements
  - Triangular/quadrilateral in shape
  - Linear/quadratic/cubic in order
  - With straight/curved edges
- Axisymmetric ring elements
  - Triangular/quadrilateral in shape
  - Linear/quadratic/cubic in order
  - With flat/curved surfaces
- Three-dimensional elements
  - Tetrahedra/hexahedra/pentahedra in shape
  - Linear/quadratic/cubic in order
  - With flat/curved faces
- Beam elements
  - Euler–Bernouli theory/shear deformation theory
  - 1-D, 2-D, 3-D beam elements
- Plate elements
  - Kirchoff theory/Mindlin theory
  - Triangular/quadrilateral shapes
  - Linear/quadratic/cubic in order
  - With straight/curved edges
- Shell elements
  - Flat shell elements/facet approximation
  - Curved shell elements: triangular/quadrilateral shapes; quadratic/cubic orders
  - Axisymmetric shell elements: with curved surfaces; linear/quadratic/cubic in order

- Special elements
  - Spring
  - Gap
  - Rigid link
  - Contact
- Crack tip elements

Some of these elements are formulated to handle large displacements, large rotations and finite strains. Some formulations use reduced integration with hourglass control.

9.3.3 Procedures library

- Linear static analysis
- Linear dynamic analysis
  - Free vibration
  - Forced vibration
  - Transient response: mode superposition
  - Transient response: direct integration
  - Acoustic excitation and response
  - Spectrum response
- Linear buckling analysis
- Non linear analysis
  - Geometric nonlinearity
  - Material nonlinearity
  - Combined geometric and material nonlinearity
  - Contact problems
- Aero-elastic analysis
  - Divergence
  - Flutter
- Design optimization (sensitivity analysis)
- Thermal analysis: computational
- Fluid dynamics: computational
- Fracture mechanics: computational
- Electromagnetics
- Electrostatics
- Magnetostatics

This allows the user to perform a wide variety of analyses. These procedures provide solutions for linear or nonlinear behaviour under static or dynamic loads. Large deformation and finite strain problems, contact problems, can also be addressed using these procedures.

9.3.4 Data processing

- Super elements
- Automated multilevel sub structuring
- Fourier analysis: axisymmetric bodies/shells under non-axisymmetric loads
- Cyclic symmetry
- Efficient numerical methods
  - Direct solver
  - Iterative solver
- Efficient computer systems
  - Super computers
  - Parallel processing systems
- Automatic adaptive mesh refinement

9.4 FEA Program: A Catalogue

A brief description of widely distributed commercial FEA programs is included here so that the reader is aware of their current capabilities.

9.4.1 MSC.Nastran

NASA Structural Analysis (Nastran) is a general-purpose program based on the finite element method developed by MacNeal Schwendler Corporation (MSC). The associated pre- and post-processor is called *MSC.Patran*. This premier FEA software is now available on the PC and runs both on DOS and Windows operating systems.

MSC.Patran provides the industry's most comprehensive and powerful tools for the creation of accurate finite element models. Backed by the world's largest CAE support organization and enhanced by continual use at some of the largest manufacturers, MSC.Patran sets the standard for finite element pre- and post-processing.

MSC.Nastran is the world standard in finite element analysis solutions. Its analysis capabilities give the user the competitive edge. With open choice of platforms from desktop PCs to supercomputers, MSC.Nastran is available where it is needed. MSC.Nastran's unique element technologies provide highly accurate results with lower modelling effort, less solution time, and reduced computer requirements. Using MSC.Nastran one can optimize designs without increasing design cycle time. MSC provides the best documentation, customer support, and user training.

Building better products lighter, stronger, safer, in less time, at less cost are the business benefits of FEA using Nastran.

Analysis procedures in MSC.Nastran include: structural statics; structural dynamics; heat transfer; aero-elastic; magnetic field; piezo electric; acoustic; and hydro-elastic.

9.4.2 NISA

Numerically Integrated Finite Elements for Systems Analysis (NISA) is a family of general-purpose finite element programs for PCs, workstations and supercomputers developed by Engineering Mechanics Research Corporation (EMRC). The associated pre- and post-processor is called DISPLAY. The distinguishing features of the NISA programs are: user-friendly documentation; excellent technical support; flexible purchase options; and best price/performance in the industry.

NISA offers independent modules for a variety of analysis: linear statics: nonlinear statics, dynamics; heat transfer; composites; optimization; fatigue and fracture; fluid dynamics; printed circuit boards; electromagnetic fields; kinematic and dynamic analysis of mechanical systems.

NISA provides an excellent library of isoparametric finite elements. A special module NISA.P ADAPT utilizes *P* elements. This program continually increases the order of the polynomial on a fixed finite element mesh until a reasonable convergence is reached. P refinement and properly designed mesh is efficient and reliable.

NISA offers interfaces to major CAD/CAM systems: pro/engineer; unigraphics; CATIA.

DISPLAY is a powerful interactive graphics pre- and post-processor, which makes complex finite element modelling and results interpretation a cinch.

These programs reflect the latest advances in CAE utilizing finite element methods.

9.4.3 MARC

The right answer for finite element analysis is the general-purpose program called MARC developed by MARC Analysis Research Corporation. Special features of this program are: fully integrated nonlinear solution; powerful automated 3-D contact; accurate, adaptive simulations; parallel processing; and multi-physics. The associated pre- and post-processor is called *Mentat*.

MARC and Mentat allow the user to perform a wide variety of structural, thermal, fluid, and coupled field analyses using finite element method. The analysis procedures provide solutions for simple to complex linear and nonlinear problems in engineering.

The capabilities in MARC include: linear; nonlinear; large deformation and finite strain; automated contact; and adaptive meshing.

MARC has an extensive library of metallic and nonmetallic material models: linear elastic; elastic plastic; elastomers; hyperelastic; rigid plastic flow; creep; viscoelastic; viscoplastic; poro-elasticity and soils; powder metallurgy; composites; and concrete.

Over 140 elements are available in MARC, which are modern, robust and accurate. They are grouped as: truss; beam; plane stress; plane strain; generalized plane strain; plate; shell; membrane; axisymmetric; 3-D solid; special elements (springs, gaps, rigid links, pipe bend, etc.); and user defined elements.

Analysis types supported by MARC are: statics; dynamics; heat transfer; thermo mechanical; fracture mechanics; fluid dynamics; hydrodynamic bearing; joule heating; acoustics; electrostatics; magnetostatics; electromagnetics; design sensitivity and optimization.

Mentat is tightly integrated with the MARC FEA program, allowing all data to be defined interactively through a powerful graphical user interface. Notable capabilities include: geometry creation; solid modelling; mesh generation; analysis support; post-processing; and advanced rendering.

Mentat's optional modules support interfaces to the leading CAD/CAM systems CATIA; pro/engineers; I-DEAS, and auto CAD.

9.4.4 LS-DYNA

LS-DYNA is a general-purpose code based on the FEM for analyzing large/elastic/inelastic deformation dynamic response of solids and structures including structures coupled to fluids. The main solution procedure is based on explicit time integration. An implicit solver is also available with somewhat limited capabilities for structural and heat transfer analysis.

A contact impact algorithm allows difficult contact problems to be easily treated with heat transfer included across the contact interfaces.

Spatial discretization is achieved by the use of four-node tetrahedral, eight-node hexahedral solid elements; two-node beam elements; three-node triangular and four-node quadrilateral shell elements; eight-noded solid shell elements; truss elements; membrane elements; discrete elements; and rigid bodies. A variety of formulations are available for each element type (solid, fluid, structural, discrete).

Specialized capabilities for modelling airbags, sensors, and seat belts have tailored LS-DYNA for applications in the automotive industry.

Adaptive meshing is available for shell elements and is widely used in sheet metal stamping simulations.

LS-DYNA currently has over two hundred material models and over ten equations of state to cover a wide range of material behavior.

LS-DYNA is operational on supercomputers, mainframes, workstations, parallel processing systems, and PCs.

The associated pre- and post-processor is called LS-TAURUS.

LS-DYNA and LS-TAURUS are developed by Livermore Software Technology Corporation.

9.4.5 ANSYS

ANSYS is an integrated design analysis tool based on the FEM developed by ANSYS, Inc. It has its own tightly integrated pre- and post-processor. The ANSYS product documentation is excellent and it includes commands reference; operations guide; modeling and meshing guide; basic analysis procedures guide; advanced analysis guide; element reference; theory reference; structural analysis guide; thermal analysis guide; electromagnetic fields analysis guide; fluid dynamics guide; and coupled field analysis guide. Taken together, these manuals provide descriptions of the procedures, commands, elements, and theoretical details needed to use the ANSYS program. All of the above manuals except the ANSYS theory reference are available online through the ANSYS help system, which can be accessed either as a standalone system or from within the ANSYS program. A brief description of the information found in each of the manuals follows.

Engineering capabilities of ANSYS products are: structural analysis (linear stress, nonlinear stress, dynamic, buckling); thermal analysis (steady state, transient, conduction, convention, radiation, and phase change); CFD analysis (steady state, transient, incompressible, compressible, laminar, turbulent); electromagnetic fields analysis (magnetostatics, electrostatics); field and coupled field analysis (acoustics, fluid–structural, fluid–thermal, magnetic–fluid, magnetic–structural, magnetic–thermal, piezoelectric, thermal–electric, thermal–structural, electric–magnetic); sub-modelling; optimization; and parametric design language.

Element library in ANSYS lists 189 finite elements. They are broadly grouped into: LINK, PLANE, BEAM, SOLID, CONTAC, COMBIN, PIPE, MASS, SHELL, FLUID, SOURCE, MATRIX, HYPER, VISCO, INFIN, INTER, SURF, etc. Under each type, different shapes and orders complete the list. Obviously, ANSYS has the best elements in its library.

Analysis procedures in ANSYS can be grouped into: static analysis; transient analysis; mode frequency analysis; harmonic response analysis; buckling analysis; sub-structuring analysis; and spectrum analysis.

In ANSYS, there are two fundamentally different types of optimization. The first is referred to as design optimization; it works entirely with the ANSYS parametric design language and is contained within its own module (ANSYS /OPT). The second is topology optimization, a form of shape optimization.

ANSYS finite element analysis software enables engineers to perform the following tasks:

- Build computer models or transfer CAD models of structures, products, components, or systems.
- Apply operating loads or other design performance conditions.
- Study physical response, such as stress levels, temperature distributions or electromagnetic fields.
- Optimize a design early in the product development process to reduce production costs.
- Do prototype testing in environments where it otherwise would be undesirable or impossible.

9.5 Closure

Spectacular advances have been made in the development, documentation, and implementation of commercial FEA programs on PCs, workstations, mainframes, and supercomputer systems. Pre-processors with graphical user interface are also available that can create finite element models of virtually all CAD models. Post-processors are capable of display and animation of the results of every finite element analysis. At present, FEA programs have been integrated in widely used CAD/CAM systems. Computer implementation of finite element procedures is not trivial; it involves hundreds of human years effort not only for development but also for updates.

It is instructive to compare and contrast the desirable features of a general purpose FEA program with the current capabilities of commercial FEA codes. This may provide directions for modifications, extensions and upgrading of commercial FEA codes.

It is recommended that the reader use one of the commercial FEA programs, not necessarily from those described here, to analyze the computational problems listed in the text. This will enable the user to acquire the skills needed to effectively use the FEM in general, and a general-purpose program in particular, in practice.

Advanced applications of the FEM, not considered so far, can be attempted using commercial FEA programs. Some of these are identified and described in the next chapter.

9.6 References

[9.1] Berbbia, C.A., editor. *Finite Element Systems: A Handbook*. Springer Verlag, Berlin, 1985.

[9.2] Krishnamoorthy, C.S. *Finite Element Analysis: Theory and Programming*. Tata McGraw Hill, New Delhi, 2nd ed., 1994.

[9.3] Reddy, J.N. *An Introduction to the Finite Element Method*. McGraw Hill, New York, 2nd ed., 1993.

[9.4] Chandrupatla, T.R. and Belegundu, A.D. *An Introduction to Finite Elements in Engineering*. Prentice Hall of India, New Delhi, 2nd ed., 2001.

10 Advanced Applications

10.1 Overview

Advanced applications of the FEM in general and commercial FEA programs in particular, not considered so far in the text are identified. They are: nonlinear analysis of solids and structures; analysis of laminated composite structures; computational fracture mechanics; structural optimization; computational heat transfer; computational fluid dynamics; electromagnetic and coupled-field analysis, and simulation of manufacturing processes. These are of interest to every engineering discipline. For each one of these topics, continuum mechanics background, finite element formulation, governing matrix equations and their solution techniques, software capabilities, and benchmark tests are described rather briefly. The available information is so vast that special volumes have been compiled and published on each one of the identified topics. An attempt is also made to identify the most relevant software product for the intended applications. Benchmark tests constitute a standard set of problems devised to critically assess the performance of a FEA code. In addition, they have been shown to be an excellent means of training novices.

10.2 Nonlinear Finite Element Analysis

There are situations where nonlinear finite element analysis must be used for a realistic assessment of structural/component response. We can identify several areas [10.1]:

- Design for both ultimate load and serviceability limit states
- Assessing structural integrity
- Establish the cause of a structural failure
- To understand basic structural behaviour
- To test the validity of proposed material models
- Simulation of manufacturing processes

There are three types of nonlinearities that deserve attention: geometric nonlinearity (large displacements, large rotations, finite strains); material nonlinearity (elastoplastic, time-dependent creep) and boundary nonlinearity (contact problems).

Nonlinear continuum mechanics [10.2] theory of plasticity [10.3] and theory of creep [10.4] provide the continuum mechanics background. Using the continuum mechanics equations, finite element equations are then formulated [10.5]. An incremental iterative solution procedure is employed to predict the nodal d.o.f for any load level and hence element strains, stresses, etc.

Commercial FEA programs have nonlinear analysis capabilities of varying sophistications. *MARC* and *Mentat* probably provide the right answer for nonlinear finite element analysis. All available FEM systems that deal with nonlinear FEA will differ in detail of input style and technical content, even though many of the underlying algorithms are common. The selection of a particular code should address the following:

- Does it contain the right pre- and post-processors?
- Are experts available to train and help?
- Does a good user's manual exist?
- Are the supporting documents pertaining to validation and quality assurance of adequate standard?

Assuming that a chosen finite element model is satisfactory, the data specific to nonlinear analysis fall into three groups:

- Discretizational parameters
- Nonlinear material data
- Control parameters

It is a good practice to perform trial runs to gain familiarization with the code.

Geometric nonlinear analysis requires a little extra data as compared to linear FEA. No special material data tables are involved. However the user should decide on large displacements or large rotations or large strains or both.

When using a commercial FEA program offering large deformation and large strain capabilities, the user should be aware of the definition of stress (Cauchy stress or Piola–Kirchoff stress) and strain (Green strain or Almansi strain) used in the code. The method of calculating these over-repeated iterations and load increments is also of concern to the user. The widely used methods are:

- Total Lagrangian
- Updated Lagrangian
- Eulerian

Since FEA codes will offer a choice, so a careful assessment of which method is best suited to the problem on hand is important.

Users of commercial FEA codes who perform computational experiments themselves will be well rewarded. NAFEMS (National Agency for Finite Element Methods and Standards, UK) benchmark tests are widely used.

Further complexities in which nonlinear behaviour is but one important aspect to be modelled are also of considerable interest. Transient dynamic analysis when combined with general nonlinear behaviour provides formidable (but not insurmountable) difficulties for the analyst. Contact problems also fall under the nonlinear analysis heading, and may be regarded as a boundary nonlinear process to complement the material and geometric nonlinearies. Nonlinear fracture mechanics is another discipline, which features here. Nonlinear behaviour in fracture mechanics analysis occurs in the form of elastoplasticity at the crack tips and also during dynamic crack propagation (a kind of reverse contact problem).

The reader is advised to refer [10.1], a special publication by NAFEMS for details.

10.3 Analysis of Laminated Composite Structures

Composite materials consisting of stiff and strong fibers (glass, carbon, kevlar, boron, etc.) embedded in compatible matrices (polymer, metal, ceramic, etc.) have emerged as engineering materials of the future. Due to their heterogeneous, anisotropic characteristics and nonlinear and/or inelastic behaviour, engineering analysis of composite parts/products/structures demand the application of FEM in general and commercial FEA programs in particular.

Mechanics of composites [10.6] provides the methodology to predict the stiffness, strength, and durability of composite laminates.

Finite elements specifically designed to model composite plates and shells should include anisotropic elastic behaviour of constituent plies, bending–stretching coupling for unbalanced laminates, and transverse shear deformation effects. Accordingly, a family of shear flexible plate and shell elements, triangular/quadrilateral in shape has been developed [10.7–10.10].

ANSYS and MSC.Nastran provide the right answer for finite element analysis of composite structures. These codes will offer a choice for failure criteria; so an assessment of which criterion is best suited to the problem on hand is important.

The test problems with target solutions presented in Ref. [10.8] can be used for performing computational experiments.

Multiple failure modes in plies (fiber breaks, matrix cracks, and shear) and onset of delaminations between plies cause progressive failure of a composite laminate. Therefore, strength and life prediction of composite structures demands a nonlinear analysis procedure.

For the purpose of analysis, the individual plies are modelled as a homogeneous, anisotropic, linear, elastic continuum characterized by measured lamina effective modulii [10.10]. For laminates, a common approach in their analysis involves the use of lamination theories [10.6]. These concepts are exclusively used in finite element analysis of composite structures. Obviously, the validity of this approach can only be verified through correlation studies, where numerical and analytical solutions (if available) are compared with reliable experimental data [10.11]. Such studies in future should focus on stress, buckling, vibration and transient response, both linear and nonlinear of composite plate and shell type structures.

Laminated composite structures can exhibit strange behaviours: a rotating annular disk can undergo asymmetric deformation [10.12]; stress concentration around a cutout in a pressurized cylindrical shell can be a strong function of material orthotropy [10.13]; and a cylindrical shell can exhibit helical modes in vibration [10.14].

Among the topics of current interest which are not covered here are:

- Theories of laminated composite plates and shells incorporating the effects of transverse normal strain and interlaminar stresses
- Modelling of three-dimensional reinforcement
- Analysis of hygro-thermal behaviour
- Analysis of bi-modulus composites
- Failure analysis including onset and growth of delamination
- Delamination buckling
- Structural optimization
- Low velocity impact damage tolerance analysis

It will be informative to critically assess the current capabilities of ANSYS and MSC.Nastran for performing the above listed advanced analysis.

Finite element analysis can be used to make more informed decisions regarding the design of test specimens and fixtures for experimental characterization of laminated composite materials. Furthermore, calibration of the tests based on a rigorous finite element analysis is also essential for a valid reduction of recorded test data [10.15].

ABAQUS is a general-purpose FEA program that can be used for modelling and analysis of laminated composite structures. Using ABAQUS, simulation of impact tests of composite plates [10.16] and prediction of nonlinear dynamic response due to lateral impact [10.17] have been successful.

10.4 Computational Fracture Mechanics

Griffith A.A. made us aware of a new phenomenon called fracture. Irwin G.R. gave flesh and blood to this phenomenon and developed fracture mechanics, which eventually lead us to the damage-tolerant design methodology.

Fracture mechanics [10.18, 10.19] provides a methodology to predict the fracture behaviour of materials, components and structures. Specifically, it enables us to quantify toughness of materials and to predict crack growth, residual strength, and fatigue life of structures/components. Fracture mechanics in general should be able to answer the following questions:

- What is the residual strength as a function of crack size?
- What is the size of crack that can be tolerated at the expected service load; i.e., what is the critical crack size?
- How long does it take for a crack to grow from a certain initial size to the critical size?
- What size of pre-existing crack can be permitted at the moment the structure component starts its service life?
- How often should the structure/component be inspected for detection, location and characterization of cracks?
- Engineering fracture mechanics today provides satisfactory answers to all these questions. The predictions made can be verified experimentally.

The elastic stress field in the vicinity of a crack tip is singular and is characterized by the mode I (opening), mode II (sliding) and mode III (tearing) components of the stress intensity factor (SIF) denoted by K_{I}, K_{II}, and K_{III}. Crack extension will occur when the SIF reaches a critical value. For mode I loading, when K_{I} reaches a critical value K_{IC}, fracture may be expected to occur; K_{IC} is a material property (fracture toughness). SIF is a widely used fracture mechanics parameter. Closely related parameters include the energy release rate denoted by G, the path independent integral denoted by J and the crack tip opening displacement denoted by δ. In linear elastic fracture mechanics all these parameters are interrelated.

The practical application of engineering fracture mechanics bears largely upon the crack tip stress intensity factor. The SIF is a function of the crack geometry (type, shape, size, etc.), component geometry, and applied loads. Much effort has been put into the determination of SIFs, and a variety of methods have been developed [10.20–10.22]. The available SIF solutions are presented in a number of handbooks [10.23, 10.24].

The application of the finite element method to determine crack tip stress intensity factors has seen rapid progress. It is fair to say that the FEM is the backbone of computational fracture mechanics [10.25]. A fairly exhaustive survey and critical assessment on finite element applications to shell structures with crack can be found in [10.26].

From the very outset the most appealing finite element model for accurate determination of crack tip stress intensity factors is that which takes explicit account of the fact that the stresses are singular at the crack tips and the order of the singularity is $1/\sqrt{\gamma}$, where γ is the radial distance measured from a crack tip. A number of crack tip elements including the singular isoparametric elements have been developed [27, 28] for this purpose. The singular elements, triangular or quadrilateral in shape, are used to model a small region in the vicinity of each crack tip while the rest of the domain under consideration is discretized using regular elements. A practical consideration in the use of singular isoparametric elements is that the crack tip SIFs have to be evaluated posteriori. K-evaluation by the crack opening displacement (COD) method is recommended for this purpose. In recent years, the evaluation of strain energy release rate G by the modified crack closure integral (MCCI) approach is advocated. For the evaluation of the J-integral, the virtual crack extension (VCE) technique represents the state-of-the art.

The use of singular elements significantly improves convergence, accuracy and computational efficiency of the FEM for the analysis of cracked bodies. In particular, it enables the development of explicit formulae for the evaluation of stress intensity factors, strain energy release rates, and the J-integral on the basis of a standard output provided by a commercial FEA software.

ANSYS provides a specific option for fracture mechanics analysis of solids. Solving a fracture mechanics problem involves performing a linear static or nonlinear (elastoplastic) static analysis and then using specialized post-processing commands or macros to calculate the desired fracture parameters. The recommended element type for a two-dimensional finite element model is PLANE 2 (six-node triangular) and PLANE 82 (eight-node quadrilateral). For a three-dimensional finite element model, the recommended element is SOLID 95 (twenty-node brick element). The first row of elements around the crack front should be singular elements, which are triangular or wedge-shaped. A special command is provided to automatically generate singular elements around the specified key point. Other fields on the command allow the user to control the radius of the first row of elements, number of elements in the circumferential direction, etc. For reasonable results, the first row of elements around the crack tip should have a radius of approximately $a/8$ or smaller, where a is the crack length. Once the static analysis procedure is completed, the user can use the general-purpose post-processor to calculate the fracture parameters namely stress intensity factors, J-integral, and energy release rate. The user is advised to refer the ANSYS user's manual for details.

FEM in general and commercial FEA programs in particular, can be applied for fracture mechanics analysis of structures (plates and shells) also [10.29, 10.30]. There is scope to improve the finite element model employed in [10.29] using singular elements around the crack tips. However, it will be necessary to derive appropriate formulae to compute the mode I and mode II components of the bending stress intensity factors $K_{\text{I}}^{(\text{b})}$, $K_{\text{II}}^{(\text{b})}$. G-evaluation by the MCCI approach and derivation of G–K relations also deserves explicit attention. It is informative to note that the finite element model proposed in [10.30], though restricted to cylindrical shells with through cracks, is applicable to laminated anisotropic as well as homogeneous isotropic materials. Therefore there is a need, if solutions to practical problems are envisaged, to extend this formulation so as to be applicable to general shell structures. It is therefore instructive to experiment with the eight-noded shell elements SHELL 99 and SHELL 91 in ANSYS. It is then possible to recover the mode I and mode II components of both membrane and bending stress intensity factors denoted by $K_{\text{I}}^{(\text{m})}$, $K_{\text{II}}^{(\text{m})}$, $K_{\text{I}}^{(\text{b})}$, and $K_{\text{II}}^{(\text{b})}$ on the basis of a standard output provided by ANSYS. Incidentally, explicit formulae for the evaluation of $K_{\text{I}}^{(\text{m})}$ and $K_{\text{II}}^{(\text{m})}$ using standard output provided by a commercial FEA program are reported in [10.31]. Similar formulae have to be developed for $K_{\text{I}}^{(\text{b})}$ and $K_{\text{II}}^{(\text{b})}$ and used.

10.5 Structural Optimization

The objective of the design process is to create a cost effective component/structure/product that meets certain requirements regarding functionality and appearance. The design is an iterative process in which the analysis of the structural behaviour leads to change in the design. The use of optimization technology can allow design changes to be derived directly from the analysis results. The design problem itself is given as a structural optimization problem. The application of structural optimization allows an efficient search for the right combination of design variables. This section gives an overview of practical applications of structural optimization methods as practiced in a CAE services company.

It has become absolutely essential to optimize a product design in order to maximize or minimize a chosen objective function. The objective function may be related to profit, cost, weight, size, efficiency, reliability, safety, quality, power output, etc. The survival of a given product is largely a function of its performance per unit cost. The optimization of the manufacturing process in order to obtain the best quality per unit cost is equally important. Therefore, optimization is of particular interest to tool designer/manufacturer, product designer/manufacturer, and process designer/developer. In the following pages we describe problem formulation, optimization methods, algorithms, software, and example applications.

An appropriate function known as the objective function and denoted by

$$U = U(X_1X_2X_3 \ldots X_n)$$

in terms of the design variables $X_1\ X_2\ X_3 \ldots X_n$ is chosen for minimization or maximization.

The constraints arising from physical limitations in the problem, maybe equalities or inequalities, are given as

$$G_i = G_i(X_1X_2X_3 \ldots X_n) = 0$$

and

$$H_j = H_j(X_1X_2X_3 \ldots X_n) \leq C_j.$$

Here, $G_i(i = 1, m)$ are the equality constraints and C_j is a given quantity corresponding to the inequality constraint $H_j(j = 1, \ell)$.

Depending on the nature of the problem, various optimization methods may be applied to find the extremum of the objective function. A sensitivity analysis may then be undertaken to determine how the objective function varies with the design variables and the operating conditions in order to choose the most appropriate, convenient, and cost effective values at or near the extremum that would optimize the system.

There are several methods that may be employed for solving the optimization problem formulated above. The choice of the method depends on whether numerical solutions are to be obtained to determine the variation of the objective function and the constraints with the design variables. Commonly used methods are: calculus methods, search methods, linear programming, dynamic programming, and geometric programming.

Today, structural optimization is an independent discipline of applied mechanics. During the past four decades, computational methods for structural optimization have seen increasing acceptance. An important prerequisite for the application of computational methods for structural optimization is the FEM. The implementation of structural optimization procedures in commercial FEA software began in the late 1980s. Layout or topology optimization based on FEM is an effective tool in automotive design. Other related developments include shape optimization, coupling of CAD systems and structural optimization programs.

The availability of commercial structural optimization software has improved during the last couple of years. What is more, the pre- and post-processing of the results has greatly improved.

In the field of linear statics and dynamics, commercial software covering the complete range from topology and sizing to shape optimization is available. The algorithms for sizing and shape optimization are mostly based on local approximation methods using semi analytical sensitivity analysis. They are fully integrated into commercial FEA codes. For topology optimization, software based on optimality criteria is available to maximize stiffness and/or natural frequencies with mass constraints. Empirical methods such as the Hard Kill Option and Soft Kill Option are also implemented for topology and shape optimization using standard FEA software. Global approximation methods are available in many codes. Such optimization software is usually interfaced with several FEA programs.

NISA.OPT is a family of computer programs available for optimum structural design. The three modules in NISA.OPT – STR OPT, SHAPE, and SEC OPT – are fully integrated with NISA/DISPLAY. State-of-the-art methods are utilized in all three programs.

In ANSYS, there are two fundamentally different types of optimization. The first is referred to as design optimization; it works entirely with the ANSYS parametric design language and is contained within its own module ANSYS/OPT. The second is topology optimization, a form of shape optimization.

Opti Struct developed by Altair computing is a finite element-based optimization program, which generates optimized topologies for structural components. It creates conceptual designs using only the following information:

- A finite element model of the available (design) space
- Static load cases, modal frequencies of interest and boundary conditions
- An estimate of available materials.

Unlike shape optimization codes, Opti Struct does not require an initial design as input. It provides an optimized shape during the conceptual design phase where the potential for improvement is the greatest. The predicted shape is thereby optimized for maximum stiffness, highest modal frequency, or a weighted combination of both under specified conditions. It is further analyzed and refined using standard FEA programs.

Topology optimization is a mathematical technique which produces the optimal shape and material distribution for a structural component within a given package space. By discretizing the given domain into a finite element mesh, Opti Struct calculates material properties for each element using a microstructural modelling method called homogenization. The Opti Struct algorithm alters the amount and orientation of material within each element to optimize the structural performance of the model. Convergence occurs when the change in compliance index between any three successive iterations is below a given tolerance.

Opti Struct/FEA developed by Altair Computing, performs basic linear static and eigen value analyses. It is a full-fledged FEA code, providing beam, shell, solid, and rigid elements. For finite element modelling of solids and structures it provides a basic evaluation of structural integrity within the hyper mesh environment. Opti Struct/FEA is the finite element solver kernel of Opti Struct.

Hyper Opt developed by Altair computing is a design optimization program that performs optimization, parametric studies and system identification. Hyper Opt performs optimization in conjunction with linear and nonlinear FEA codes such as ABAQUS, LS-DYNA, PAM CRASH, etc. Hyper

Opt allows the choice of design variables so that the user can perform both size (shell wall thickness, beam section, and material properties) and shape (grid point locations) optimization.

Example applications can be found in the user's manual for NISA, ANSYS, and Hyper works. The reader is advised to refer [10.32 to 10.35] for algorithmic details, software capabilities, and applications.

10.6 Computational Heat Transfer

The computer is now the preferred vehicle for solution of many heat transfer problems, and finite element method is the preferred numerical method for computational heat transfer. Regardless of the computer and FEA program used, there is always a need for an understanding of basic equations and computational algorithms. This information is readily available in Holman [10.36] and Baker [10.37]. Our objective in this section is to discuss the application of the FEM to heat transfer analysis. It is therefore instructive to first recall governing differential equations and boundary/initial conditions. This information is the basis for design of finite elements leading to derivation of element equations and computation of element matrices and vectors.

Using a weak statement or a weighted residual approach, finite element theory simply requires that the approximation error be a minimum! All combinations of appropriate boundary conditions become naturally included in the process. The end result is a set of algebraic equations, written on each finite element, which are then collected together using a procedure called assembly to form the global matrix statement. The nodal d.o.f are then determined via solution of this matrix statement using any linear algebra technique. The finite element algorithm implementation in one, two, and three dimensions employs the Galerkin criterion for the weak statement for optimal accuracy. Full details of finite element formulation for heat transfer analysis are available in Bathe [10.38] and Baker [10.37]. For the solution of both linear and nonlinear, steady state and transient problems, incremental equilibrium equations are developed in [10.38]. The effects of phase changes and latent heat generation and of bodies radiating onto each other can also be included in the analysis.

The solution of the global matrix statement is a function of time is also outlined in [10.38]. Steady state heat transfer analysis by FEM produces a global matrix statement for which the global rank of the coefficient matrix can become large compared to the current PC memory capacity. While a system of linear algebraic equations can be solved directly in theory, on a PC environment it may be preferable and perhaps mandatory to use a numerically approximate solution technique wherein the large global matrix is not equation solved. There are a number of candidates such as Jacobi iteration, Gauss-Seidel, Successive Over Relaxation (SOR), Strongly Implicit (SIP), and Alternating Direction (ADI), to name only a few. Regarding direct methods, no algorithm for equation solving involves fewer arithmetic operations counts than Gauss elimination, which is a two-step process. The first involves factorization and the second back substitution. Most commercial FEA codes replace Gauss elimination with variations of the Cholesky method, which yields more efficient storage and faster computational speed.

Thermal analysis means primarily the calculation of temperatures within a solid body. A by-product of temperature calculation is the information about the magnitude and direction of heat flow in the body. Temperature and temperature gradients are important causes of stress. Heat is transferred to or from a body by convection and radiation. There may be internal heat generation, produced by electric current, dielectric heating, or other sources. At some points on the boundary or within, temperatures may be prescribed. Heat moves within the body by conduction. Thermal conductivity and other properties may depend on temperature. Thus the problem is inherently nonlinear if there is radiation heat transfer, because the heat fluxes across the boundary depends on the differences of fourth powers of absolute temperatures. If steady state conditions do not prevail, a thermal mass matrix and a vector of nodal rates of change of temperature are included.

ANSYS has a thermal analysis option and can handle the three modes of heat transfer, i.e., conduction, convection and radiation. The basis for thermal analysis is a heat balance equation obtained from the principle of conservation of energy. The principle of virtual temperatures is used to derive finite element equations while the Newton–Raphson method is used to solve the global matrix statement. Provision for defining material properties like conductivity, specific heat, and density, etc., as varying with temperature is available. A library of finite elements is included in ANSYS for thermal analysis as well as coupled thermal structural analysis.

NISA/DISPLAY has an independent module for heat transfer analysis. Analysis features are: linear and nonlinear/steady state and transient problems; constant or variable time step sizes; time and/or temperature dependent loading and boundary conditions; view factor calculations for surface radiation; and time integration techniques. Nonlinear capabilities include: temperature-dependent material properties; time and/or temperature-dependent boundary conditions (convection, radiation); phase change (melting and freezing); and surface radiation. The material properties can be isotropic, orthotropic, time and/or temperature-dependent. The loading is concentrated heat source, distributed heat flux, or internal heat generation (node or element). Boundary conditions include specified nodal temperature, convection, and radiation. The output includes nodal temperature time history; temperature distribution at specified time step; nodal heat values; and temperature gradients in shells. A very good library of two-dimensional, axisymmetric, three-dimensional and plate/shell elements is implemented for thermal analysis.

Finite element modelling considerations for thermal analysis and an example application featuring thermal analysis and subsequent stress analysis can be found in Ref. [10.39].

The subject of computational heat transfer is not static. New developments occur quite regularly, and better numerical methods and experimental data are continuously made available. A huge amount of information is available in research literature. For the serious student these publications offer an open door to the literature on computational heat transfer.

10.7 Computational Fluid Dynamics

The study of fluid mechanics [10.40] is of equal importance to mechanical, civil, chemical, aerospace and environmental engineers and is to be found as a core course in engineering education. The techniques available to enable students to solve fluid flow problems have been revolutionized by the readily available computing facilities. The use of computers to aid fluid flow analysis through the provision of numerical simulations, including the use of commercial codes, has lead us to Computational Fluid Dynamics (CFD).

Fluids can exist either as gases or liquids. The field of fluid mechanics is concerned with the motion of fluids and the conditions affecting the motion. It is therefore necessary to review the basic concepts of fluid mechanics and derive the governing equations. The basic equations of fluid mechanics are derived from the global laws of conservation of mass, momentum, and energy. Conservation of mass gives continuity equation, while conservation of momentum results in the equations of motion.

Conservation of energy leads to energy balance equation. Often, these equations are supplemented by the equation of state.

An ideal fluid is one that has zero viscosity and is incompressible. A real fluid is one with finite viscosity, and may or may not be incompressible. A viscous fluid is Newtonian if its viscosity coefficient is independent of the velocity gradient. Non-Newtonian fluids are those for which the viscosity is a function of the velocity gradient.

Two different viewpoints are used in the analytical description of the continuum mechanics equations governing fluid motion. The Eulerian description is the one that is most commonly used in fluid mechanics. The other one known as Lagrangian description is widely used in solid mechanics.

In the Eulerian description of the basic equations, one considers the motion of all matter passing through a fixed spatial location. Here we are concerned with changes in velocity, pressure, density, temperature, etc., that are taking place in the matter which happens to occupy the fixed spatial location. This is the one commonly used in fluid mechanics. In the Lagrangian description of continuum mechanics equations, one considers the motion of a fixed set of material particles (i.e., fixed matter), irrespective of its spatial location. The viewpoint is used in solid/structural mechanics.

A frequently followed approach in the analysis of fluid mechanics problems is to derive the governing differential equations for specific flow, geometry, and boundary conditions under consideration, and then solve these equations numerically using the Finite Difference Method (FDM).

However, during the last decade, much progress has been made toward the analysis of fluid flow problems using finite element procedures. At present very complex fluid flows are simulated. Inviscid and viscous, compressible and incompressible fluid flows with or without heat transfer are analyzed, including the coupling between structural and fluid response.

The objective in this section is to discuss briefly how the FEM presented earlier can also be applied in the analysis of fluid flow problems and identify CFD codes for the analysis of practical problems.

The continuum mechanics equations for laminar flow of viscous, incompressible fluid including heat transfer are summarized in Bathe [10.38]. It is effective to use an Eulerian formulation in fluid flow analysis. The essence of the formulation is that we focus attention on a stationary control volume. However, if the solution domain changes, such as in free surface flow problems, a pure Eulerian formulation would require the creation of new control volumes, as it is more effective to use an arbitrary Lagrangian–Eulerian formulation. This approach allows modelling of general free surfaces, and interaction between fluid flow and structural response. However, the Lagrangian formulation is widely used in solid/structural mechanics.

The finite element solution of the continuum mechanics equations governing the fluid flow is then obtained by establishing a *weak statement* of the equations using the weighted residual approach. Specifically, the momentum equations are weighted with velocities, the continuity equation is weighted with pressure, and the heat transfer equation is weighted with temperature. Integrating over the domain of interest and using the divergence theorem to lower the order of the derivatives in the expressions and to incorporate the natural boundary conditions as forcing terms gives the variational equations.

Assume that the finite element descritization of the variational equations is performed with any one of the continuum elements, with velocity and pressure as variables and temperature as an additional variable at all nodes. The governing matrix equations for a single element can then be derived [10.38]. The resulting finite element equations are highly nonlinear.

A major difficulty in the FEA of fluid flow problems is that the number of solution variables is generally very large because to obtain a realistic resolution of fluid response, very fine meshes are needed and the coefficient matrices are non-symmetric. Hence, explicit schemes that do not require the solution of a set of equations and iterative methods for the solution of equations are very attractive.

NISA/3-D-FLUID is a general-purpose FEM-based CFD code for solving fluid flow and heat transfer problems in arbitrary and complex geometries. It is integrated with DISPLAY for FE model generation, application of boundary conditions, graphical interpretation and presentation of results. A specialized utility, FLUTL, is provided for computing lift and drag forces, and heat transfer coefficients. Analysis features include compressible fluid flow, incompressible fluid flow, internal and external flows in 2-D, 3-D and axisymmetric domains, advanced turbulence modelling, flow through porous media, particle tracing, and fluid–structure interaction.

CF Design/I-DEAS is a FEM-based CFD code for solving complex laminar and turbulent fluid flow and heat transfer problems. It is based on a proven CFD solver integrated into I-DEAS master series environment, providing a complete CAD system for performing fluid flow and heat transfer analyses. All of the geometric modelling features coupled with the finite element modelling and post-processing features available in I-DEAS are accessible to CF Design/I-DEAS users. The capabilities include:

- Incompressible, transonic, and supersonic compressible flows
- Laminar and turbulent flows
- Natural, forced, mixed convection and radiative heat transfer
- Conjugate heat transfer (fluid convection/solid conduction)
- Newtonian and non-Newtonian fluids
- Distributed flow resistances/porous media
- 2-D quadrilateral and triangular elements (plane and axisymmetric)
- 3-D tetrahedral (4 and 10 nodes), wedge (6 nodes) and hexahedral (8 nodes) elements
- Turbulence models

CF Design/I-DEAS allows engineers access to complex mathematical models and numerical algorithms necessary to simulate fluid flow and thermal conditions. These features combined with the robust solution procedures provide a highly efficient tool for the engineer. In addition, since CF Design/I-DEAS is based on finite element technology, which has been accepted as the de facto standard for engineering analysis, calculated values from fluid flow and heat transfer analyses can be utilized for other analyses.

ANSYS/FLOTRAN offers comprehensive tools for analysis of two-dimensional and three-dimensional fluid flow fields. Using the FLUID 141 and FLUID 142 elements, the following are the example problems solved:

- Lift and drag on airfoils
- Flow in supersonic nozzles
- Complex, 3-D flow patterns in pipe bends
- Calculating gas pressure and temperature distributions in engine exhaust manifolds
- Studying thermal stratification and break-up in piping systems
- Flow mixing studies to evaluate potential for thermal shock
- Conducting heat exchanger studies involving different fluids separated by solid regions

ANSYS/FLOTRAN analysis capabilities are:

- Laminar and turbulent flow analysis
- Thermal or adiabatic analysis

- Free surface analysis
- Compressible or incompressible flow analysis
- Newtonian or non-Newtonian fluid flow analysis
- Multiple species transport analysis

CFX computational fluid dynamics software combines advanced mathematical models and numerical methods with powerful pre- and post-processing capabilities to undertake simple or complex fluid flow analysis. CFX capabilities include:

- Any combination of tetrahedral, hexahedral, prism and/or pyramid elements
- Incompressible/compressible subsonic, transonic, supersonic
- Steady state/transient
- Laminar/turbulent
- Rotating or stationary frame of reference
- Turbulence models
- Heat transfer
- Newtonian and non-Newtonian fluids
- Species transport
- Multi-component fluids
- Coupled Eulerian multi-phase flow
- Pre-processor
- Boundary conditions
- Mesh generation conservative finite element based control volume method
- Implicit, pressure based algorithms
- Advection modelling
- Parallel computing
- Post-processing

FLUENT is a general purpose CFD software based on the finite volume method. Example applications are:

- Internal and external automotive flows
- High speed aerodynamics
- Rocket flows
- Turbo machinery
- Reactor vessels
- Cyclones
- Bubble columns
- Mixing tanks
- Fluidized beds

Gambit is a single, integrated pre-processor for CFD analysis using FLUENT. The capabilities include:

- Geometry creation
- Mesh generation
- Mesh quality examination
- Boundary zone assignment

The mathematical models in FLUENT include:

- Fluid flow and heat transfer
- Turbulence
- Chemical species transport and reacting flows
- Multi-phase flows
- Flows involving moving parts
- User defined scalar transport

Field variables (stored at cell centers) must be interpolated to the faces of the control volume in the finite volume method. FLUENT offers a number of interpolation schemes. Additional interpolation options are available for calculating face pressure.

FLUENT is based on the finite volume method where the domain is discretized into a finite set of control volumes or cells. Continuum mechanics equation for mass, momentum, energy, etc., is applied to each cell. Each transport equation is discretized into algebraic form, for each cell. Equation sets are solved iteratively using explicit or implicit coupled or segregated solver.

Heat transfer modelling options available in FLUENT include:

- Species diffusion heat source
- Combustion heat source
- Conjugate heat transfer
- Natural convection
- Radiation
- Periodic heat transfer

FLUENT is also the code of choice for combustion modeling. A wide range of reacting flow applications can be addressed by the combustion models in FLUENT.

10.8 Electromagnetic and Coupled-Field Analyses

More and more modern technologies are dependent on electromagnetic analysis. From microwave ovens to mobile telephones, from radars to digital radio, electromagnetic based systems are now truly ubiquitous. Computational electromagnetics is therefore becoming vital in civil and commercial engineering situations.

Unlike traditional finite element analysis, there are many algorithms and solvers in computational electromagnetics, each suitable for a particular physical situation. Each of these requires a particular kind of mesh to be built, some of which mix surface boundary elements and finite elements in the same mesh. Efforts are on to make available a versatile general electromagnetic solver for providing accurate solutions of computational electromagnetic problems.

NISA/EMAG is a finite element based electromagnetic analysis program to determine electric and magnetic field distributions in electromagnetic devices. The program solves numerically the appropriate Maxwell's formulations. EMAG is integrated with DISPLAY for pre-and post-processing.

The capabilities of EMAG include:

- Two and three-dimensional problems
- Linear, nonlinear, and anisotropic materials
- Steady and/or time-varying escalations

The analysis features in EMAG are:

- Electrostatic analysis
- Magnetostatic analysis
- Transient magnetic field analysis
- Magnetodynamic analysis
- Coupled thermal electromagnetic field analysis
- High frequency electromagnetic analysis

The outputs provided by EMAG are:

- Colour contour plots of electric field, magnetic field, flux densities, magnetic vector potential, voltage, and current distribution
- Arrow plots of all vector parameters
- Stored electric energy
- Stored magnetic energy and co-energy
- Power losses
- Capacitance, inductance, and conductance of the device
- Induced eddy currents
- Ampere's force in current-carrying region
- Temperature distribution
- Force and torque calculations using either virtual work method or Maxwell's stress tensor approach

ANSYS/EMAG also provides state-of-the-art capabilities for electromagnetic analysis based on the finite element method.

A *coupled-field analysis* is a finite element analysis that takes into account the interaction (coupling) between two or more disciplines (fields) of engineering. A piezoelectric analysis, for example, handles the interaction between the deformation and electric fields. It solves for the voltage distribution due to applied displacements, or vice versa. Other examples of coupled-field analysis are thermal stress analysis, thermal electric analysis, and fluid structure analysis. Some of the applications in which coupled-field analysis by the FEM may be required are: pressure vessels (thermal stress analysis), induction heating (magnetic thermal analysis), ultrasonic transducers (piezoelectric analysis), magnetic forming (magneto-structural analysis), and micro, electromechanical systems (MEMS).

The procedure for a coupled field analysis depends on which fields are being coupled, but two distinct methods can be identified: sequential and direct. The sequential method involves two or more sequential analyses, each belonging to a different field. Two fields are coupled by applying results from the first analysis as loads for the second analysis. The direct method usually involves just one analysis that uses a coupled field element type containing all necessary terms.

The ANSYS program is the right answer for coupled-field analyses. The ANSYS program performs sequentially coupled analyses using the concept of physics environment. The direct method of coupled-field analyses uses a coupled-field element. ANSYS provides a list of coupled-field elements. ANSYS/ multi-physics uses the direct method. The reader is advised to see the ANSYS theory reference for full details. The ANSYS verification manual consists of test cases demonstrating the coupled-field analysis capabilities of ANSYS program.

10.9 Simulation of Manufacturing Processes

The FEM in general and commercial FEA programs in particular play a key role in the numerical simulation of manufacturing processes [10.41]. This is illustrated through case studies.

In the *metal casting process*, the solidification front can be studied using transient thermal analysis including phase change. After solidification, subsequent cooling of the solid can also be simulated to predict shrinkage. ANSYS is recommended for this application.

Nonlinear finite element analyses can be used to simulate sheet metal forming, rolling of thick plates and forging processes. These problems involve geometric nonlinearity (large displacements, large rotations, finite strains), material nonlinearity (plasticity) and boundary nonlinearity (closing/ opening of gaps, 3-D contact, follower force), etc. MARC is identified as ideal for this application.

In the *sheet metal forming* simulation, the blank is modelled using triangular/quadrilateral shell elements. The punch and/or the die are usually considered perfectly rigid, and the same are not modelled using finite elements. Instead, they are retained as geometrical surfaces defined by simple entities such as line segments, circular arcs, and splines since MARC allows geometric entities itself to be defined as rigid contact bodies. Further, the rigid surfaces are offset from the middle surface of the bank as the contact algorithm in MARC takes the shell thickness into account.

Friction between the blank and the punch and/or die needs to be considered. Numerical modelling of the friction has been simplified to two idealistic models. The first is the adhesive friction or Coulomb friction model, while the second is the cohesive shear based model.

Taking advantage of the options available in MARC, an appropriate friction model has to be chosen and the same is modelled along all contact interfaces.

Coming to material properties of the blank, Young's modulus of elasticity, Poisson's ratio, yield stress and the stress–strain curve have to be defined.

The component is formed by displacing the punch downward with a velocity. The velocity with respect to time is entered as a curve. The punch slowly descends on the blank and forces it to take the relevant shape defined by its end by the time it reaches the bottom of the die. As the punch draws the sheet into the die cavity the sidewalls increase in thickness while there is a slight thinning of the shell at the bottom corners. MARC offers a number of options to post-process the results.

Rolling of steel plates between rigid rollers is another metal forming process demanding nonlinear finite element analysis for its numerical simulation. The objectives are to study the thickness reduction in the plate apart from determining the deformations strains and stresses in the plate. Finite element computations are increasingly used to determine the elongation and spread of the material during rolling.

No better solver than MARC could be identified for this application, as it has the ability to handle large deformations, plasticity, and contact in addition to rezoning and adaptive meshing.

In manufacturing simulations the goal is to plastically deform the workpiece from some initial (simple) shape to a final, often complex shape. The deformation of the material, results in mesh distortion. For this reason it is often required to perform a rezoning/remeshing step. At this point the nonlinear analysis is stopped, a new mesh is created, the current state of deformation, strains, stresses etc., is transferred to the new mesh, and the analysis continued. Contact between a deformable body (workpiece) and a rigid body (rollers) is insured such that the nodes do not penetrate the rigid surface. One can request that the mesh be adaptively refined due to contact. In such a case, where a node on the workpiece comes into contact with the roller, the elements associated with that node are refined. This results in a larger number of elements and nodes on the exterior region where contact occurs. This can lead to a substantial improvement in the accuracy of the solution.

Using MARC, rolling in two or more passes can be simulated. In addition, 3-D rolling wherein the rollers are placed in two different positions can be performed.

Forging is still the best way to produce a strong part like turbine disk. During the forging set-up, a block of metal, which may be cylindrical or any other shape, is transformed to a shape close to that of a disk in one or several operations. It is now possible to optimize the forging process using numerical simulation and geometry modelling. The nonlinear analysis capabilities in MARC are useful here also.

10.10 Closure

Advanced applications presented in this chapter are critical in the competitive business strategies of many industries—automotive, aerospace, nuclear energy, thermal power, appliances, consumer electronics, etc. In these businesses, physical prototypes can cost a fortune and take months to build. So companies across the board are implementing numerical simulation wherever possible to refine designs, catch errors and evaluate part and assembly configurations as early as possible in the product development, when changes are most easily made.

Finite elements and analysis procedures implemented in a commercial FEA software may or may not handle all of the advanced applications identified in this chapter. The user therefore has the responsibility to make the right choice. The technical information included here has remained introductory, hence the reader should now approach the literature with confidence. The skills required to use a chosen FEA program have to be acquired. Benchmark tests devised to critically assess the performance of a code are found to be an excellent means of training novices.

This draws to a close the topical coverage in this text. It is our hope you have found the contents and the presentation format, of significant help in establishing a feeling of comfort with FEM. While there are several methods of engineering analysis, the most comprehensive is the finite element method.

10.11 References

[10.1] Hinton, E., editor. *Introduction to Nonlinear Finite Element Analysis*. NAFEMS, National Engineering Laboratory, Glasgow, UK, 1992.

[10.2] Bonet, J. and Wood, R.D. *Nonlinear Continuum Mechanics for Finite Element Analysis*. Cambridge University Press, Cambridge, 1997.

[10.3] Owen, D.R.J. and Hinton, E. *Finite Elements in Plasticity—Theory and Practice*. Pineridge Press, Swan Sea, 1980.

[10.4] Gittus, J. *Creep, Visco Elasticity, and Creep Fracture in Solids*. Applied Science Publishers, London, 1975.

[10.5] Crisfield, M.A. *Nonlinear Finite Element Analysis of solids and Structures—Volume 1: Essentials*. John Wiley, Chichester, UK, 1991.

[10.6] Whitney, J.M. *Structural Analysis of Laminated Anisotropic Plates*. Technomic Publishers, West Port, 1987.

[10.7] Noor, A.K. and Mathers, M.D. Shear flexible finite element models of laminated composite plates and shells. *NASA TN D*, page 8044, 1975.

[10.8] Lakshminarayana, H.V. and Sridhara Murthy, S. A shear flexible triangular finite element model for laminated composite plates. *Int J. for Numer. Methods in Engng*, 20:591–623, 1984.

[10.9] Lakshminarayana, H.V. and Kailash, K. A shear deformable curved shell element of quadrilateral shape. *Computers & Structures*, 33:987–1001, 1989.

[10.10] Carlson, L.A. and Pipes, R.B. *Experimental Characterization of Advanced Composite Materials.* Prentice Hall, Englewood Cliffs, N.J., 1987.

[10.11] Lakshminarayana, H.V. and Viswanath, S. A correlation study of finite element modeling for the stress analysis of composite material laminates. *J. Strain Analysis*, 13:205–211, 1978.

[10.12] Lakshminarayana, H.V. Finite element analysis of rotating laminated composite annular discs. *Int. J. of Composites*, 17:42–47, 1986.

[10.13] Lakshminarayana, H.V. and Viswanath, S. Influence of material orthotropy on the state of stress around a circular hole in a cylindrical shell. *ASME J. Pressure Vessel Technology*, 99:454–458, 1977.

[10.14] Lakshminarayana, H.V. and Dwarakanath, K. Free vibration characteristics of cylindrical shells made of composite materials. *Journal of Sound and Vibration*, 154:431–439, 1992.

[10.15] Lakshminarayana, H.V. A symmetric rail shear test for mode ii fracture toughness (G_{IIC}) of composite materials—finite element analysis. *Journal of Composite Materials*, 18:227–238, 1984.

[10.16] Lakshminarayana, H.V., Boukhili, R, and Gauvin, R. Finite element simulation of impact tests of laminated composite plates. *Composite Structures*, 28:47–59, 1994.

[10.17] Lakshminarayana, H.V., Boukhili, R, and Gauvin, R. Impact response of laminated composite plates: Prediction and verification. *Composite Structures*, 28:61–72, 1994.

[10.18] Broek, D. *Elementary Engineering Fracture Mechanics.* Sijthoff & Noordhoff, 1978.

[10.19] Broek, D. *The Practical Uses of Fracture Mechanics.* Kluwer Academic Publishers, Dordrecht, 1988.

[10.20] Sih, G.C., editor. *Methods of Analyses and Solutions of Crack Problems.* Noordhoff, 1973.

[10.21] Newman, J.C. and Raju, I.S. Methods for analyses of cracks in three dimensional solids. *Journal of the Aeronautical Society of India*, 36:153–172, 1984.

[10.22] Smith, C.W. Experimental methods for the determination of the stress intensity factor. *J. Aero. Soc. India*, 36:173–193, 1984.

[10.23] Rooke, D.P. and Cartwright, D.J. *Compendium of Stress Intensity Factors.* Her Majesty Stationary Office, London, 1976.

[10.24] Murakami et al., editors. *Stress Intensity Factors Hand Book*, volume 1 and 2. Pergamon Press, Oxford, 1987.

[10.25] Liebowitz, H. and Moyer, E.T. Finite element method in fracture mechanics. *Computers & Structures*, 31:1–9, 1989.

[10.26] Lakshminarayana, H.V., Murthy, M.V.V., and Rao, A.K. Finite element analysis of shell structures with cut-outs and cracks. In J.W. Bull, editor, *Finite Element Applications to Thin walled Structures.* Elsevier Applied Science Publishers, 1990.

[10.27] Barsoum, R.S. On the use of isoparametric finite elements in linear elastic fracture mechanics. *Int. J. Numer. Meth. Engng*, 10:25–37, 1976.

[10.28] Barsoum, R.S. A degenerated solid element for linear fracture analysis of plate bending and general shells. *Int. J. Numer. Method Engg*, 10:551–564, 1076.

[10.29] Viswanath, S., Lakshminarayana, H.V., and Ravindranath, D.D. A modified crack closure integral method for calculating stress intensity factors for cracked plates subjected to bending loads. *Int. J. Fracture*, 41, 1989.

[10.30] Lakshminarayana, H.V., Murthy, M.V.V., and Srinath, L.S. On a finite element model for the analysis of through cracks in laminated anisotropic cylindrical shells. *Engineering Fracture Mechanics*, 19:257–275, 1982.

[10.31] Lakshminarayana, H.V., Bernard, M., and Bui–Quoc, T. Finite element study of a penny shaped crack along the interface in a bi-material cylinder. *Engineering Fracture Mechanics*, 43:981–991, 1992.

[10.32] Vanderplaats, G.N. Structural optimization—past, present and future. *AIAA Journal*, 20:922–1000, 1982.

[10.33] Morris, A.J., editor. *Foundations of Structural Optimization: A Unified Approach*. Wiley & Sons, New York, 1982.

[10.34] Haftka, R.T., Gurdal, Z., and Kamath, M.P. *Elements of Structural Optimization*. Kluwer Academic Publishers, Dortrecht, The Netherlands, 1990.

[10.35] Herskovits, editor. *Advances in Structural Optimization*. Kluwer Academic Publishers, Dortrecht, The Netherlands, 1995.

[10.36] Holman, J.P. *Heat Transfer*. McGraw Hill, New York, 8th ed., 1997.

[10.37] Baker, A.J. and Pepper, D.W. *Finite Elements 1-2-3*. McGraw Hill, New York, 1991.

[10.38] Bathe, K.J. *Finite Element Procedures*. Prentice Hall of India, New Delhi, 1996.

[10.39] Cook, R.D. *Finite Element Modeling For Stress Analysis*. John Wiley, New York, 1995.

[10.40] Douglas, J.F., Gasiorek, J.M., and Swaffield, J.A. *Fluid Mechanics*. Addison Wesley Longman, Essex, UK, third edition, 1995.

[10.41] Kobayashi, S. *Metal Forming and The Finite Element Method*. Oxford Univ. Press, New York, 1989.

Index